Computational Techniques in Precision Agriculture

This book discusses the components and foundations of precision agriculture, followed by smart farming practices, sustainable development in precision agriculture, and the use of artificial intelligence techniques in the field of agriculture.

Key Features:

- Discusses remote sensing techniques, global positioning systems, robotics, and unmanned aerial vehicles in precision agriculture.
- Covers sustainable development in precision agriculture, data management and analytics, and variable rate technology.
- Highlights the implementation of artificial intelligence techniques and the use of prediction algorithms in improving crop efficiency.
- Presents machine learning, deep learning, and computer vision applications in precision agriculture.
- Showcases case studies on cotton crops, tea pests, and crop recommendations.

This book is primarily written for senior undergraduates, graduate students, and academic researchers in the fields of electrical engineering, electronics and communication engineering, smart grids, environmental engineering, agricultural engineering, and computer science and engineering.

Bhoomika Batra is a founder of a research and data analytics academy called EdBhoomi Learning and a visiting professor at Alliance University and Wellingkar Institute of Management, Bengaluru, India. Her research interests are in the areas of agriculture, blockchain, neural networks, finance, corporate governance, and digitalization. She is an expert on data analysis and predictive analytics having over three years of experience using tools like IBM SPSS, E-views, STATA, and SMART_PLS.

Vinod Kumar Shukla is an associate professor and head at the School of Engineering, Architecture, and Interior Design, Amity University, Dubai, UAE. He completed his PhD in the field of Semantic Web and Ontology. He is an active member of IEEE and has published research papers in various reputed journals and conferences. He holds international patents in the broader field of information technology. His areas of research include the Internet of Things, semantic web, computer networking, and network security.

Vaibhav Bhatnagar is an assistant professor in the Department of Computer Applications at Manipal University, Jaipur, India. He has published research articles in journals indexed in Scopus and Web of Science. His areas of research include fog computing, Internet of Things, data analytics, and machine learning.

Computational and Intelligent Systems

Series Editor: Ramesh Chandra Poonia
CHRIST (Deemed to be University), Bangalore, Karnataka, INDIA - 560029

About the Series:

In todays' world, the systems that integrate intelligence into machine-based applications are known as intelligent systems. In order to simplify the man-machine interaction, intelligent systems play an important role. The books under the proposed series will explain the fundamentals of intelligent systems, review the computational techniques and also offer step-by-step solutions to the practical problems. Aimed at senior undergraduate students, graduate students, academic researchers and professionals, the proposed series will focus on broad topics including artificial intelligence, deep learning, image processing, cyber physical systems, wireless security, mechatronics, cognitive computing, and industry 4.0.

Application of Soft Computing Techniques in Mechanical Engineering
Amar Patnaik, Vikas Kukshal, Pankaj Agarwal, Ankush Sharma, and Mahavir Choudhary

Computational Intelligence Based Optimization of Manufacturing Process for Sustainable Materials
Deepak Sinwar, Kamalakanta Muduli, Vijaypal Singh Dhaka, and Vijander Singh

Advanced Mathematical Techniques in Computational and Intelligent Systems
Sandeep Singh, Aliakbar Montazer Haghighi, and Sandeep Dalal

Natural Language Processing and Information Retrieval: Principles and Applications
Edited By Muskan Garg, Sandeep Kumar, and Abdul Khader Jilani Saudagar

Soft Robotics for Medical and Healthcare Applications
Edited by Shaik Himam Saheb, Tharakeshwar Appala, and Mohammad S. Khan

Smart Cyber Physical Systems: Innovations and Real-World Implications
Edited by Ramesh Chandra Poonia, Kamal Upreti, and Mohammad S Khan

Cognitive Cloud Computing: Building Intelligent Systems for Tomorrow
Edited by Fr. Jossy George, Kamal Upreti, Ramesh Chandra Poonia, Ankit Gautam, and Danish Nadeem

Computational Techniques in Precision Agriculture: Algorithms and Applications
Edited by Bhoomika Batra, Vinod Kumar Shukla, and Vaibhav Bhatnagar

For more information on this series page, please visit our website link:
https://www.routledge.com/Computational-and-Intelligent-Systems/book-series/CRCCIS

Computational Techniques in Precision Agriculture

Advances and Applications

Edited by

Bhoomika Batra, Vinod Kumar Shukla,
and Vaibhav Bhatnagar

CRC Press
Taylor & Francis Group
Boca Raton London New York

CRC Press is an imprint of the
Taylor & Francis Group, an **informa** business

Designed cover image: © Shutterstock

MATLAB® and Simulink® are trademarks of The MathWorks, Inc. and are used with permission. The MathWorks does not warrant the accuracy of the text or exercises in this book. This book's use or discussion of MATLAB® or Simulink® software or related products does not constitute endorsement or sponsorship by The MathWorks of a particular pedagogical approach or particular use of the MATLAB® and Simulink® software.

First edition published 2026
by CRC Press
2385 NW Executive Center Drive, Suite 320, Boca Raton FL 33431

and by CRC Press
4 Park Square, Milton Park, Abingdon, Oxon, OX14 4RN

CRC Press is an imprint of Taylor & Francis Group, LLC

ISBN: 9781032859774 (hbk)
ISBN: 9781032859767 (pbk)
ISBN: 9781003520733 (ebk)

DOI: 10.1201/9781003520733

Typeset in Sabon
by codeMantra

Contents

Part II
Precision Agriculture and Machine Learning

Part III
Precision Agriculture, Technology, Integration, and Food Security

Preface

Agriculture stands at the threshold of a technological revolution. As the global population grows and climate change continues to affect food systems, there is an urgent need for smarter, data-driven, and sustainable farming practices. Precision agriculture (PA), powered by computational techniques, is transforming traditional farming into a high-tech, efficient, and environmentally conscious endeavor.

This book, *Computational Techniques in Precision Agriculture: Advances and Applications*, provides a comprehensive exploration of how computational tools—ranging from artificial intelligence and machine learning to blockchain, big data, and edge computing—are reshaping agricultural systems. It serves as both a reference for researchers and a guide for practitioners, students, and policymakers aiming to harness digital innovation for agricultural sustainability and productivity.

Divided into four thematic parts, this book reflects a progressive journey from foundational understanding to futuristic applications:

Part I: Exploring Precision Agriculture introduces the technological backbone of modern farming. It discusses core concepts such as automated irrigation using IoT and AI, drone-based aerial analytics, remote sensing with GIS, and integrated smart farming systems. These chapters highlight the evolution of PA into a multidisciplinary field that merges hardware, software, and environmental science.

Part II: Precision Agriculture and Machine Learning delves into the integration of intelligent algorithms to enhance agricultural decision-making. Here, techniques like Random Forest Regression, K-Nearest Neighbors, and predictive modeling are applied to problems such as crop classification, yield prediction, and environmental optimization. The regional focus, including a case study on India, underscores the global relevance and adaptability of these technologies.

Part III: Precision Agriculture, Technology, Integration, and Food Security explores the broader ecosystem of technologies that enable precision agriculture. Topics include blockchain for transparency, big data for real-time decision-making, climate-smart approaches for resilience, and variable rate technology (VRT) for soil and yield management. This section links technological innovation to global food security and sustainable development goals.

Part IV: Precision Agriculture: Future Direction looks ahead to the digital transformation of farming. Emerging technologies such as edge computing, digital twins, and renewable energy integration are examined for their potential to revolutionize agricultural landscapes. This section concludes with visionary perspectives on using PA to address global challenges and prepare for an intelligent farming future.

This book is the outcome of collaborative efforts by researchers and professionals from diverse backgrounds, unified by the goal of enhancing agricultural efficiency and sustainability through computational innovation. Whether you are a researcher, educator, agronomist, technologist, or policymaker, we hope this work serves as a valuable resource in your exploration of PA.

We would like to thank all the contributing authors for their insightful chapters and the reviewers for their feedback and guidance. May this book inspire continued innovation and collaboration in the pursuit of smarter agriculture for a sustainable world.

Acknowledgments

We extend our heartfelt gratitude to all those who contributed to the successful completion of this book, *Computational Techniques in Precision Agriculture: Advances and Applications*.

First and foremost, we sincerely thank all the authors whose dedication, expertise, and scholarly contributions form the foundation of this volume. Their thoughtful research, innovative perspectives, and commitment to advancing the field of precision agriculture have made this compilation both timely and impactful.

We are equally grateful to the reviewers, whose valuable insights and constructive feedback have significantly enhanced the quality and clarity of the chapters. Their rigorous evaluations and academic diligence ensured that the content meets high standards of relevance, accuracy, and scholarly excellence.

A special note of appreciation goes to our publishers (Apple Academic Press, an exclusive co-publishing partnership with CRC Press, a member of the Taylor & Francis Group), whose unwavering support, professional guidance, and commitment to academic dissemination have been instrumental in bringing this book to life. Their encouragement and resources played a vital role throughout the publication process.

We would also like to express our sincere thanks to our friends and family members for their constant encouragement, patience, and understanding. Their support and belief in our work provided the strength and motivation we needed to see this project through to completion.

Finally, we acknowledge the broader academic and research community, whose continued efforts and collaborative spirit continue to inspire innovation in precision agriculture and computational technologies.

Contributors

Bhaskar Arora
Department of Commerce
Manipal University
Jaipur, Rajasthan, India

Nidhi Arora
Department of Economics
Manipal University
Jaipur, Rajasthan, India

Shaik Salma Banu
Acharya N.G. Ranga Agricultural University
Lam Guntur, Andhra Pradesh, India

Anjana Barola
Dr. D. Y. Patil Institute of Pharmaceutical Sciences and Research
Pimpri, Pune, India

S. Bathrinath
Department of Mechanical Engineering
Kalasalingam Academy of Research and Education
Krishnankoil, Tamil Nadu, India

Devershi Pallavi Bhatt
Department of Computer Applications
Manipal University
Jaipur, Rajasthan, India

G. Bhuvaneswari
CSE Department of Computer Science and Engineering
Saveetha Engineering College
Chennai, Tamil Nadu, India

Seema Darekar
Information Technology
Dr. D. Y. Patil Institute of Technology
Pimpri, Pune, India
Soumyakant DashGlobal Business School and Research Centre
Dr. D. Y. Patil Vidyapeeth, Pune
Tathawade, Pune, Pimpri-Chinchwad, Maharashtra, India

Komal Dhanraj Deshmukh
Dr. D. Y. Patil Institute of Pharmaceutical Sciences and Research
Pimpri, Pune, India

Amol M. Dhepe
Department of Computer Engineering
Dr. D. Y. Patil Institute of Technology
Pimpri, Pune, India

Ifeanyichukwu Eze
A Diversity and Inclusion Specialist (Freelancing)

Amit Gangopadhyay
Department of Electronics and Communication Engineering
Mohan Babu University (Erstwhile Sree Vidyanikethan Engineering College)
Tirupati, Andhra Pradesh, India

Pragya Gaur
Department of Commerce and Management
University of Kota
Kota, Rajasthan, India

Santosh Gore
Sai Info Solution
Nashik, Maharashtra, India

Tejashree Jadha
Dr. D. Y. Patil School of Science and Technology
Dr. D. Y. Patil Vidyapeeth, Pune
Tathawade, Pune, India

Babasaheb Jadhav
Global Business School and Research Centre
Dr. D. Y. Patil Vidyapeeth, Pune
Tathawade, Pune, Pimpri-Chinchwad, Maharashtra, India

Pranali Anandrao Jadhav
Dr. D. Y. Patil Institute of Pharmaceutical Sciences and Research
Pimpri, Pune, India

Neelu Jain
Centre for Distance and Online Education
Manipal University
Jaipur, Rajasthan, India

Rekha Jain
Department of Computer Applications
Manipal University
Jaipur, Rajasthan, India

Aarti Mohan Jathar
Dr. D. Y. Patil Institute of Technology
Pimpri, Pune, India
Shripad JoshiGlobal Business School and Research Centre
Dr. D. Y. Patil Vidyapeeth, Pune
Tathawade, Pune, Pimpri-Chinchwad, Maharashtra, India

P. Jyothi
CSE Department
VNR Vignana Jyothi Institute of Engineering and Technology
Hyderabad, Telangana, India

Madhuri Mohanrao Karad
Dr. D. Y. Patil Institute of Technology
Pimpri, Pune, India

Tejasvini Rahul Katkar
Department of E&TC (First Year Engineering)
Dr. D. Y. Patil Institute of Technology
Pimpri, Pune, India

Ritvik Khurana
Independent Researcher

Midhun Mathew Kizhakethil
Dr. D. Y. Patil Institute of Pharmaceutical Sciences and Research
Pimpri, Pune, India

Anish Kumar
Amity Institute of Information and Technology
Amity University
Noida, Uttar Pradesh, India

Pardeep Kumar
Department of Electronics and Communication Engineering
SGT University
Gurugram, Haryana, India

P. Lakshmi Supriya
Department of Electrical and Electronics Engineering
Mahatma Gandhi Institute of Technology (A)
Gandipet, Hyderabad, Telangana, India

Timothy Malche
Department of Computer Applications
Manipal University
Jaipur, Rajasthan, India

Chaitalee Naresh Mali
Department of Chemistry (First Year Engineering)
Dr. D. Y. Patil Institute of Technology
Pimpri, Pune, India

G. Manikandan
Department of Artificial Intelligence and Data Science
R.M.K. Engineering College
Kavaraipettai, Tiruvallur District, Tamil Nadu, India

Harsh Mishra
Amity Institute of Information and Technology
Amity University
Noida, Uttar Pradesh, India

Poonam Mishra
Department of Economics
Manipal University
Jaipur, Rajasthan, India

Monika
School of Computer Science Engineering and Technology
Bennett University
Greater Noida, Uttar Pradesh, India

Priyanshi V. Mulwani
Dr. D. Y. Patil School of Science and Technology
Dr. D. Y. Patil Vidyapeeth, Pune
Tathawade, Pune, India

Rakhi Mutha
Department of Information Technology
Amity University
Jaipur, Rajasthan, India

Shraddha Amit Pathak
Dr. D. Y. Patil Institute of Pharmaceutical Sciences and Research
Pimpri, Pune, India

Kamakshi Patil
Dr. D. Y. Patil School of Science and Technology
Dr. D. Y. Patil Vidyapeeth, Pune
Tathawade, Pune, India

Sonali Lalasaheb Patil (Shinde)
Dr. D. Y. Patil Institute of Pharmaceutical Sciences and Research
Pimpri, Pune, India

Sonali V. Patil
Dr. D. Y. Patil School of Science and Technology
Dr. D. Y. Patil Vidyapeeth, Pune
Tathawade, Pune, India

Hemlata Yashwant Pawar
Dr. D. Y. Patil Institute of Technology
Pimpri, Pune, India

Minal Pawar
Dr. D. Y. Patil School of Science and Technology
Dr. D. Y. Patil Vidyapeeth, Pune
Tathawade, Pune, India

Kiran Kumar Reddy Penubaka
Department of CSE-AIML
MLR Institute of Technology
Hyderabad, Telangana, India

Madhu Priya
Department of Computer Applications
Manipal University
Jaipur, Rajasthan, India

Sharmila Pudaruth
University of Mauritius
Réduit, Mauritius

P. Ram Kishore Kumar Reddy
Department of Electrical and Electronics Engineering
Mahatma Gandhi Institute of Technology
Ranga Reddy, Hyderabad, Telangana, India

Mallu Rama Prasada Reddy
EEE Department
G. Pullaiah College of Engineering and Technology
Kurnool, Andhra Pradesh, India

Sreejith Ramachandran
Department of Business Administration
Rajagiri College of Social Sciences (Autonomous)
Kochi, Kerala, India

E. R. Ramdas
Department of Processing and Food Engineering
Kelappaji College of Agricultural Engineering and Food Technology
Tavanur, Kerala, India

Pratiksha Dadasaheb Sabale
Dr. D. Y. Patil Institute of Pharmaceutical Sciences and Research
Pimpri, Pune, India
Abid SalatiGlobal Business School and Research Centre
Dr. D. Y. Patil Vidyapeeth, Pune
Tathawade, Pune, Pimpri-Chinchwad, Maharashtra, India

Rajesh Sannegadu
University of Mauritius
Réduit, Mauritius

Deepika Ghanasham Sarode
Dr. D. Y. Patil Institute of Technology
Pimpri, Pune, India

S. Sathish Kumar
Department of Artificial Intelligence and Machine Learning
JB Institute of Engineering and Technology
Hyderabad, Telangana, India

Komal Saxena
Amity Institute of Information and Technology
Amity University
Noida, Uttar Pradesh, India

Vibha Manish Saxena
Dr. D. Y. Patil Institute of Pharmaceutical Sciences and Research
Pimpri, Pune, India

S. Selvakanmani
Department of Information Technology
R.M.K. Engineering College
Kavaraipettai, Tiruvallur District, Tamil Nadu, India

R. Senthilkumar
Department of Processing and Food Engineering
Kelappaji College of Agricultural Engineering and Food Technology
Tavanur, Kerala, India

Nazeer Shaik
Department of Computer Science and Engineering
Srinivasa Ramanujan Institute of Technology (SRIT)-Autonomous
Anantapur, Andhra Pradesh, India

Naim Shaikh
Global Business School and Research Centre
Dr. D. Y. Patil Vidyapeeth, Pune
Tathawade, Pune, Pimpri-Chinchwad, Maharashtra, India

Manoj Sharma
Department of Electronics and Communication Engineering
Giani Zail Singh Campus College of Engineering and Technology, MRSPTU
Bathinda, Punjab, India

Parveen Kumar Sharma
DIT University
Dehradun, Uttarakhand, India

Shivangi Sharma
Department of Social Science
B.K Birla Public School Kalyan
Mumbai, India

Swati Sharma
Dr. D. Y. Patil School of Science and Technology
Dr. D. Y. Patil Vidyapeeth, Pune
Tathawade, Pune, India

Shubhangi Ravindra Shelar
Dr. D. Y. Patil Institute of Pharmaceutical Sciences and Research
Pimpri, Pune, India

Pragya Singh
Symbiosis Centre for Management Studies NOIDA
Symbiosis International University
Pune, Maharashtra, India

Jalari Somasekar
Faculty of Engineering and Technology, Department of Computer Science and Engineering
Jain (Deemed-to-be University)
Bangalore, Karnataka, India

M. Srivani
Department of Computer Science and Engineering (DS)
CVR College of Engineering
Ibrahimpatnam, Hyderabad, Telangana, India

P. S. Sruthi
Department of Processing and Food Engineering
Kelappaji College of Agricultural Engineering and Food Technology
Tavanur, Kerala, India

P. B. Sruthy
Department of Processing and Food Engineering
Kelappaji College of Agricultural Engineering and Food Technology
Tavanur, Kerala, India

Subramanya Sarma S
Ramachandra College of Engineering
Eluru, Andhra Pradesh, India

Ajith Sundaram
Amrita School of Business
Amrita Vishwa Vidyapeetham, Kochi, Kerala, India

Manyam Thaile
Department of Information Technology
School of Engineering, Anurag University
Hyderabad, Telangana, India

M. Tholkapiyan
Department of Civil Engineering
Chennai Institute of Technology
Chennai, Tamil Nadu, India

Amit Verma
Department of Management
Symbiosis International (Deemed University)
Pune, Maharashtra, India
and
Symbiosis International University
Dubai, United Arab Emirates

Part I

Exploring Precision Agriculture

Chapter 1

Precision agriculture

Leveraging technology for efficient farming

Rekha Jain and Rakhi Mutha

1.1 INTRODUCTION TO PRECISION AGRICULTURE

In the 21st century, agriculture faces challenges such as climate change, limited natural resources, and an increasing global population. Traditional farming methods, while effective in the past, are no longer sufficient to meet modern demands for food security, sustainability, and economic viability. This is where precision agriculture (PA) steps in. PA involves the use of advanced technologies to manage crops and livestock by optimizing inputs such as water, fertilizers, and pesticides, improving yield, efficiency, and sustainability.

In this chapter, we explore the key aspects of precision agriculture, its core technologies, and how it is transforming modern farming practices.

PA, also known as satellite farming or site-specific crop management, is a farming management concept that uses data analysis and technology to ensure crops and soil receive exactly what they need for optimum health and productivity. This method relies on GPS, remote sensing, and Internet of Things (IoT) to monitor and apply data-driven insights to farming, allowing farmers to make precise, real-time decisions to improve yield and sustainability [1].

1.1.1 Key components of precision agriculture

Soil Monitoring: Soil conditions vary across a field, affecting crop growth. PA employs soil sensors to monitor pH, moisture, and nutrient levels. These insights allow farmers to apply fertilizers or water only where needed, reducing waste and costs. For example: In a 100-acre wheat field, soil sensors may reveal that a section has low nitrogen levels. Instead of spreading nitrogen uniformly, the farmer applies it only to that specific section, conserving resources.

Remote Sensing and Imagery: Drones, satellites, and aircraft collect images in infrared and visible light spectrums, detecting plant stress, pest infestations, or water deficiencies before they are visible to the human eye. For example: A vineyard may use drones to capture aerial images, identifying areas where vines are under stress due to a lack of water [2]. Irrigation can then be directed to those areas, ensuring uniform grape quality without wasting water on well-irrigated sections.

Yield Monitoring and Mapping: Harvesters are equipped with sensors that measure crop yield in real time. This data helps farmers create yield maps, showing which areas of a field produce more and which less, helping inform planting and fertilizing strategies for the next season. For example: A corn farmer uses a yield monitor to determine that one part of his field consistently produces lower yields. By adjusting seeding rates or applying more nutrients to this area in future plantings, the farmer can potentially boost productivity.

DOI: 10.1201/9781003520733-2

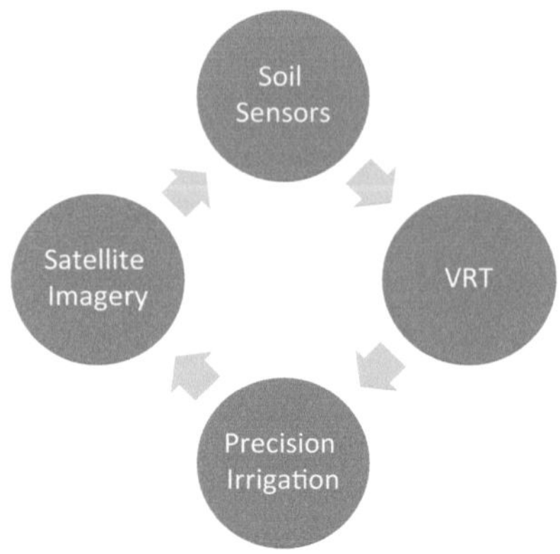

Figure 1.1 Precision agriculture implementation.

Variable Rate Technology (VRT): VRT allows the application of inputs—such as seeds, fertilizers, and pesticides—in varying amounts across a field. This ensures that resources are used only where necessary, reducing costs and environmental impact. For example: On a cotton farm, sensors and GPS data might show that certain areas need more pesticide application due to higher pest density, while other areas need less or none. VRT allows pesticides to be applied only where needed, reducing chemical usage and costs.

Precision Irrigation: Instead of watering entire fields uniformly, precision irrigation uses sensors and drip systems to target areas with the exact water volume needed [3]. This conserves water and ensures that plants do not get overwatered, which can lead to disease or root issues. For example: In a greenhouse, tomatoes in different rows may require different amounts of water. Precision irrigation systems control the amount each row receives based on plant needs, conserving water and ensuring healthy growth.

PA is revolutionizing the farming industry (Figure 1.1) by making it more data-driven and sustainable, paving the way for a more efficient and resilient agricultural future [4].

1.2 SOIL SENSORS

Soil sensors are devices used in PA to monitor soil conditions, such as moisture, temperature, pH, nutrient levels, and EC [5]. These sensors help farmers understand the variability in their fields and make informed, data-driven decisions. Soil sensors allow for more precise irrigation, fertilization, and planting, saving water and nutrients while boosting crop yield and health.

1.2.1 Types of soil sensors and their functions

Moisture Sensors: Measure the volumetric water content in soil, indicating if irrigation is needed or if there is enough moisture. This data helps farmers avoid both under- and over-watering crops, conserving water and preventing waterlogging. For example: A moisture sensor in a vegetable farm might indicate that the soil in one area is dry, prompting the farmer to irrigate only that area instead of the entire field. This saves water and prevents overwatering in already wet zones.

Nutrient Sensors: Detect the presence of key nutrients like nitrogen (N), phosphorus (P), and potassium (K), which are essential for crop growth. This allows farmers to apply fertilizers only where necessary, reducing costs and environmental impact [6]. For example: On a corn farm, nutrient sensors reveal that nitrogen levels are low in specific field sections. The farmer applies nitrogen fertilizer only to those areas, ensuring healthy crop growth while reducing fertilizer use and costs.

pH Sensors: Measure the acidity or alkalinity of the soil, which is crucial since most crops thrive within specific pH ranges. Knowing the soil pH helps farmers amend the soil with lime or sulfur as needed to optimize conditions for crop growth. For example: A farmer growing tomatoes might use pH sensors to find acidic zones in their field. Adding lime to these areas helps bring the pH into a range favorable for tomato growth, leading to better yield and crop health.

Temperature Sensors: Monitor soil temperature, important for germination and growth. Different crops require specific soil temperatures to thrive, so this information can guide planting schedules. For example: A rice farmer might wait to plant until soil sensors indicate that the temperature has reached an optimal range, ensuring rapid germination and uniform plant growth.

Electrical Conductivity (EC) Sensors: Measure the soil's ability to conduct electricity, which correlates with soil salinity and nutrient availability. High salinity or nutrient concentration affects water uptake and can lead to toxicity for plants, so monitoring EC helps manage soil amendments effectively. For example: An EC sensor on a wheat farm might show high salinity in certain areas. The farmer can then treat these areas by leaching salts or applying gypsum, ensuring healthier crop growth and preventing reduced yields.

Various studies have been conducted in the field of soil sensors [7] and PA [8], highlighting the importance of utilizing sensor technology to enhance agricultural practices.

1.3 SATELLITE IMAGERY

Satellite imagery is a core tool in PA, offering farmers a comprehensive, real-time view of their fields from space. This technology allows for the monitoring of crop health, soil condition, water stress, and potential pest issues across large areas without physically visiting every part of the field. By analyzing multispectral and hyperspectral images, farmers gain actionable insights that can optimize input use, reduce costs, and improve yield.

1.3.1 Key aspects of satellite imagery in precision agriculture

Multispectral and Hyperspectral Imaging: Satellites capture images in multiple wavelengths, not just the visible spectrum. Multispectral images typically include near-infrared (NIR) and red-edge bands, which help detect chlorophyll levels in plants, a key indicator of health and stress. Hyperspectral imaging, with more bands, provides even more detailed information on crop health and soil conditions. For example: A corn farmer analyzes NIR data, which highlights chlorophyll-rich areas. If certain sections show lower chlorophyll levels, indicating stressed plants, the farmer can target those areas with extra care, such as adding water or fertilizer.

Vegetation Indices (VIs): Satellite imagery is often processed into indices like the Normalized Difference Vegetation Index (NDVI), which measures plant health based on NIR and red-light reflection. High NDVI values indicate healthy vegetation, while lower values suggest stress or pest infestations. For example: An NDVI map of a wheat field reveals areas with low values, suggesting nutrient deficiencies or water stress. The farmer can then investigate these specific areas and apply corrective measures only where needed [9].

Water Stress Detection: Water availability is crucial for crop health, and satellite imagery helps detect areas with water stress. By analyzing thermal data (capturing surface temperature), farmers can identify zones where plants may be struggling due to insufficient moisture. For example: A vineyard manager uses thermal images to find hotter, drier patches within the field, indicating water stress. The manager then targets these areas for additional irrigation, improving grape quality and conserving water.

Soil Condition Analysis: Using data from specific wavelengths, satellite imagery can help evaluate soil moisture levels, texture, organic matter, and nutrient availability. This information is invaluable for pre-planting decisions and ongoing management of crop health. For example: Before planting a rice crop, a farmer studies soil moisture data from satellite imagery, choosing areas with the highest moisture retention for rice, which thrives in wet soil. This ensures a good crop start without unnecessary water input [10].

Pest and Disease Monitoring: Satellite imagery can identify areas where crops exhibit stress due to pests or diseases. Changes in color and reflectance patterns alert farmers to potential problems before they are visible to the naked eye, allowing for early intervention. For example: In a soybean field, satellite images reveal unusual reflectance in certain patches, indicating pest activity. This information helps the farmer apply pest control measures in these spots, saving the rest of the field from infestation.

Yield Prediction: By analyzing patterns in VIs throughout the growing season, farmers can use satellite data to forecast yield potential [11]. This helps with supply chain planning and managing expectations for harvest quantities. For example: A farmer growing barley observes high NDVI values in most areas of the field. This suggests a robust crop, allowing the farmer to plan for storage and marketing strategies ahead of harvest [12].

1.3.2 Advantages of satellite imagery in precision agriculture

- **Large-Scale Monitoring:** Provides an overview of large fields in one image, enabling efficient management without on-the-ground labor.
- **Real-Time Updates:** Near real-time data helps farmers respond quickly to issues like water stress, nutrient deficiencies, or pest outbreaks [13].
- **Cost Savings:** Reduces unnecessary applications of fertilizers, pesticides, and water by targeting resources only where needed.
- **Environmental Benefits:** Minimizes chemical runoff, conserves water, and reduces the carbon footprint by using fewer resources.

Satellite imagery is thus a powerful tool in PA, enabling a more data-driven, efficient, and sustainable approach to farming.

1.4 VARIABLE RATE TECHNOLOGY (VRT)

VRT is a key component of PA that allows farmers to apply inputs—such as seeds, fertilizers, pesticides, and water—at varying rates across different parts of a field. Instead of applying inputs uniformly, VRT uses data-driven insights to deliver precise amounts only where needed, optimizing resource use, reducing costs, and improving crop productivity [14]. VRT relies on GPS, sensors, mapping software, and application equipment to adjust input rates automatically, adapting to the specific needs of each zone within a field.

1.4.1 Components and types of VRT

Variable Rate Seeding (VRS): Adjusts the seed planting rate according to soil and environmental conditions. In areas with high productivity potential, a higher seeding rate can be used, while in areas with lower productivity, the seeding rate is reduced. For example: A farmer planting corn in a 50-acre field with soil variability might use VRS to plant more seeds per acre in rich, fertile sections while planting fewer seeds in areas with poorer soil, ensuring optimal growth and reducing seed costs.

Variable Rate Fertilization (VRF): Fertilizer application rates are adjusted based on nutrient availability in different zones. This helps avoid over-fertilization, which can harm the environment, and under-fertilization, which can reduce yields. For example: On a wheat farm, soil nutrient maps created through sensors or soil testing reveal nitrogen levels in different areas. Using VRF, the farmer applies more nitrogen to areas with low levels, while reducing it in zones where nitrogen is sufficient, balancing crop health and reducing waste.

Variable Rate Irrigation (VRI): This type of VRT allows farmers to apply water variably based on soil moisture, crop type, and weather forecasts [15]. VRI is especially beneficial in water-scarce areas where water conservation is essential. For example: In a vineyard, moisture sensors detect that some areas of the field are drier than others. Using VRI, the farmer irrigates only those areas, ensuring that vines receive adequate water without over-irrigating and wasting water in already wet sections.

Variable Rate Pesticide Application: Pesticides are applied selectively to control pests only where they are detected or likely to develop, reducing chemical usage and minimizing environmental impact. For example: A soybean farmer receives data from pest monitoring sensors showing a higher concentration of aphids in certain areas [16]. The VRT sprayer applies pesticides only in these areas, conserving chemicals and reducing environmental impact.

1.4.2 How VRT works

VRT systems require several components to function effectively, such as:

- **Data Collection:** Information on soil health, moisture levels, nutrient distribution, crop condition, and pest presence is gathered using sensors, satellite imagery, soil testing, and yield monitors. This data is mapped to show variability across the field [17].
- **Mapping and Zoning:** Fields are divided into management zones based on the data collected. Each zone represents areas with similar characteristics, such as soil type, nutrient levels, or moisture content.
- **Prescription Maps:** Farmers and agronomists use mapping software to create "prescription maps" based on the zones. These maps guide the VRT equipment, telling it the exact amount of input to apply in each area.
- **VRT Equipment:** Specialized equipment, like VRT-enabled sprayers, seeders, and irrigation systems, interprets the prescription map in real time. This machinery adjusts the input rate as it moves across the field, following GPS guidance to apply inputs accurately.

1.4.3 Advantages of VRT

- **Cost Savings:** By applying inputs only where needed, VRT reduces the amount of seeds, fertilizers, pesticides, and water used, saving money.
- **Increased Yield:** By matching input levels to crop requirements, VRT maximizes crop health and productivity, leading to higher yields.
- **Environmental Benefits:** VRT minimizes chemical runoff, nutrient leaching, and water overuse, leading to more sustainable farming practices.
- **Efficient Resource Use:** VRT optimizes input application, ensuring each part of the field receives exactly what it needs, reducing waste.

1.4.4 Challenges of VRT

- **High Initial Cost:** VRT systems, including mapping software, sensors, and specialized equipment, can be expensive.
- **Data Management Complexity:** VRT relies on large amounts of data, which require analysis and skilled interpretation.
- **Technical Skill Requirements:** Farmers may need training to operate VRT equipment effectively and to understand data-driven decision-making.

VRT plays a critical role in modern PA, making farming more efficient, sustainable, and profitable by aligning input application with field variability.

1.5 KEY TECHNOLOGIES IN PRECISION AGRICULTURE

The success of PA lies in its integration of multiple cutting-edge technologies [18]. These tools provide farmers with actionable data and enable precise management [19]. Some of the core technologies include:

1.5.1 Geographic Information System (GIS) and GPS in precision agriculture

GIS and Global Positioning System (GPS) play an essential role in modern PA by helping farmers collect, map, and analyze highly detailed data about their fields. These technologies enable farmers to make decisions based on precise, location-specific data, leading to more efficient use of resources and improved crop yield.

1.5.1.1 Geographic Information System (GIS)

GIS allows farmers to gather and visualize data about various parameters across their fields. Here is how it works in agriculture:

- **Field Mapping:** GIS helps farmers create detailed, layered maps of their fields. These maps can represent a wide range of data, such as soil type, elevation, water flow patterns, and crop health. For example, a soil map could show areas with varying nutrient levels, while a crop health map might highlight regions experiencing stress due to pests or lack of nutrients.
- **Data Layering:** Farmers can combine multiple layers of information, such as soil pH, moisture levels, and historical yield data, to get a complete view of field conditions [20]. This allows for targeted interventions, such as applying fertilizer only where it is most needed, reducing waste and costs.
- **Decision-Making:** GIS enables farmers to see patterns and correlations in their fields. For instance, they may discover that certain areas have consistently low yields and require specific improvements, like drainage or additional nutrients.

1.5.1.2 Global Positioning System (GPS)

GPS technology provides farmers with real-time, location-specific tracking of machinery and equipment. This helps in several ways:

- **Precise Equipment Tracking:** GPS-equipped tractors and other farming machinery can navigate fields with pinpoint accuracy, reducing human error. This precision is especially useful in large fields where manually tracking every detail would be challenging.
- **Optimized Planting and Fertilizing:** GPS enables automated guidance systems to control the movement of tractors, seeders, and fertilizer spreaders, ensuring consistent spacing and alignment. This minimizes overlap and gaps, meaning seeds and fertilizers are distributed more uniformly across the field.
- **Efficient Harvesting:** GPS also aids in the efficient harvesting of crops by helping harvesters track previously covered ground, ensuring that no sections are missed or harvested twice. This reduces fuel costs, labor time, and wear on equipment.

1.5.1.3 Combined benefits of GIS and GPS in precision agriculture

Together, GIS and GPS allow for site-specific farming, where every part of the field is managed according to its unique conditions [21]. This data-driven approach improves resource utilization, reduces waste, and can lead to healthier crops and higher yields by ensuring each area of the field receives precisely what it needs.

In essence, GIS and GPS make PA feasible, supporting farmers in making data-informed decisions that drive sustainability and profitability.

1.5.2 Remote sensing and drones in precision agriculture

Remote sensing and drone technologies have become transformative tools in PA, allowing farmers to monitor crop health, soil conditions, and pest issues with unprecedented detail and efficiency [22]. These technologies provide valuable insights into crop conditions, enabling proactive and targeted interventions.

1.5.2.1 Drones in precision agriculture

Drones, or unmanned aerial vehicles, are equipped with cameras and sensors to capture high-resolution images and data from above. They provide real-time information about field conditions that would otherwise be difficult or time-consuming to collect manually.

- **High-Resolution Imaging:** Drones can fly over fields and capture images in extremely high resolution. These images show detailed variations in crop health, soil conditions, and moisture levels. For example, drones can detect areas of crop stress, which might indicate insufficient water, nutrient deficiency, or pest damage.
- **Multispectral and Thermal Imaging:** Equipped with multispectral or thermal cameras, drones can capture data that is not visible to the naked eye. Multispectral imaging can identify variations in leaf color and chlorophyll levels, which often signal plant health and vigor [23]. Thermal imaging helps monitor soil and plant temperatures, which can indicate moisture content and even highlight areas of water stress.
- **Pest and Disease Detection:** Drones can quickly cover large areas to locate signs of pest infestations or diseases. Early detection enables farmers to address problems promptly, applying pest control measures only where needed, thus reducing pesticide use and improving crop health.
- **Cost-Effectiveness and Accessibility:** Drone flights are more affordable than traditional aerial methods, like manned flights, allowing for regular monitoring. They provide timely information, enabling farmers to respond quickly to changes in the field.

1.5.2.2 Remote sensing in precision agriculture

Remote sensing uses data collected from satellites or aircraft to gather insights over large agricultural areas, providing broad, consistent coverage over time.

- **Vegetation Indices**: Remote sensing measures VIs, such as the NDVI, to assess plant health. NDVI readings show the density and vigor of plant growth across a field by capturing differences in light absorption and reflection [24]. This helps farmers assess overall crop health and detect any stressed areas.
- **Soil Moisture and Temperature Monitoring**: Satellite-based remote sensing can measure soil moisture and temperature across large fields, indicating where irrigation may be necessary or highlighting areas vulnerable to drought. Knowing which areas are dry or prone to water stress helps farmers optimize irrigation, using water more efficiently.
- **Scalability and Long-Term Monitoring**: Unlike drones, which are effective for frequent, smaller-scale monitoring, remote sensing allows for consistent observation over vast areas, making it valuable for large farms or regional assessments. It also provides a long-term view, capturing trends over time that can inform crop rotation, field management, and other strategic decisions.

1.5.2.3 Benefits of combining drones and remote sensing

Together, drones and remote sensing technologies allow for multiscale monitoring of agricultural fields, where both immediate, detailed data and larger-scale trends are accessible. This dual approach gives farmers the ability to

- **Detect Problems Early**: Both drones and remote sensing help catch issues like crop stress, water deficiency, or pest outbreaks early, allowing farmers to take corrective measures before the problem spreads.
- **Optimize Inputs**: These technologies support efficient use of resources by guiding where fertilizers, pesticides, and water are needed, reducing waste and improving cost-effectiveness.
- **Increase Yields and Quality**: With timely, accurate data from both drones and remote sensing, farmers can implement precise interventions, improving crop yields and quality, leading to better profitability and sustainability.

In summary, drones and remote sensing are essential for modern PA, offering real-time, detailed insights that enhance crop management, resource efficiency, and productivity. Together, they enable a proactive approach to farming that is responsive to the specific needs of each field.

1.5.3 IoT and smart sensors in precision agriculture

The IoT has become integral to PA by connecting various smart devices and sensors across the farm, providing farmers with real-time data for more accurate decision-making [24]. These sensors capture key metrics related to soil, weather, and livestock, which enables precise resource management, improved crop yields, and healthier livestock.

1.5.3.1 Soil and crop sensors

Soil and crop sensors play a vital role in monitoring field conditions to optimize crop growth and resource application.

- **Soil pH and Nutrient Monitoring:** Soil pH sensors measure the acidity or alkalinity of the soil, which affects nutrient availability to crops. When combined with nutrient sensors, these devices allow farmers to detect nutrient deficiencies in real time and apply targeted fertilizers only where needed [25].
- **Moisture and Temperature Sensors:** Soil moisture sensors track water content, ensuring that plants receive adequate hydration without over-irrigating, which can lead to root rot or nutrient leaching. Temperature sensors, meanwhile, help farmers monitor soil and air temperatures, which influence plant growth cycles. By integrating these readings, farmers can adjust irrigation schedules to match weather patterns and soil needs, improving water efficiency and promoting healthy crop growth.
- **Smart Fertilization and Irrigation:** By gathering data on soil and crop conditions, IoT sensors enable variable-rate fertilization and irrigation. This means farmers can apply just the right amount of fertilizer and water to each area of the field, reducing waste, minimizing environmental impact, and enhancing crop quality [26].

1.5.3.2 Weather stations

On-farm weather stations equipped with IoT sensors provide farmers with precise, hyper-local weather data, essential for timing and adjusting farm operations according to real-time conditions.

- **Wind Speed and Direction Monitoring:** Wind speed data helps farmers determine safe times for pesticide application to avoid drift and improve effectiveness. It is also important for greenhouse management and irrigation, as high winds can accelerate soil moisture loss.
- **Rainfall Tracking:** Rainfall sensors provide insight into recent precipitation levels, helping farmers make informed decisions on irrigation. After a heavy rain, irrigation schedules may be adjusted to avoid over-watering, which conserves water and prevents soil erosion.
- **Humidity and Temperature Readings:** Humidity sensors allow farmers to anticipate dew points and the likelihood of fungal diseases, which are more likely to spread in moist conditions [27]. Temperature readings inform planting schedules and enable better pest and disease management by indicating when certain insects may be more active.
- **Customized Microclimate Data:** These weather stations capture data specific to a farm's microclimate rather than relying on regional weather reports, which can often miss subtle climate variations within a single farm. Farmers can use this data to better plan planting, harvesting, and crop rotation.

1.5.3.3 Livestock monitoring sensors

IoT sensors are also widely used in livestock farming to monitor animal health, improve welfare, and streamline farm operations.

- **Animal Health Monitoring:** Sensors attached to animals, like collars or ear tags, can track vital signs, body temperature, and movement patterns, allowing farmers to detect early signs of illness [28]. This helps in minimizing disease spread and reducing veterinary costs by allowing for early intervention.
- **Tracking Movement and Activity:** Movement sensors help monitor grazing patterns and identify animals that may be under stress or experiencing illness, as less active animals may be sick. This data assists in managing grazing schedules and maintaining pasture health.

- **Feeding and Watering Patterns:** IoT sensors can also monitor feeding and watering behavior, ensuring that each animal is consuming adequate food and water. Deviations from normal patterns can signal health issues, while tracking feed intake helps farmers reduce waste and optimize feed costs.

1.5.3.4 Benefits of IoT and smart sensors in precision agriculture

Together, IoT and smart sensors enable an automated, data-driven approach to managing farms, from optimizing crop growth to ensuring animal welfare [14]. Key benefits include

- **Resource Efficiency:** With real-time data, farmers can optimize water, fertilizer, and pesticide use, conserving resources and reducing environmental impact.
- **Reduced Labor:** Automated systems for monitoring soil, crops, and livestock reduce the need for manual labor, freeing up time for farmers to focus on strategic planning.
- **Enhanced Yield and Quality:** By adjusting farming practices to specific conditions in each field or barn, farmers can increase crop yields, improve product quality, and maintain healthy livestock [29].

IoT and smart sensors allow for a more informed, responsive, and efficient approach to farming, ultimately leading to higher productivity and sustainability in agriculture [30,31].

1.5.4 Data analytics and artificial intelligence (AI) in precision agriculture

Data analytics and AI have become essential components in modern PA, helping farmers make data-driven decisions by analyzing the extensive data generated by sensors, drones, and other IoT devices on the farm. These technologies allow farmers to uncover valuable insights from complex data, improving productivity, sustainability, and profitability [32,33].

1.5.4.1 Predictive analytics in precision agriculture

Predictive analytics uses both historical data and real-time inputs to forecast future agricultural trends and optimize farm management practices. This approach leverages statistical algorithms and machine learning (ML) models to anticipate various factors critical to farm operations.

- **Crop Yield Forecasting:** Predictive models analyze historical yield data along with current conditions like soil fertility, moisture levels, and climate patterns to estimate future crop yields. This information allows farmers to adjust input levels and make early planning decisions, such as choosing the best time for harvesting to maximize output.
- **Pest and Disease Outbreak Prediction:** Predictive analytics can help detect early signs of pest infestations or disease outbreaks by analyzing environmental factors like temperature, humidity, and plant stress indicators. By forecasting these outbreaks, farmers can implement preventive measures or apply treatments more efficiently, reducing the need for extensive pesticide use and protecting crop health [26].
- **Optimal Planting and Harvesting Times:** By analyzing weather patterns, soil conditions, and crop development stages, predictive models suggest the most favorable times for planting and harvesting. This optimizes crop growth and yield potential, ensuring that crops are harvested when they are at peak quality and nutritional value.

1.5.4.2 *Machine learning in precision agriculture*

ML, a subset of AI, enables computers to learn from data without explicit programming, making it well-suited for analyzing the diverse and often complex datasets found in agriculture. In PA, ML models continuously improve by learning from new data, refining their ability to identify trends and make recommendations.

- **Pattern Recognition:** ML models analyze data from multiple sources, like soil sensors, drones, and weather stations, to detect patterns that may be difficult for humans to identify. For example, ML can uncover correlations between soil moisture and crop growth rates, guiding optimal irrigation strategies for maximum efficiency.
- **Efficient Irrigation Scheduling:** ML can analyze real-time data, such as soil moisture levels, plant growth stages, and weather conditions, to recommend precise irrigation schedules. This ensures that crops receive the right amount of water when needed, reducing water waste and avoiding over-irrigation, which can damage crops and reduce yield.
- **Pest Control Optimization:** By identifying patterns in pest activity, ML models can suggest targeted pest control measures based on pest location, type, and life cycle. This data-driven approach minimizes pesticide use, which reduces costs and environmental impact while enhancing crop quality.

1.5.4.3 *Benefits of data analytics and AI in agriculture*

The integration of data analytics and AI offers numerous advantages for farmers, allowing them to optimize farming operations, reduce costs, and improve yields through precision and insight:

- **Enhanced Decision-Making:** AI and data analytics provide farmers with timely, actionable insights, supporting better decision-making for critical aspects of crop and livestock management. Instead of relying on intuition, farmers can use data to back their decisions, increasing accuracy and efficiency.
- **Proactive Farm Management:** Predictive models help farmers anticipate challenges, such as drought or disease outbreaks, and take preventive actions rather than reactive ones. This proactive approach reduces risks and keeps farm operations more consistent and resilient.
- **Cost and Resource Efficiency:** By analyzing data on input usage, weather patterns, and field conditions, AI helps farmers allocate resources—such as water, fertilizers, and labor—where they are most needed. This results in optimized use of resources, lowering input costs and boosting farm sustainability.
- **Real-Time Monitoring and Adjustments:** AI-driven systems monitor crop conditions, soil health, and other variables in real time, enabling adjustments to be made immediately [34]. For example, a sudden change in weather may trigger adjustments in irrigation or fertilization, helping to protect crops and maintain growth rates.

1.5.4.4 *How data analytics and AI work in practice*

The effective use of data analytics and AI in agriculture typically follows these steps:

- **Data Collection:** Sensors, drones, and satellite imagery gather data on soil conditions, crop health, weather, and other variables.

- **Data Processing:** This raw data is cleaned, organized, and stored in a structured format so it can be analyzed.
- **Analysis and Modeling:** ML and statistical algorithms analyze the data, identifying patterns, correlations, and trends.
- **Recommendations and Decision Support:** Based on the analysis, AI models provide recommendations on actions like fertilizing, watering, pest control, and harvesting, optimizing farm practices.
- **Continuous Improvement:** As more data is collected, the models learn and improve, refining predictions and recommendations over time.

Data analytics and AI bring powerful tools to PA, turning vast datasets into meaningful insights that enhance farm management. By enabling farmers to predict outcomes, optimize inputs, and respond to real-time data, these technologies make modern farming more efficient, sustainable, and profitable.

1.6 APPLICATIONS OF PRECISION AGRICULTURE

PA applies advanced technology to enhance crop productivity, efficiency, and sustainability. By using GPS, IoT sensors, drones, data analytics, and AI, farmers can monitor soil health, optimize irrigation, control pests, and manage crop health with precision. These technologies allow targeted application of fertilizers, pesticides, and water, reducing waste and environmental impact while maximizing yields. PA also includes livestock management, where sensors monitor animal health and behavior. Together, these applications enable data-driven decision-making, helping farmers adapt to changing conditions, improve resource use, and increase profitability in a sustainable way.

1.6.1 Precision irrigation in agriculture

Precision irrigation is a key component of PA, designed to optimize water usage in farming by delivering the exact amount of water needed for crops at the right time. Given that water is one of the most essential and limited resources in agriculture, precision irrigation aims to improve water-use efficiency, reduce waste, and ensure healthy crop growth, especially in water-scarce regions.

1.6.1.1 Drip and micro-irrigation systems

Drip and micro-irrigation are advanced techniques in precision irrigation that allow water to be delivered directly to the plant's root zone, minimizing water loss through evaporation and runoff.

- **Drip Irrigation:** In drip irrigation, a network of pipes with small holes or emitters is laid close to the plants' root zones. Water drips slowly from these emitters, ensuring that each plant receives adequate moisture directly at the root level. By targeting only the plant's base, drip irrigation prevents water from spreading across unnecessary areas, reducing evaporation and runoff, which is especially beneficial in dry climates.
- **Micro-Irrigation:** This method includes various micro-sprinklers or mini-sprayers that release water in a controlled, low-volume manner. Micro-irrigation is often used for crops that require broader water coverage, as it maintains a balance between water efficiency and coverage by delivering small amounts to specific areas within the root zone.

Both systems are known for their efficiency in water conservation, as they use less water than traditional sprinkler or flood irrigation. They also support nutrient application (fertigation) through irrigation lines, delivering both water and nutrients simultaneously to improve plant growth.

1.6.1.2 Automated irrigation systems

Automated irrigation systems leverage real-time data to precisely control the timing and volume of water applied to crops. This automation can adjust water distribution based on soil moisture, weather conditions, and crop growth stages, ensuring that crops receive the optimal amount of water when they need it most.

- **Real-Time Data Integration:** Automated irrigation systems are equipped with **soil moisture sensors** that monitor the water content in the soil. When soil moisture drops below a certain threshold, the system automatically triggers irrigation to maintain optimal hydration levels [35]. This reduces water stress on crops, which can stunt growth and decrease yields.
- **Weather Data Utilization:** Some automated systems integrate weather forecasts and local climate data to adjust irrigation schedules. For instance, if rain is predicted, the system can hold off on watering, preventing over-irrigation and water wastage. Additionally, automated systems adjust for factors like temperature and humidity, providing more water during hot, dry periods and less during cooler, humid times.
- **Customized Watering for Crop Stages:** As crops progress through various growth stages, their water requirements change. Automated systems can be programmed to deliver different water quantities based on growth stages, ensuring efficient resource use. For example, certain plants may require more water during flowering or fruiting stages and less during early vegetative growth.

1.6.1.3 Benefits of precision irrigation

Precision irrigation offers numerous benefits, making it a critical practice in sustainable agriculture:

- **Water Conservation:** By only delivering water where and when it is needed, precision irrigation reduces water wastage and lowers overall water consumption, especially in arid regions.
- **Increased Crop Yield and Quality:** Ensuring that crops receive adequate water improves their growth and productivity, leading to higher-quality yields and healthier plants.
- **Reduced Soil Erosion and Nutrient Runoff:** Controlled water application minimizes the risk of soil erosion and nutrient leaching, which can occur with excessive water flow in traditional irrigation systems.
- **Lower Costs:** Although initial setup costs can be higher, precision irrigation reduces long-term water and nutrient costs by preventing overuse.
- **Sustainability:** By optimizing water use and reducing environmental impact, precision irrigation supports more sustainable farming practices that are resilient to climate variability.

Precision irrigation, through drip and micro-irrigation systems and automated real-time monitoring, revolutionizes water management in agriculture [36]. By targeting specific water needs and adjusting to real-time data, it supports water conservation, enhances crop health, and helps farmers achieve higher yields with less environmental impact.

1.6.2 Precision planting in agriculture

Precision planting is a critical practice in PA that involves planting seeds at the ideal depth, spacing, and rate to maximize crop yield potential. By controlling these variables, precision planting helps farmers ensure that each seed has the optimal conditions for germination, growth, and resource access. This technique uses advanced equipment and data analytics to enhance planting accuracy, minimize seed waste, and improve overall field productivity.

1.6.2.1 Seed mapping

Seed mapping uses GPS-enabled planters and advanced software to guide the precise placement of each seed. This technology ensures that seeds are planted in exact locations with accurate depth and spacing, which is essential for achieving high yield and crop uniformity.

- **GPS Guidance:** With GPS technology, planters can follow a specific route mapped out for planting. This allows seeds to be spaced perfectly within rows, reducing competition for nutrients, water, and sunlight among plants. GPS guidance minimizes the likelihood of double-planting (where two seeds are placed in one spot) or gaps (where no seeds are planted), both of which can negatively impact crop yield.
- **Correct Depth and Spacing:** Seed mapping ensures that each seed is placed at the proper depth and distance from others. **Depth** affects how well a seed absorbs moisture and resists environmental stress, while **spacing** impacts how much space each plant has to grow. For instance, seeds planted too close may compete for resources, leading to smaller plants and lower yields, while seeds planted too deeply may struggle to germinate.
- **Minimizing Competition:** Seed mapping prevents plants from competing for the same resources by ensuring consistent spacing between each seed. For crops like corn, soybeans, and wheat, where plant density directly impacts productivity, precise spacing is crucial for maximizing yields and achieving uniform crop development.

1.6.2.2 Automated planters

Automated planters use data-driven insights to adjust planting patterns and settings in real time, optimizing seed placement according to soil conditions and yield goals [37]. These advanced machines are equipped with sensors, data analytics tools, and GPS technology, making them highly adaptable to changing field conditions.

- **Soil-Based Adjustments:** Automated planters analyze soil data, such as soil type, moisture level, and compaction, to make adjustments during planting. For example, in areas with compacted soil, the planter might increase planting depth to ensure seeds reach less compacted soil layers for better germination. Conversely, in looser, nutrient-rich soil, seeds may be placed at a shallower depth, promoting faster growth and efficient nutrient uptake.
- **Variable Seeding Rates:** Automated planters can vary the seed rate based on different sections of the field. In highly fertile zones, the machine may plant seeds at a higher density to take advantage of the soil's nutrient-rich properties, increasing yield. In less fertile areas, it may reduce seed density to prevent overcrowding and reduce input waste. This technique, known as variable-rate seeding, maximizes productivity across varying field conditions.

- **Real-Time Adaptation:** With real-time monitoring capabilities, automated planters can respond to unforeseen obstacles or changes in soil conditions during planting. For instance, if sensors detect a sudden increase in moisture, the planter can adjust its depth to prevent seeds from sitting in overly wet soil, which can lead to poor germination.
- **Reduced Seed Waste:** Automated planters ensure that seeds are planted only where they are most likely to thrive, reducing seed waste. By planting according to optimal conditions, farmers can save on seed costs, improve germination rates, and minimize the resources spent on seeds that might not contribute to yield.

1.6.2.3 Benefits of precision planting

Precision planting offers numerous advantages that make it a valuable practice in modern agriculture:

- **Improved Germination and Crop Uniformity:** By planting seeds at the ideal depth and spacing, precision planting promotes higher germination rates and ensures uniform crop development. Uniform crops are easier to manage and harvest, and they tend to produce higher yields.
- **Increased Yield Potential:** With each seed positioned for optimal growth, plants can reach their full potential, leading to higher yields. Precision planting also helps farmers achieve better land use efficiency by utilizing each part of the field to its best advantage.
- **Reduced Seed and Input Costs:** By ensuring that only the necessary amount of seed is used, precision planting minimizes waste and reduces the cost of seeds and inputs. Farmers can allocate resources more effectively, lowering overall production costs.
- **Enhanced Resource Use Efficiency:** Precision planting aligns seed placement with soil conditions, allowing for more efficient use of water, nutrients, and sunlight. By preventing overcrowding, it reduces the strain on resources, leading to more sustainable farming practices.
- **Data-Driven Decision-Making:** Precision planting uses data to guide planting decisions, providing farmers with insights that can be used for future planting seasons. The data collected can also help farmers monitor field performance, assess soil health, and improve crop management strategies over time [38].

1.6.2.4 How precision planting works in practice

The precision planting process generally involves the following steps:

- **Data Collection and Analysis:** Data on soil health, moisture, and field topography is collected using sensors and GPS mapping. This data is analyzed to identify optimal planting patterns for different areas of the field.
- **Mapping and Route Planning:** Based on the analysis, a planting map is created that guides the automated planter. The map outlines seed placement, depth, and spacing, optimizing each factor for different field zones.
- **Planting Execution:** Automated planters follow the GPS-guided map, adjusting their depth, spacing, and rate in real time to match field conditions. This ensures that each seed is planted in the ideal location.
- **Post-Planting Monitoring:** Once seeds are planted, the system continues to monitor field conditions, enabling farmers to make any necessary adjustments during the crop's growth stages.

Precision planting, through seed mapping and automated planters, maximizes the efficiency of planting operations by using GPS guidance and real-time data to ensure that each seed has optimal conditions to thrive. This practice not only improves crop yields and uniformity but also reduces costs and resource usage, making it a highly effective approach to sustainable and profitable agriculture.

1.6.3 Crop health monitoring in precision agriculture

Crop health monitoring is a key aspect of PA that allows farmers to continuously assess the condition of their crops and take corrective actions when issues arise. By leveraging technologies such as remote sensing, drones, and data analytics, farmers can detect problems early—such as diseases, nutrient deficiencies, and pest infestations—before they significantly impact crop yields. Continuous crop health monitoring leads to better-targeted treatments, more efficient resource use, and healthier, more productive plants.

1.6.3.1 Disease detection

Early detection of diseases is crucial to maintaining crop health and avoiding yield loss. PA uses advanced imaging technologies, such as remote sensing and drone imagery, to detect subtle signs of crop stress that are often invisible to the naked eye.

- **Remote Sensing Technology:** Remote sensing involves the use of satellite or aerial imagery to detect changes in crop health across vast fields. Through multispectral and **hyperspectral imaging**, remote sensors can capture data across different wavelengths, revealing variations in crop health that may indicate disease. For example, areas where plants show reduced chlorophyll levels may appear different in infrared imaging, signaling early stages of a disease before visible symptoms appear.
- **Drone-Based Imaging:** Drones equipped with high-resolution cameras and specialized sensors can capture detailed images of fields, covering large areas quickly and efficiently. Drones fly at low altitudes, providing clear, close-up views of crops. They can identify early symptoms of diseases such as leaf blight, mildew, or rust, allowing farmers to treat affected areas before the disease spreads further. Drones can also perform regular flyovers to monitor crop health over time, making it easier to detect trends and assess the effectiveness of any treatments applied.
- **Preventative Action:** By identifying disease hotspots early, farmers can apply targeted treatments, reducing the risk of widespread infection and minimizing the need for broad-spectrum pesticides or fungicides. Targeted action helps prevent disease spread, protects healthy plants, and reduces overall chemical usage, contributing to more sustainable farming practices.

1.6.3.2 Pest control

PA technologies also support efficient pest management by identifying and targeting pest hotspots. Rather than applying pesticides across an entire field, precision tools allow for localized treatment, reducing chemical usage and environmental impact.

- **Early Detection of Pest Infestations:** Remote sensing and drone imagery can also detect the early stages of pest infestation. By analyzing crop stress patterns, imaging can reveal areas where pest activity may be occurring. For instance, certain pest infestations cause distinctive patterns of plant damage, such as leaf wilting, discoloration, or holes. By spotting these signs early, farmers can respond promptly to limit the infestation's spread.

- **Targeted Pesticide Application:** With precision pest control, pesticides are applied only where necessary, reducing chemical exposure to healthy plants, soil, and nearby ecosystems. For example, if pest activity is identified in a specific section of the field, the farmer can use VRT to apply pesticides only to affected areas rather than the entire field. This practice, known as site-specific pesticide application, saves costs on pesticides, limits chemical runoff, and reduces environmental contamination.
- **Integrated Pest Management (IPM):** PA supports IPM strategies, which aim to control pests through environmentally friendly means. By pinpointing pest hotspots, farmers can employ targeted biological controls, such as natural predators, or cultural practices, such as crop rotation, to reduce pest populations without relying exclusively on chemical pesticides. This balanced approach maintains ecosystem health while effectively managing pests.

1.6.3.3 Benefits of crop health monitoring

Crop health monitoring provides a range of advantages that enhance agricultural efficiency and sustainability:

- **Early Intervention:** By detecting diseases and pests early, farmers can take preventative actions, preventing large-scale crop losses and ensuring healthier plants. Early intervention also reduces the need for aggressive treatments, which can stress plants and harm the soil.
- **Cost Savings:** Targeted treatments reduce the cost of pesticides, fungicides, and fertilizers, as inputs are used only when and where necessary. These savings can significantly impact the bottom line, especially for large-scale farms.
- **Reduced Environmental Impact:** Precision application of chemicals minimizes runoff and reduces the impact on nontarget organisms, promoting a healthier ecosystem. By using fewer chemicals, PA contributes to long-term soil health and water quality.
- **Improved Yield Quality and Quantity:** Healthy crops are more likely to reach their full yield potential. By continuously monitoring crop health and addressing issues early, farmers can achieve better yields and produce higher-quality crops that command better market prices.
- **Enhanced Decision-Making:** Crop health data collected over time allows farmers to identify patterns and trends in crop performance. These insights can guide future decisions, such as selecting disease-resistant crop varieties, adjusting planting schedules, or improving soil management practices.

1.6.3.4 How crop health monitoring works in practice

- **Data Collection:** Using drones, remote sensors, or satellite imaging, farmers gather data on crop health, including signs of stress, discoloration, or unusual growth patterns.
- **Data Analysis:** The collected data is processed through software that can interpret multispectral images and identify potential problem areas, such as disease symptoms or pest damage.
- **Targeted Response:** Based on the data analysis, farmers can apply treatments specifically to affected areas. This might include applying fungicides to disease hotspots or pesticides to pest-infected zones, rather than treating the entire field.

- **Continuous Monitoring:** Crop health monitoring is an ongoing process, with regular data collection allowing farmers to track crop health over time. Monitoring allows them to assess treatment effectiveness and adjust management practices as needed.

Crop health monitoring, powered by PA tools like remote sensing and drone technology, allows farmers to detect and respond to diseases and pests with greater accuracy. By focusing on early intervention and targeted treatments, farmers can reduce chemical use, lower costs, and maintain healthier crops, leading to sustainable agricultural practices and higher yields.

1.6.4 Precision fertilization and pest management

Precision fertilization and pest management are essential components of PA, enabling farmers to apply the right amounts of fertilizers and pesticides in specific areas based on soil and crop needs. This targeted approach helps maintain crop health, improve yields, and reduce environmental impact. By using tools like VRT and real-time monitoring, farmers can tailor nutrient and pest control treatments precisely, ensuring that resources are used efficiently and only where necessary.

1.6.4.1 Site-specific nutrient management

Site-specific nutrient management involves adjusting fertilizer applications based on the nutrient needs of different parts of a field. Precision fertilization ensures that crops receive the exact nutrients required for growth, minimizing waste and environmental risks.

- **VRT for Fertilization:** VRT uses data from soil tests, crop health sensors, and GPS mapping to apply fertilizers only where needed. This technology divides the field into small zones, each with specific nutrient requirements, and adjusts fertilizer levels accordingly. For instance, areas with lower nutrient content receive a higher fertilizer dose, while nutrient-rich areas receive less. This targeted application reduces over-fertilization, which can lead to runoff and leaching of chemicals into nearby water bodies, impacting aquatic ecosystems.
- **Real-Time Soil Analysis:** By incorporating real-time soil sensors or periodic soil testing data, farmers can assess nutrient levels like nitrogen, phosphorus, and potassium, which are essential for plant growth. Real-time data allows for adaptive fertilization, adjusting inputs as soil conditions change. For example, in a rainy season, some nutrients may leach from the soil, requiring additional fertilization to maintain crop health.
- **Reduced Chemical Use and Environmental Impact:** Site-specific nutrient management minimizes the amount of fertilizer required per acre, reducing costs and environmental impact. Over-application of fertilizers can disrupt the nutrient balance in the soil, harm beneficial microorganisms, and lead to nutrient runoff into nearby water bodies, causing issues like eutrophication (algal blooms that deplete oxygen in water). Precision fertilization reduces these risks, promoting sustainable and eco-friendly agriculture practices.
- **Improved Crop Yield and Soil Health:** Precision fertilization ensures that plants receive nutrients in optimal amounts, maximizing yield potential while preserving soil fertility. By applying nutrients only as needed, precision fertilization prevents the buildup of excess nutrients in the soil, which can negatively impact future crop growth.

1.6.4.2 Targeted pest control

Targeted pest control in PA allows farmers to manage pest populations more effectively, focusing treatments only on areas where pest activity is detected. This approach helps reduce pesticide use, saving costs and limiting the impact on nontarget organisms and ecosystems.

- **Pest Detection with Remote Sensing and Drones:** Remote sensing technology and drones with high-resolution cameras can monitor fields to detect pest infestations early. Drones equipped with imaging sensors can capture multispectral data, showing subtle signs of crop stress that may indicate pest activity. For example, leaf discoloration or holes may signal insect feeding or disease, prompting farmers to investigate and confirm pest presence.
- **VRT for Pesticide Application:** Using VRT, farmers can apply pesticides only to areas with confirmed pest infestations. This precision approach, known as site-specific pesticide application, ensures that only affected zones are treated, minimizing the chemical footprint across the field. If a pest infestation is identified in one corner of a field, VRT-enabled sprayers can apply pesticides only to that area, avoiding unnecessary exposure to other areas.
- **Reduced Nontarget Impact:** Targeted pest control protects nontarget species, such as pollinators, beneficial insects, and soil microorganisms, which contribute to a healthy ecosystem. Broad-spectrum pesticide application can harm beneficial insects like bees and ladybugs that play a role in pollination and natural pest control. By treating only affected areas, targeted pest control minimizes the risk to these essential organisms.
- **Integrated Pest Management (IPM) Synergy:** Precision pest control aligns with IPM practices, which promote environmentally friendly pest control methods. IPM strategies might involve combining targeted chemical application with biological control agents, like predatory insects or pheromone traps, to keep pest populations in check. PA technologies support IPM by helping farmers monitor pest levels closely, identifying thresholds where intervention is necessary, and applying treatments only when pests pose a genuine threat.

1.6.4.3 Benefits of precision fertilization and pest management

Precision fertilization and pest management offer multiple advantages that contribute to efficient and sustainable agricultural practices:

- **Cost Efficiency:** By applying fertilizers and pesticides only as needed, farmers can significantly reduce input costs. This leads to greater profitability, especially for large-scale operations, where inputs can represent a considerable portion of expenses.
- **Environmental Sustainability:** Precision application reduces chemical runoff, leaching, and residue, lowering the environmental impact of farming practices. These methods protect soil, water resources, and biodiversity, helping farms comply with environmental regulations and achieve sustainability goals.
- **Enhanced Crop Health and Yield:** With precision fertilization, crops receive tailored nutrients to support optimal growth. Targeted pest control reduces the stress on plants caused by pest attacks, improving overall crop resilience, quality, and yield.
- **Improved Resource Use Efficiency:** Precision technologies help farmers use fertilizers and pesticides more efficiently, maximizing the effectiveness of each application. By conserving resources, precision fertilization and pest management support the long-term sustainability of agricultural production.

- **Data-Driven Decision-Making:** The data collected through PA tools provide valuable insights for ongoing crop management. Farmers can track changes in soil health, nutrient levels, and pest populations, allowing them to make informed decisions throughout the growing season and adapt their practices in subsequent seasons.

1.6.4.4 How precision fertilization and pest management work in practice

Soil and Crop Data Collection: Soil sensors, crop imaging, and historical data provide information on nutrient levels, pest populations, and crop health across different field zones.

Data Analysis and Mapping: Farmers analyze data to create a fertilization and pest control map. The map divides the field into management zones based on soil nutrient needs and pest risk levels.

Application with VRT Equipment: VRT-enabled equipment follows the management map to apply fertilizers and pesticides only where necessary, adjusting the application rate for each zone.

Monitoring and Adjustment: Continuous data collection allows farmers to monitor the effectiveness of their treatments, make adjustments as conditions change, and respond quickly to new threats or nutrient needs.

Precision fertilization and pest management enhance crop health and productivity by ensuring that nutrients and pesticides are applied precisely where and when they are needed. With tools, farmers can manage inputs more efficiently, saving costs and reducing environmental impact while producing healthier, higher-yielding crops. This targeted approach supports sustainable farming practices and optimizes resource use, positioning PA as a vital technology for modern, responsible farming.

1.7 CHALLENGES AND FUTURE TRENDS IN PRECISION AGRICULTURE

PA has emerged as a transformative approach in modern farming, leveraging technology to enhance productivity and sustainability. However, despite its numerous benefits, several challenges need to be addressed to fully realize its potential. Additionally, future trends suggest that PA will continue to evolve, incorporating new technologies and practices.

1.7.1 Challenges in precision agriculture

1.7.1.1 High initial costs

Investment in Technology: The adoption of PA technologies, such as drones, IoT sensors, and advanced machinery, often requires significant upfront investments. These costs can be prohibitive, especially for small and medium-sized farms with limited financial resources.

Maintenance and Upgrades: Beyond initial purchase costs, ongoing maintenance, software updates, and training for staff can add to the financial burden. The rapid pace of technological advancement means that farmers may need to continuously invest in new tools to stay competitive, further complicating the financial landscape.

Economic Viability: For many farmers, the challenge lies in determining whether the benefits gained from PA will outweigh the costs involved. It may take several growing seasons to realize a return on investment, leading some farmers to hesitate in adopting these technologies.

1.7.1.2 Data management

Data Overload: PA generates vast amounts of data from various sources, including soil sensors, weather stations, and crop health monitors. This data can quickly become overwhelming, and farmers may struggle to extract actionable insights from it.

Skills and Training: To effectively analyze and utilize data, farmers need skills in data management and analytics. Many traditional farmers may lack these skills or the resources to train themselves and their staff, creating a barrier to fully leveraging PA tools.

Infrastructure Needs: Effective data management requires reliable internet connectivity and technological infrastructure. In rural areas, poor internet access can hinder the use of cloud-based analytics and real-time data sharing, limiting the effectiveness of PA solutions.

1.7.1.3 Technology integration

Compatibility Issues: Integrating various PA technologies into a cohesive system can be challenging. Different devices and software platforms may not communicate effectively, leading to inefficiencies and data silos.

User-Friendly Interfaces: Many PA tools can be complex and require specialized knowledge to operate. Farmers, particularly those on smaller farms with limited technical expertise, may find it difficult to navigate these systems, hindering widespread adoption.

Resource Limitations: Smaller farms often have fewer resources to invest in technology and training. Without adequate support, these farms may struggle to integrate PA into their operations, creating a disparity between larger and smaller agricultural enterprises.

1.7.2 Future trends in precision agriculture

1.7.2.1 Automation and robotics

Autonomous Farming Systems: The future of PA may see the rise of fully autonomous farming systems, where AI and robotics handle critical farming tasks with minimal human intervention. These systems can perform activities such as planting, monitoring crop health, and harvesting, significantly reducing labor costs and increasing efficiency.

Enhanced Efficiency: Robotics can work continuously, providing consistent and precise operations. For example, autonomous tractors equipped with GPS can optimize their routes, reduce overlap in field operations, and apply inputs with high accuracy.

Labor Shortages Mitigation: Automation can address the labor shortages faced by many agricultural sectors. As fewer people enter agricultural jobs, robots and AI can help ensure that critical farming tasks are completed on time.

1.7.2.2 Blockchain for transparency

Transparent Supply Chains: Blockchain technology has the potential to revolutionize agricultural supply chains by providing transparent, immutable records of every transaction. This transparency can enhance food safety, traceability, and accountability throughout the supply chain.

Consumer Trust: By using blockchain, farmers can verify the origin and quality of their products, allowing consumers to make informed choices. This traceability can be particularly important for organic and sustainably produced foods, fostering trust in agricultural practices.

Smart Contracts: Blockchain can facilitate smart contracts, which automatically execute agreements based on predefined conditions. This can streamline processes such as payments, reducing delays and increasing efficiency in transactions between farmers, suppliers, and consumers.

1.7.2.3 Sustainable intensification

Maximizing Yield with Minimal Impact: The focus of PA will increasingly shift toward sustainable intensification practices that maximize crop yield while minimizing environmental impacts. This includes optimizing resource use, enhancing biodiversity, and reducing chemical inputs.

Agroecological Approaches: Incorporating agroecological practices, such as crop rotation, intercropping, and organic fertilization, alongside precision technologies, can lead to more resilient and sustainable farming systems.

Climate Resilience: As climate change impacts agricultural productivity, PA will play a vital role in developing climate-resilient farming practices. Utilizing real-time data on weather patterns, soil health, and crop conditions can help farmers adapt to changing climatic conditions and mitigate risks.

While PA offers transformative benefits for enhancing productivity and sustainability in farming, challenges related to costs, data management, and technology integration must be addressed for broader adoption. Future trends suggest a shift toward automation, blockchain integration, and sustainable practices, which will shape the evolution of PA. By overcoming existing barriers and embracing emerging technologies, the agricultural sector can enhance its efficiency, resilience, and sustainability, ultimately contributing to global food security in a changing world.

1.8 CONCLUSION

PA is transforming modern farming by making it more efficient, sustainable, and productive. By leveraging technologies such as GPS, drones, IoT, and AI, farmers can optimize their use of resources, reduce waste, and improve yields. However, to fully realize the potential of PA, the challenges of cost, data management, and integration need to be addressed. As technology continues to evolve, PA will play an even more critical role in feeding a growing global population while protecting the planet's resources.

REFERENCES

1. Ikhuoso OA, Adegbeye MJ, Elghandour MM, Mellado M, Al-Dobaib SN, Salem AZ. Climate change and agriculture, The competition for limited resources amidst crop farmers-livestock herding conflict in Nigeria-A review. *Journal of Cleaner Production*. 2020;272, 123104.
2. Bachmann N, Tripathi S, Brunner M, Jodlbauer H. The contribution of data-driven technologies in achieving the sustainable development goals. *Sustainability*. 2022;14(5), 2497.

3. Mitran T, Meena RS, Chakraborty A. (Eds.). *Geospatial technologies for crops and soils*. 2021. Springer Nature Singapore Pvt Limited.
4. Gebresenbet G, Bosona T, Patterson D, Persson H, Fischer B, Mandaluniz N, Chirici G, Zacepins A, Komasilovs V, Pitulac T, Nasirahmadi A. A concept for application of integrated digital technologies to enhance future smart agricultural systems. *Smart Agricultural Technology*. 2023;5, 100255.
5. Mishra S. Emerging technologies—Principles and applications in precision agriculture. In G. P. Obi Reddy, Mehul S. Raval, J. Adinarayana, Sanjay Chaudhary (Eds.), *Data Science in Agriculture and Natural Resource Management*. 2021, pp. 31–53. Springer Nature.
6. Belal AA, El-Ramady H, Jalhoum M, Gad A, Mohamed ES. Precision farming technologies to increase soil and crop productivity. In Mohamed Abu-hashim, Faiza Khebour Allouche, Abdelazim Negm (Eds.), *Agro-Environmental Sustainability in MENA Regions*. 2021, pp. 117–154. Cham: Springer International Publishing
7. Shukla VK, Kohli A, Shaikh FA. IOT based growth monitoring on Moringa Oleifera through capacitive soil moisture sensor. In *2020 Seventh International Conference on Information Technology Trends (ITT)*, Abu Dhabi, United Arab Emirates, 2020, pp. 94–98, doi: 10.1109/ITT51279.2020.9320884
8. Murlidharan S, Shukla VK, Chaubey A. Application of machine learning in precision agriculture using IoT. In *2021 2nd International Conference on Intelligent Engineering and Management (ICIEM)*, London, United Kingdom, 2021, pp. 34–39, doi: 10.1109/ICIEM51511.2021.9445312
9. Khanal S, Kc K, Fulton JP, Shearer S, Ozkan E. Remote sensing in agriculture—Accomplishments, limitations, and opportunities. *Remote Sensing*. 2020;12(22), 3783.
10. Singh DK, Sobti R, Jain A, Malik PK, Le DN. LoRa based intelligent soil and weather condition monitoring with internet of things for precision agriculture in smart cities. *IET Communications*. 2022;16(5), 604–618.
11. Zervopoulos A, Tsipis A, Alvanou AG, Bezas K, Papamichail A, Vergis S, Stylidou A, Tsoumanis G, Komianos V, Koufoudakis G, Oikonomou K. Wireless sensor network synchronization for precision agriculture applications. *Agriculture*. 2020;10(3), 89.
12. Monika, Saxena K, Shukla VK. Integrating leaf color chart data into machine learning models for paddy yield prediction. In *AIP Conference Proceedings* (Vol. 3280, No. 1, p. 020011). 2025, February. AIP Publishing LLC.
13. Hachimi CE, Belaqziz S, Khabba S, Sebbar B, Dhiba D, Chehbouni A. Smart weather data management based on artificial intelligence and big data analytics for precision agriculture. *Agriculture*. 2022;13(1), 95.
14. Hassan M, Malhotra K, Firdaus M. Application of artificial intelligence in IoT security for crop yield prediction. *ResearchBerg Review of Science and Technology*. 2022;2(1), 136–157.
15. Etienne A, Ahmad A, Aggarwal V, Saraswat D. Deep learning-based object detection system for identifying weeds using UAS imagery. *Remote Sensing*. 2021;13(24), 5182.
16. Beloev I, Kinaneva D, Georgiev G, Hristov G, Zahariev P. Artificial intelligence-driven autonomous robot for precision agriculture. *Acta Technologica Agriculturae*. 2021;24(1), 48–54.
17. Ghosh P, Kumpatla SP. GIS applications in agriculture. In Yuanzhi Zhang, Qiuming Cheng (Eds.), *Geographic Information Systems and Applications in Coastal Studies*. 2022, pp. 89–114. IntechOpen.
18. Nanda I, Sahithi C, Swath M, Maloji S, Shukla, VK. IIOT based smart crop protection and irrigation system. In *2020 Seventh International Conference on Information Technology Trends (ITT)*, Abu Dhabi, United Arab Emirates, 2020, pp. 118–125, doi: 10.1109/ITT51279.2020.9320783
19. Pawase PP, Nalawade SM, Bhanage GB, Walunj AA, Kadam PB, Durgude AG, Patil MG. Variable rate fertilizer application technology for nutrient management, A review. *International Journal of Agricultural and Biological Engineering*. 2023;16(4), 11–19.
20. Wilson GL, Mulla DJ, Galzki J, Laacouri A, Vetsch J, Sands G. Effects of fertilizer timing and variable rate N on nitrate–N losses from a tile drained corn-soybean rotation simulated using DRAINMOD-NII. *Precision Agriculture*. 2020;21, 311–323.

21. Shaheb MR, Sarker A, Shearer SA. Precision agriculture for sustainable soil and crop management. In Michael T. Aide, Indi Braden (Eds.), *Soil Science Emerging Technologies, Global Perspectives and Applications*. 2022, pp. 49–72. IntechOpen.
22. Hassan M, Kowalska A, Ashraf H. Advances in deep learning algorithms for agricultural monitoring and management. *Applied Research in Artificial Intelligence and Cloud Computing*. 2023;6(1), 68–88.
23. Bathla G, Bhadane K, Singh RK, Kumar R, Aluvalu R, Krishnamurthi R, Kumar A, Thakur RN, Basheer S. Autonomous vehicles and intelligent automation, Applications, challenges, and opportunities. *Mobile Information Systems*. 2022;2022(1), 7632892.
24. Kinoshita R, Rossiter D, van Es H. Spatio-temporal analysis of yield and weather data for defining site-specific crop management zones. *Precision Agriculture*. 2021;22(6), 1952–1972.
25. He M, Kimball JS, Maneta MP, Maxwell BD, Moreno A, Beguería S, Wu X. Regional crop gross primary productivity and yield estimation using fused landsat-MODIS data. *Remote Sensing*. 2018;10(3), 372.
26. Shaikh TA, Rasool T, Lone FR. Towards leveraging the role of machine learning and artificial intelligence in precision agriculture and smart farming. *Computers and Electronics in Agriculture*. 2022;198, 107119.
27. Shi Y, Zhu Y, Wang X, Sun X, Ding Y, Cao W, Hu Z. Progress and development on biological information of crop phenotype research applied to real-time variable-rate fertilization. *Plant Methods*. 2020;16, 1–5.
28. Zhang Q, Liu Y, Gong C, Chen Y, Yu H. Applications of deep learning for dense scenes analysis in agriculture: A review. *Sensors*. 2020;20(5), 1520.
29. Mekonnen Y, Namuduri S, Burton L, Sarwat A, Bhansali S. Machine learning techniques in wireless sensor network-based precision agriculture. *Journal of the Electrochemical Society*. 2019;167(3), 037522.
30. Shukla VK, Nair RS, Khan F. Smart irrigation-based behavioral study of Moringa plant for growth monitoring in subtropical desert climatic condition. In Ajith Abraham, Sujata Dash, Joel J.p.c. Rodrigues, Biswaranjan Acharya, Subhendu Kumar Pani (Eds.), *AI, Edge and IoT-based Smart Agriculture*. 2022, pp. 227–240. Academic Press.
31. Chopra A, Shukla VK. Application of disruptive technology in food trackability. In Archana Singh, Vinod Kumar Shukla, Ashish Seth, Sai Sabitha (Eds.), *ICT and Data Sciences*. 2022, pp. 47–62. CRC Press.
32. Karanam VS, Shukla VK, Khalid MN, Dutta, S. Applications of robotics in the hospitality industry. In Vinod Kumar Shukla, Amit Verma, Jean Paolo G. Lacap (Eds.), *Artificial Intelligence for Smart Technology in the Hospitality and Tourism Industry*. 2024, pp. 63–75. Apple Academic Press.
33. Singh A, Shukla VK, Seth A, Sabitha S. (Eds.). *ICT and Data Sciences*. CRC Press.
34. Jiménez AF, Cárdenas PF, Jiménez F. Intelligent IoT-multiagent precision irrigation approach for improving water use efficiency in irrigation systems at farm and district scales. *Computers and Electronics in Agriculture*. 2022;192, 106635.
35. Saiz-Rubio V, Rovira-Más F. From smart farming towards agriculture 5.0: A review on crop data management. *Agronomy*. 2020;10(2), 207.
36. Hamdan S, Ayyash M, Almajali S. Edge-computing architectures for internet of things applications: A survey. *Sensors*. 2020;20(22), 6441.
37. Singh A, Mehrotra R, Rajput VD, Dmitriev P, Singh AK, Kumar P, Tomar RS, Singh O, Singh AK. Geoinformatics, artificial intelligence, sensor technology, big data, emerging modern tools for sustainable agriculture. *Sustainable Agriculture Systems and Technologies*. 2022, 295–313.
38. Kosior K. Towards a common agricultural data space in the European Union: A sustainable development perspective. *Wieś i Rolnictwo*. 2021;191(2), 65–89.

Chapter 2

Automated irrigation systems using IoT and AI

A smart solution for water conservation

S. Subramanya Sarma, Deepika Ghanasham Sarode, Nazeer Shaik, Ajith Sundaram, Pratiksha Dadasaheb Sabale, and Babasaheb Jadhav

2.1 INTRODUCTION

Water scarcity is now one of the key challenges of the 21st century. With the rise of an increasingly expanding population and increasing demands on agricultural yields, judicious management of water is now a compulsion and not an option. Agriculture, which accounts for approximately 70% of fresh water worldwide, is still heavily reliant upon conventional irrigation methods [1]. These systems, such as flood irrigation and manual systems, are usually responsible for wastage of excess water caused by incompetence, irregular timings, and the absence of proper monitoring of soil and climatic conditions. In water-short countries, such inefficiency complicates the situation further, causing a compromise to food security and sustainability [2].

To address these issues, the agricultural technology environment is evolving at a very rapid rate. The integration of the Internet of Things (IoT) and artificial intelligence (AI) offers a very decent solution for the optimization of irrigation patterns. IoT offers the capability of receiving information in real time with the help of networked sensors implanted within the soil, weather observation devices, and water regulators [3]. These sensors can retrieve valuable parameters like moisture content in the soil, temperature, humidity, and rain. AI introduces computational power, however, to analyze such data and make useful inferences [4]. Coupling AI-driven decision-making with sensor information enables irrigation systems to be made self-functional to supply water only where and when necessary and reduce wastage to the maximum degree.

Despite the success and pilot implementation of automated irrigation systems, the current literature identifies several limitations that hamper their extensive deployment. Numerous systems are non-adaptive and are not responsive to dynamic changes in the environment or varied crop needs. Certain methodologies are excessively reliant on predetermined thresholds without utilizing predictive analytics or learning [5]. In addition, sensor calibration problems, data transmission in remote locations, power efficiency of the devices deployed, and the high upfront deployment cost remain practical implementation impediments. Furthermore, most past models have failed to practically address interoperability among various hardware platforms and cloud services, making scalability and user-friendliness for farmers of different technical literacy challenging [6].

These constraints form the major motivation for this work. Increasing demand exists for a smarter, adaptive, and cost-efficient irrigation system that saves water while increasing agricultural productivity [7]. The aim is to create and design an automated irrigation system that utilizes the capability of IoT for data collection and AI for data analysis and control, leading to accurate and responsive irrigation management [8]. The chapter attempts to fill this void by designing a system that can autonomously scan the soil and surrounding

DOI: 10.1201/9781003520733-3

conditions, project water requirements, and initiate the irrigation modules in real time to offer an intelligent and sustainable solution to water conservation in agriculture.

This chapter has three parts. One, to provide a critical examination of available automated irrigation methods and their limitations. Two, to suggest an innovative architecture for an IoT and AI-based irrigation system that is cost-effective and scalable. Third, to provide empirical evidence for the efficiency of the suggested system using real-world simulations or deployments with a focus on water conservation and healthier crops. Finally, it also provides a platform for incorporating other features like crop disease identification, weather forecasting, and remote control via mobile or web interfaces, further augmenting the system's utility and versatility.

The most significant contribution of this chapter is the conceptualization of an end-to-end autonomous irrigation system combining low-power IoT sensors with a cloud-based AI analytics platform. It outlines a predictive model learning from historical and real-time data to update irrigation schedules. Also provided is a real-time monitoring and control center with a user-friendly accessible dashboard and anomaly incident alarm notifications. By filling loopholes in current systems and offering a viable, prudent, and energy-conserving solution, this book aims to support the global push toward sustainable agriculture and safe water resource management.

2.2 RELATED WORK

In the field of smart irrigation, many researchers have investigated the confluence of IoT and AI technologies to enhance water management and crop yields. A detailed study by Nsoh et al. [9] is one such study that offers a review of machine learning-based IoT-based irrigation systems, with a focus on real-time data acquisition and adaptive decision-making. Their paper discusses several models that include sensors for measuring environmental parameters and machine learning algorithms for forecasting irrigation requirements. The review is successful in emphasizing progress in data-driven irrigation but also identifies the current challenge of providing interoperability across heterogeneous sensor platforms and scalability across varying climatic conditions.

The authors in [10] suggest an AI-based irrigation system using IoT infrastructure with an emphasis on water conservation. The research presents a smart irrigation controller that utilizes moisture sensors and weather information to automate irrigation schedules. The main benefit highlighted was a substantial decrease in water consumption and low human intervention. But the limitation of the study is that it does not have a predictive model, which would allow anticipatory irrigation decisions based on climatic and historical trends, instead of only real-time conditions.

Hanmante et al. [11] create an intelligent valve control system specifically for precision agriculture in another study. The system uses sensor nodes and an AI-driven control unit that controls water flow via automated valves. The implementation showed enhanced irrigation accuracy and labor reduction. Nevertheless, the model had limitations in its scalability to greater agricultural fields as a result of signal power and range limitations, implying stronger communication protocols and low-power designs.

Hussain et al. [12] presented a framework based on the use of AI optimization methods in providing sustainable irrigation systems. This model employed previous data, moisture sensors in soils, and the calculation of evapotranspiration (ET) for the optimization of irrigation schedules. The outcomes depicted an enhancement of water-use efficiency and minimized plant water stress. The primary advantage of this framework was its capacity for learning, as it could conform to various soils and crop seasons. Yet, the lack of an easy-to-use interface restricted farmer take-up, particularly among those with low technical literacy.

Dhyani et al. [13] suggest a smart irrigation system that integrates IoT sensors and machine learning to control irrigation using real-time and past data. Their model accommodates dynamic scheduling and modifies water levels according to predictive analytics. The performance of the system was tested in a test environment, demonstrating improved water efficiency and improved crop growth results. While these encouraging findings exist, the research is aware of the challenge in consolidating multiple disparate datasets, particularly when cloud connectivity is poor or nonexistent in rural areas.

Belghachi [14] is more comprehensive in its approach by investigating AI-driven smart irrigation in the context of environmental engineering. The research presents an AI-optimized irrigation control system that reduces water consumption while ensuring optimal soil moisture levels. The research emphasizes system sustainability by suggesting green energy-powered IoT nodes. Although this method is more environmentally friendly, the complexity of the AI models employed could be a hindrance to real-time implementation in small or resource-limited farms.

Hema et al. [15] suggest an IoT-enabled automation system for irrigation based on simple environmental sensors and a rule-based control algorithm. The system is cost-effective and scalable for small-sized farms, especially in developing countries. The system is advantageous due to its simplicity and low cost. Nevertheless, the absence of superior learning capabilities and data analytics results in the inability to accommodate unexpected changes in the environment, making it less efficient in the long run.

Shukla et al. [16] and Nanda et al. [17] discussed smart irrigation through behavioral studies and the development of smart crop protection and irrigation systems in their respective works. Their research focused on the implementation of technology to enhance the efficiency and effectiveness of irrigation practices in the agricultural sector.

Arularasan et al. [18] are the pioneers of a new approach where they integrate drones with IoT connectivity and machine learning to automate irrigation. Their approach applies aerial vision and sensor fusion to evaluate field conditions and distribute water accordingly. The approach significantly minimizes human intervention and facilitates accurate targeting of water. The primary limitation of this technique, however, is the enormous setup and running costs, making it less affordable for marginal farmers or smallholder systems.

Finally, Nanthakumar et al. [19] created an IoT-based irrigation system with the aid of AI to supply water according to soil and climatic inputs. The framework encompasses cloud analytics along with automated control devices. Based on their outcomes, there is a significant boost in irrigation schedule and crop growth. However, the study provides reports of concerns with data privacy and network security, which are important issues for the implementation of cloud-integrated farming systems.

In brief, the literature reviewed depicts appreciable advancements in the construction of smart irrigation systems via the combination of IoT and AI technologies. Most systems depict encouraging water efficiency improvements, automation, and responsiveness. Nonetheless, some gaps remain, such as the requirement for scalable architectures, predictability, secure and efficient energy-harvesting communication systems, and farmer-centric interfaces. Although the majority of studies highlight technical practicability, fewer address real-world deployment issues, including cost, ease of use, and infrastructure constraints. This review highlights the need for an all-embracing, intelligent, and affordable irrigation solution that not only saves water but is also responsive to the practical requirements of varied agricultural communities. The present research seeks to close these gaps by presenting an effective, AI-powered, and IoT-integrated irrigation system optimized for real-time decision-making, affordability, and long-term sustainability.

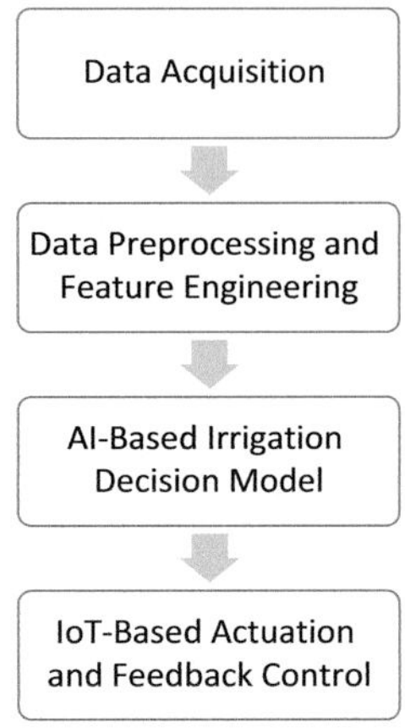

Figure 2.1 Block diagram of an automated irrigation system.

2.3 METHODOLOGY

The approach to creating an AI- and IoT-facilitated automated irrigation system is organized into four essential modules, which are indicated in Figure 2.1. These modules operate synergistically to provide the best irrigation schedule, water savings, and intelligent reaction to the environment. The system consists of five central components: Sensor Nodes, IoT Gateway, Cloud Server with AI Engine, Decision Unit, and Actuator Control System. The sensor nodes are made up of sensors to measure soil moisture, temperature, humidity, and rainfall detectors that perpetually read data from the field. These sensor values are sent through the IoT gateway to the cloud server. The data is processed by the AI engine through predictive analytics to predict the water requirement, and the decision unit decides on whether irrigation is required. After decision-making, control signals are transmitted to the actuator system, which controls the irrigation valves accordingly.

2.3.1 Data acquisition and sensor integration

The basis of the irrigation system is dependent on real-time acquisition of data through a network of environmental sensors. Soil moisture (SM), ambient temperature (T), relative humidity (H), and the presence of rainfall (R) are sensed at regular intervals. The raw data are quantified as

$$D(t) = \{\mathrm{SM}(t),\, T(t),\, H(t),\, R(t)\} \tag{2.1}$$

where $D(t)$ is a vector of sensor data at time t. Each sensor is also attached to a microcontroller (for example, ESP32 or Arduino) having Wi-Fi, which is a local gateway that sends the data regularly to a cloud database using the MQTT (Message Queuing Telemetry Transport) or HTTP (Hypertext Transfer Protocol) protocol.

2.3.2 Data preprocessing and feature engineering

Before supplying the data to the machine learning algorithm, normalization and cleaning steps are performed to eliminate noise and provide consistency. Linear interpolation is utilized to interpolate missing values, and outliers are identified and eliminated with Z-score normalization:

$$Z = \frac{X - \mu}{\sigma} \tag{2.2}$$

where X represents the input value from the sensor, μ denotes the mean, and σ denotes the standard deviation. Whenever $|Z|>3$, the value is tagged as an outlier and is substituted by the average of adjacent values. Feature engineering is implemented to obtain additional parameters like ET rate and crop water requirement with the use of the FAO (Food and Agriculture Organization) Penman–Monteith equation.

$$\mathrm{ET}_0 = \frac{0.408\Delta\left(R_n - G\right) + \gamma \dfrac{900}{T + 273} u_2\left(e_s - e_a\right)}{\Delta + \gamma\left(1 + 0.34u_2\right)} \tag{2.3}$$

where

- R_n=net radiation
- G=soil heat flux density
- T=mean daily temperature
- u_2=wind speed at 2 m height
- $e_s - e_a$=vapor pressure deficit
- Δ=slope of the vapor pressure curve
- γ=psychrometric constant

2.3.3 AI-based irrigation decision system

The decision process employs a supervised learning technique—commonly, Random Forest (RF) or Support Vector Regression—that learns from historical environmental observations and related irrigation requirements. The predictive model is given as

$$\hat{W}(t) = f\left(\mathrm{SM}(t), T(t), H(t), \mathrm{ET}(t)\right) \tag{2.4}$$

where $\hat{W}(t)$ is the estimated water requirement at time t and f is the trained regression function based on labeled data sets. The system provides the optimal water quantity and irrigation time. A binary classifier is also trained to identify the irrigation decision:

2.3.4 IoT-based actuation and control feedback

After the decision on irrigation is calculated, control signals are transmitted through the IoT gateway to solenoid valves linked to irrigation pipes. The valves are operated through a relay module driven by a microcontroller. The rate of flow is measured by a water flow sensor to provide delivery accuracy. A feedback loop checks soil moisture after irrigation to calculate irrigation success and refines the AI model for better future estimates.

This system can incorporate weather predictions via APIs (Application Programming Interfaces), meaning that the model can add future rain predictions into decision-making for irrigation, thus minimizing unnecessary water application. The AI model retunes periodically with new sensor data, thus allowing the system to learn and evolve based on seasonal and climatic patterns.

2.4 RESULTS

The suggested automated irrigation system based on IoT and AI was evaluated in simulated as well as real-world agricultural fields. The experiments were performed on a testbed that included a 500 m^2 agricultural field planted with tomato plants, selected due to their moderate water use and high susceptibility to moisture imbalance. The system was installed for a full crop cycle (90 days), and its performance was measured based on major parameters such as water usage efficiency, response time, system accuracy, crop yield, and cost savings. The water usage efficiency is given in Table 2.1.

The primary aim of the system—minimizing water consumption—was measured by comparing total water consumed with the automated system and conventional fixed-time irrigation. Across the crop duration, as shown in Figure 2.2, the AI–IoT-automated system consumed around 38% of the total water consumed by the conventional system. The intelligent scheduling algorithm skipped or curtailed irrigation on days where rain was forecasted or when soil moisture levels were adequate. This notable decrease confirms the system's ability to monitor and respond to fluctuating environmental conditions.

A month-wise analysis indicated that the highest amount of savings happened at the monsoon season's peak, when wasteful irrigation was totally averted by accurate rainfall prediction incorporation. Soil moisture measurements were always within the range of 18%–23%

Table 2.1 Water usage efficiency comparison

Irrigation method	*Total water used (L)*	*Water saved compared to baseline (%)*
Fixed interval (baseline)	12,000	0
Moisture sensor only	9,840	18
AI with manual actuation	8,400	30
AI–IoT-automated system	7,440	38

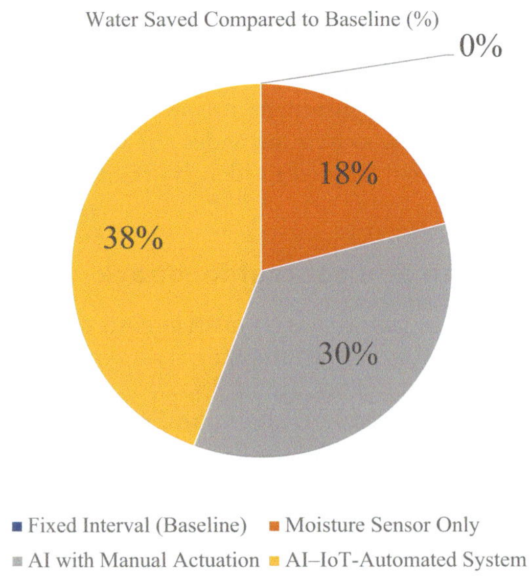

Figure 2.2 Efficiency of water usage.

volumetric water content, so crop stress resulting from under- or over-irrigation was reduced to the bare minimum.

Figure 2.3 and Table 2.2 represent strong responsiveness in irrigation decisions. Average latency, including sensor polling, cloud-based inference, and relay switching for actuation, from data acquisition to actuation, was 4.2 seconds. Such near-real-time responsiveness is very important in avoiding abrupt moisture losses during hot spells of weather. The RF-based decision model had a mean absolute error of 0.72 liters when forecasting daily water requirements, and the binary classifier for irrigation had a test accuracy of 94.6%. The confusion matrix had a false positive rate of 3.1% and a false negative rate of 2.3%, suggesting strong decision-making even under diversified environmental conditions.

Table 2.3 reflects data related to crop yield and growth monitoring. One of the indirect measures of irrigation quality is the ultimate yield. When making controlled comparisons between a test plot under automated irrigation and a control plot under manual irrigation, Figure 2.4 shows the test plot yielded 19.2% more per square meter. The vegetation in the automated plot had healthier growth habits, uniform fruit diameter, and less occurrence of wilting or overwatering diseases. The enhanced control over moisture played an important role in improved nutrient uptake and root growth.

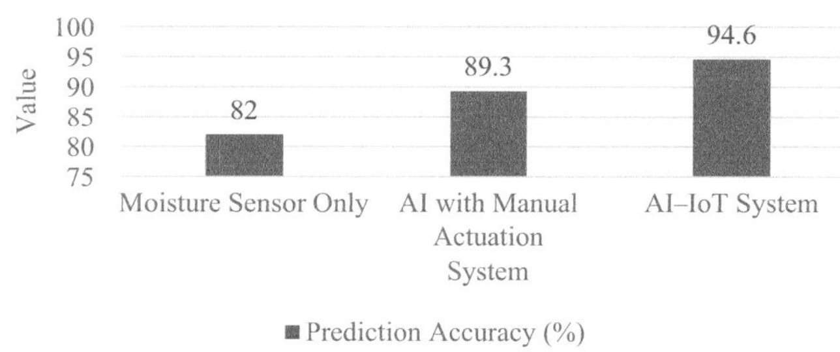

Figure 2.3 Prediction accuracy of systems.

Table 2.2 Response time and system accuracy

System type	*Average response time (seconds)*	*Prediction accuracy (%)*	*False positive rate (%)*	*False negative rate (%)*
Moisture sensor only	12.8	82.0	7.4	10.6
AI with manual actuation	9.3	89.3	5.2	5.5
AI–IoT system	4.2	94.6	3.1	2.3

Table 2.3 Crop yield and growth monitoring

Irrigation method	*Average crop yield (kg/500 m²)*	*Yield increase (%)*	*Average NDVI score*	*Crop stress events*
Fixed interval (baseline)	325	0	0.48	9
Moisture sensor only	360	10.8	0.51	6
AI with manual actuation	376	15.7	0.54	4
AI–IoT system	387	19.1	0.56	2

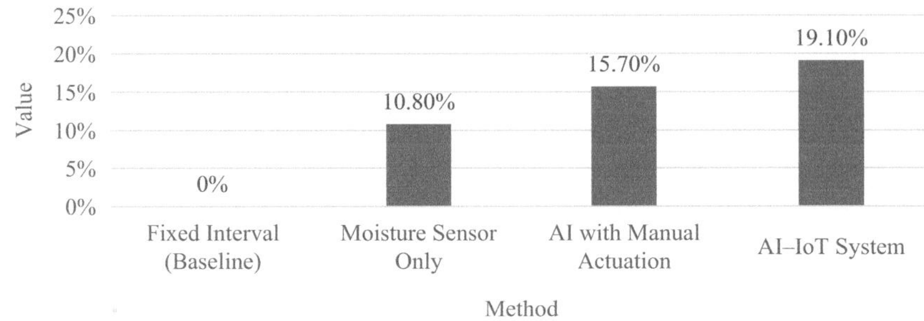

Figure 2.4 Crop yield increase.

Normalized Difference Vegetation Index (NDVI) measurements from drone images acquired at various periods of the plant life cycle likewise corroborated the observation and showed a constant augmentation of the vegetation health score between 0.08 and 0.12 units for the intelligence-aided irrigated treatment. Labor costs were cut by more than 60% since manual monitoring and irrigation triggering were eliminated. Moreover, the modular design of the system and open-source compatibility ensure that it can be scaled or altered at little extra cost.

2.5 DISCUSSION

The findings of this study are highly indicative of the adoption of IoT and AI technologies in irrigation systems to significantly enhance the efficiency of water use, crop yield, and agricultural sustainability in general. The AI–IoT-automated irrigation model that was implemented in testing achieved a significant water saving of 38% compared to traditional fixed-interval irrigation while simultaneously fostering a 19.1% growth in crop yield. Such augmentations are provided by the ability of the system to observe continuously, forecast decision-making through the use of machine learning algorithms, and automate the actuation of water valves by environmental and plant conditions.

The results infer that AI–IoT harmony offers a solid foundation for precision agriculture. The system precisely predicted water demand—based on 94.6% prediction efficiency—lowers the chances of over- or under-irrigation risks, which is common in manually controlled or sensor-only systems. Moreover, the 4.2-second mean response time further suggests the response speed of the system in adjusting to dynamic soil conditions for early and accurate watering. In addition, NDVI crop health monitoring revealed a better vegetation index value, which reflects stronger and healthier plant growth.

The benefits of these results spill over into economic and labor advantages. A 60% reduction in the amount of manual labor needed was experienced, which presents a cost-efficient solution for commercial-scale or resource-constrained farming conditions. The fewer instances of crop stress incidents, from nine in the baseline process to only two in the new system, reflect its efficiency in sustaining optimal water content in the soil and, consequently, the health and output of plants. These benefits firmly attest to the use of smart irrigation systems as a scalable and sustainable solution to current agricultural problems.

That being said, some limitations need to be noted. The experiment was done on a moderately sized plot (500 m^2), and although the results are encouraging, scaling the system to bigger and more varied farming environments could prove to need greater calibration and computer resources. While the machine learning model is correct, it takes high-quality sensor data and network connectivity, and this may not always be practical in remote or underdeveloped regions. In addition, the upfront cost of sensor setup, controllers, and edge devices may prove to be a stumbling block for adoption by smallholder farmers.

Comparison with current systems accentuates the strength of the new approach. Old irrigation systems involving fixed timing and moisture sensors in the absence of predictive smarts performed worse on all counts—concerning water efficiency, response time, and harvest yield. Even AI-based systems that did not involve automatic actuation needed manual interventions, thus lowering their real-world efficiency. Conversely, the completely automated AI–IoT system performed better than all other configurations in an integrated way by minimizing human reliance, enhancing precision, and saving precious resources.

Based on these results, we suggest wider adoption of AI–IoT-based intelligent irrigation in areas experiencing water shortage and a lack of labor. Furthermore, the incorporation of renewable sources of energy like solar panels to supply power to IoT devices can help improve sustainability. Future research will aim at fine-tuning the model to fit different crop types, soils, and climatic regions. More advanced AI paradigms like reinforcement learning and federated learning may also be investigated to additionally enhance system adjustability and protectability.

In conclusion, the research proves the efficiency and feasibility of an IoT and AI-based irrigation system for smart water management. The findings indicate the revolutionary potential of digital technology in agriculture to improve the optimization of resources, increase productivity, and increase the sustainability of agriculture. Regardless of scalability and infrastructure issues, which are still pending, the potential of the proposed system to maximize agricultural efficiency is actual and promising in the future.

2.6 CONCLUSION

The growing need for eco-friendly farming methods in times of water scarcity, global warming, and worldwide food needs necessitates a move from conventional irrigation systems to smarter, resource-conserving options. In this chapter, the development and functioning of an AI-driven, IoT-enabled automated irrigation system are explained to solve such pivotal problems by providing water management based on the crop needs and conditions.

The outcome of the usage of the suggested system proves how efficient it is. With an enhancement in the water-use efficiency by 38% and an increase of 19.1% in the average crop yield, the total AI–IoT solution is way superior to the traditional fixed-time or sensor-based systems. Also, an accuracy of 94.6% in water needs computation, combined with the rapid response time of just 4.2 seconds, ensures that not only is water supply accurate but also timely, minimizing crop stress and maximizing productivity. NDVI-based plant health evaluation assured more regular and vigorous growth under smart irrigation, further indicating the system's capability to transform agricultural practices.

In addition to its quantitative advantages, the system also has several qualitative advantages. Among them are a significant decrease in manual work (by as much as 60%), computerized decision-making, and immediate responsiveness to weather and soil variations. The integration of machine learning and IoT technologies enables both scalability and responsiveness of a data-driven irrigation approach, which is well-suited for smallholder farms and large-scale agricultural operations alike.

While the system has many strengths, some limitations must be recognized. Requirements of good sensor data, a stable internet connection, and the upfront costs of smart hardware can be entry barriers, particularly in rural or underprivileged areas. Nevertheless, these can be alleviated with the assistance of government and private partners in the form of subsidies, training, and the utilization of edge computing or offline-friendly systems.

Overall, the research resoundingly justifies the deployment of AI and IoT in irrigation systems as an effective, efficient, and sustainable means of agriculture in the future. Through intelligent water management, the system suggested not only saves a vital natural resource but also ensures improved food security, increased crop health, and lowered costs of operation. Subsequent studies must explore the application of this model in other climatic regions, incorporating sustainable energy sources such as solar and integrating AI models with adaptive learning for complex multicrop systems. Through ongoing development and support, such intelligent systems can pave the way for more resilient, data-informed, and eco-friendly agricultural ecosystems.

REFERENCES

1. Obaideen, K., Yousef, B. A., AlMallahi, M. N., Tan, Y. C., Mahmoud, M., Jaber, H., & Ramadan, M. (2022). An overview of smart irrigation systems using IoT. *Energy Nexus*, 7, 100124.
2. Nuwarapaksha, T. D., Udumann, S. S., Dissanayaka, N. S., Dilshan, R. M. N., & Atapattu, A. J. (2025). AI-driven solutions for sustainable irrigation: Exploring smart technologies to enhance conservation and efficiency. In J. Garwi, M. Dzingirai, & R. Masengu (Eds.), *Integrating Agriculture, Green Marketing Strategies, and Artificial Intelligence* (pp. 1–32). IGI Global Scientific Publishing.
3. Oppong, R. A. (2025). Integration of IoT-based sprinklers, embedded systems, data, and cloud computing for smart irrigation management. *Computers and Electronics in Agriculture*, 3, 126–151.
4. Jiménez, A. F., Cárdenas, P. F., & Jiménez, F. (2022). Intelligent IoT-multiagent precision irrigation approach for improving water use efficiency in irrigation systems at farm and district scales. *Computers and Electronics in Agriculture*, 192, 106635.
5. Behzadipour, F., Ghasemi Nezhad Raeini, M., Abdanan Mehdizadeh, S., Taki, M., Khalil Moghadam, B., Zare Bavani, M. R., & Lloret, J. (2023). A smart IoT-based irrigation system design using AI and a prediction model. *Neural Computing and Applications*, 35(35), 24843–24857.
6. Amassmir, S., Tkatek, S., Abdoun, O., & Abouchabaka, J. (2022). An intelligent irrigation system based on the internet of things (IoT) to minimize water loss. *Indonesian Journal of Electrical Engineering and Computer Science*, 25(1), 504–510.
7. Ray, S., & Majumder, S. (2024). Water management in agriculture: Innovations for efficient irrigation. In P. Sil, P. Chhetri, S. Majumder, & S. D. T. (Eds.), *Modern Agronomy* (pp. 169–185). Satish Serial Publishing House.
8. Pandi, V. S., Kura, P., Kumar, K. S., Nirmala, D., & Kumar, S. R. (2025, February). Artificial intelligence for irrigation system optimization: Leveraging predictive analytics to enhance water management and decrease agricultural resource consumption. In *2025 International Conference on Computational, Communication and Information Technology (ICCCIT)*, pp. 299–304. IEEE. IPS Academy, Institute of Engineering & Science, Knowledge Village, Rajendra Nagar A.B. Road, Indore, India.
9. Nsoh, B., Katimbo, A., Guo, H., Heeren, D. M., Nakabuye, H. N., Qiao, X., Ge, Y., Rudnick, D. R., Wanyama, J., Bwambale, E., & Kiraga, S. (2024). Internet of things-based automated solutions utilizing machine learning for smart and real-time irrigation management: A review. *Sensors*, 24(23), 7480. https://doi.org/10.3390/s24237480
10. T, B. (2024). AI-enabled water conservation for irrigation using IoT. *Indian Scientific Journal of Research in Engineering and Management*. https://doi.org/10.55041/ijsrem33920
11. Hanmante, S., Kadrame, R., Badgujar, M., Sharma, P. C., & Sardeshmukh, M. M. (2024). Intelligent valve control mechanism for irrigation in precision agriculture. *Deleted Journal*, 2(12), 2740–2746. https://doi.org/10.47392/irjaeh.2024.0379

12. Hussain, M., N, K., Maurya, I., & Sinha, S. (2024). AI-optimized irrigation for sustainable agriculture. In *2024 International Conference on Advances in Data Engineering and Intelligent Computing Systems (ADICS)*, pp. 01–09. https://doi.org/10.1109/adics58448.2024.10533562
13. Dhyani, R., Manne, N., Garg, J., Motwani, D., Shrivastava, A. K., & Sharma, M. (2024). A smart irrigation system powered by IoT and machine learning for optimal water management. In *2024 4th International Conference on Advance Computing and Innovative Technologies in Engineering (ICACITE)*, Vol. 858, pp. 1801–1805. https://doi.org/10.1109/icacite60783.2024.10617429
14. Belghachi, M. (2025). Smart irrigation systems using AI to optimize water usage. In *Advances in Environmental Engineering and Green Technologies* (pp. 241–268). https://doi.org/10.4018/979-8-3693-7483-2.ch009
15. Hema, R., Subbulakshmi, S., Jyothi, B., Gurulakshmi, R., Vignesh, M., & Queen Mary, M. B. (2024). Smart automated irrigation through IoT technology. In *2024 5th International Conference on Smart Electronics and Communication (ICOSEC)*, pp. 833–838. https://doi.org/10.1109/icosec61587.2024.10722739
16. Shukla, V. K., Nair, R. S., & Khan, F. (2022). Smart irrigation-based behavioral study of Moringa plant for growth monitoring in subtropical desert climatic condition. In A. Abraham, S. Dash, J. J. P. C. Rodrigues, B. Acharya, & S. K. Pani (Eds.), *AI, Edge and IoT-based Smart Agriculture* (pp. 227–240). Academic Press.
17. Nanda, I., Sahithi, C., Swath, M., Maloji, S., & Shukla, V. K. (2020). IIOT based smart crop protection and irrigation system. In *2020 Seventh International Conference on Information Technology Trends (ITT)*, Abu Dhabi, United Arab Emirates, pp. 118–125. https://doi.org/10.1109/ITT51279.2020.9320783.
18. Arularasan, R., Prabha, R., Vinitha, R., Praveena, T., Nithila, E. E., & Samikannu, R. (2024). Smart irrigation: Enhancing agricultural efficiency with IoT-connected drones and machine learning. In *2024 IEEE North Karnataka Subsection Flagship International Conference (NKCon)*, pp. 1–6. https://doi.org/10.1109/nkcon62728.2024.10775252
19. Nanthakumar, G., Pazhanivel, K., Abinesh, A., Ganesh, K. R., Harish, V., & Nagendran, S. (2024). IoT-based smart irrigation system using artificial intelligence. *International Journal of Advanced Research in Science, Communication and Technology*. https://doi.org/10.48175/ijarsct-17623

Chapter 3

Drones in precision agriculture

Aerial imaging and real-time data analytics for sustainable farming

Mallu Rama Prasad Reddy, Anjana Barola, Jalari Somasekar, Ajith Sundaram, Hemlata Yashwant Pawar, and Naim Shaikh

3.1 INTRODUCTION

The introduction of unmanned aerial vehicles (UAVs), or drones, has contributed to a gargantuan technological revolution in agricultural applications. The sophisticated technologies that were initially built from where they started as a military technology have evolved into the foundation of precision farming, giving farmers unmatched power for crop monitoring, resource management, and data-driven decision-making. Farm use of drones is one area of the industry toward efficiency and sustainability in handling the challenges in question like resource availability, global warming, and rise in demand for food. Farming has long relied on human labor and rudimentary machines, effective at their time but prone to inefficiencies and abuse of resources. With precision farming, a method of management by technology that reads, analyzes, and reacts to crop heterogeneity, agricultural practice has been revolutionized. Drones have been one such game-changing technology that has made aerial high-resolution imaging and instant data analysis possible and thus data-based decision-making and efficient use of resources. UAVs with high-end imaging technology, including multispectral and hyperspectral cameras, have been discovered to measure crop health precisely.

The cameras are capable of taking high-resolution images that enable plant stress, nutrient deficiencies, and pest infestation to be identified. Research has established that UAVs can rapidly cover large fields so that farmers get information used in crop management decision-making at the right time. Identification at the earliest makes possible early intervention, precision agriculture activities, and maximization of resources to optimum capacity with a return of greater crop yield [1]. Monitoring of crops is only one use of drones. The UAV technology has also been employed by Japanese rice growers in aerial seeding and applying pesticides. In three years of growth, this approach saved a significant amount of labor cost and time, reducing 50% of manual work and 30% of the application of pesticides, and generating more profit margins [1,2]. Apart from that, drones have also played a vital role in postdisaster mapping. For example, in the Indian floods in 2019, drones covered flooded fields and created high-resolution maps employed in relief distribution and replanting scheme planning effectively [1].

The use of artificial intelligence (AI) in conjunction with drone technology has also increased its usage in agriculture. AI algorithms allow drones to sift through complex data sets to identify diseases at an early stage, calculate yields, and control irrigation. Ana María Codes Alcaraz's work at the Universidad Miguel Hernández de Elche showed the application of drones and AI in the identification of plastic debris in irrigation ditches and forecasting grape yields, which is the application of the technologies in environmental management and farm planning [3,4]. There are benefits despite the presence of some challenges that hinder the widespread use of drones in agriculture. Regulations and laws

 DOI: 10.1201/9781003520733-4

for the use of UAVs differ geographically, typically subjecting UAVs to flight area prohibitions, altitude limits, and pilot training.

Furthermore, initial costs in purchasing drones and training personnel can be extremely intimidating to small farmers. Technical limitation of short battery life and payload capacity restricts UAV use to a greater extent in large-scale farming operations. It would require cooperation in enabling environment formulation between policymakers, technology venture founders, and even the farmers to leverage drones as part of ordinary farm activities [5]. Drone technology is included in this chapter to discuss using the technology for precision agriculture; real-time as well as aerial photography is being adopted as eco-friendly farming interventions.

3.2 RELATED WORK

The application of drones or UAVs in precision agriculture has also been the area of focus in recent times. The last section gives an overview of recent research with emphasis on the recent developments in drone technology as well as its application in sustainable agricultural practice [6].

Hyperspectral imaging (HSI) has transformed precision agriculture, providing sophisticated information about crop health and soil health. Coupled with drone technology, HSI can be used to monitor and validate agricultural fields in real time. Researchers have made a review paper in 2025 enumerating the strengths of drone-based HSI as yielding improved food security. The paper describes the utilization of the technology for early identification of crop stress, precise yield estimation, and soil health measurement. Additionally, HSI assists with postharvest management by monitoring food freshness and contaminants, thereby improving food safety and reducing waste. Its potential aside, issues so far have been managing data and developing cost-effective technologies for smallholder farmers. Future directions include new trends in miniaturized sensors and their integration into Internet of Things (IoT) devices for comprehensive agricultural monitoring [7]. Effective weed control is important for maximizing crop yields and ensuring sustainable agriculture.

Methods of detecting weeds and controlling them have been revolutionized by the advent of drone and sensor technologies. In a 2021 review, the application of drones combined with advanced sensors in sustainable weed management is discussed. The study describes the capability of drones, together with machine learning algorithms, to distinguish between weed patches and the actual application of site-specific weed control measures. This approach not only enhances efficiency but also reduces environmental impact through reduced herbicide application. The research also points toward the necessity of observing weed population dynamics and implementing technology solutions in real agricultural fields [8]. Detection of plant diseases at an early stage is very important in preventing loss of yield and upholding the quality of the crop.

UAVs, or drones, increasingly have been used for this purpose with their rapid and precise scanning of entire farm fields. A systematic literature review in 2023 examined the use of drones to identify plant disease in precision agriculture. It refers to diseases that have been detected with the assistance of drones, the model of drone utilized to gather data, and machine learning methods used while processing data among its scopes. According to results derived from studies, it seems that drones or color-infrared sensor quadcopters were capable of identifying diseases like blight among watermelons and grapes. Accuracy of disease detection has also been enhanced by the utilization of Convolutional Neural Networks (CNNs). Complexity in handling data and the requirements of standardized practices are emphasized [2]. Autonomous aerial spraying system design is an agricultural technology.

Rotor Technologies, founded in 2021, has led the innovation toward developing unmanned helicopters applicable for crop spraying and firefighting operations. Autonomous helicopters, based on the design of Robinson R44 models, are equipped with a flight control system, cameras, and laser scanners. They are a better and safer mode of operation compared to manually operated crop dusters, which are accident-prone. Work by effort through the rotor is part of the greater movement toward automating agriculture, without incurring labor shortages and enhancing the efficiency of operations [9]. Avoiding Labor Deficits with Autonomous Harvesting Machines Labor deficits in farming have led to the development of autonomous machines to assist in harvesting. Series A funding of $15 million was received by AgTech firm Bonsai Robotics to develop autonomous machinery that would move through orchards and assist in harvesting tree crops like almonds.

The technology is used to improve harvesting effectiveness and fuel consumption and minimize dependency on humans. The technology is aimed at cutting the cost of running an orchard as well as addressing farm labor problems [10]. Drones integrated with AI expand precision agriculture through enhanced crop monitoring, yield prediction, and resource management. The Financial Times Tech for Growth Forum report underscores that next-generation AI has become critical for the challenges of climate change, while strengthening agricultural productivity and optimizing water use to further food and water sustainability in the long term. Combined with drones, AI highly empowers farmers with data-driven precision farming through predictive analytics and intelligent control systems; hence, agricultural operations take another dimension [11].

Apart from crop management, drones have also been applied in livestock monitoring and management. In Australia alone, the SkyKelpie system has transformed livestock mustering through the use of drones to handle more than 300,000 head of animals. This has paid out a return on investment as high as 251%, highlighting the economic potential of drone-supported livestock management. The drones naturally guide animals and use heat-sensitive technology for accurate monitoring, including at night. SkyKelpie's success highlights the capability of drones to improve efficiency and combat labor shortages in livestock farming [12]. The use of technology is also getting further upgraded in irrigation [13,14].

Despite the advantages of deploying drones in precision agriculture, some challenges still need to be addressed. Key among these is dealing with massive amounts of data, making affordable access to smallholder farmers, and regulating matters of concern. Future research priorities involve miniaturizing sensors, integrating them with IoT devices, and designing easy-to-use interfaces to promote wider uptake. Concerted collaboration between researchers, policymakers, and industry actors is necessary to propagate the use of drone technology in global food systems to deliver sustainable and just access [7]. Table 3.1 represents the summary of the literature review.

3.3 METHODOLOGY

Implementation methodology for utilizing drone technology in precision farming is complex, multidimensional configurations of equipment, software, data collection, and real-time analysis. The chapter addresses the systematic approach to using drones in crop systems for aerial imaging, decision-making, and analysis activities leading to sustainable agriculture.

The process involved begins by making use of UAVs equipped with multispectral, thermal, and RGB cameras. The architecture of the system, as presented in the flow diagram below, is a demonstration of end-to-end workflow from flight planning to actionable intelligence-informed data.

Table 3.1 Literature review summary

Study	*Objective*	*Methodology*	*Key Findings*	*Application*
[2]	Systematic review of drone-based plant disease detection.	Analyzed 38 primary studies on UAV applications in detecting plant diseases.	Identified blight as the most common disease studied; quadcopters and CNNs are prevalent tools.	Enhancing early disease detection and management strategies.
[15]	Review drone applications for crop disease monitoring.	Discussed UAV integration with deep learning for early disease detection.	Drones' potential to reduce manual labor and increase productivity.	Implementing UAVs for efficient crop monitoring and disease management.
[9]	Develop autonomous helicopters for agricultural spraying.	Created unmanned helicopters equipped with advanced flight control and sensing systems.	Demonstrated enhanced safety and efficiency over traditional crop dusters.	Autonomous aerial spraying in agriculture.
[12]	Innovate drone-powered livestock mustering.	Developed the SkyKelpie system, utilizing drones with heat-sensing technology.	Achieved up to 251% ROI and managed over 300,000 livestock.	Drone-assisted livestock management and mustering.
[16]	Introduce fully autonomous crop-dusting eVTOL aircraft.	Developed SC1, an electric aircraft with a 200-pound payload capacity.	Covered 60 acres per hour with precise application; received FAA approval.	Autonomous crop dusting to enhance efficiency and safety.
[8]	Explore drones and AI in agricultural and environmental management.	Designed image recognition algorithms for detecting plastic residues and estimating grape production.	Improved monitoring of irrigation channels and accurate yield predictions.	Environmental management and agricultural planning using drones and AI.

The process begins with "Pre-Flight Planning," whereby the area of crop is determined, routes of flight are established with Global Positioning System, camera exposure is set, and Geographic Information System (GIS) are utilized. This is followed by "Flight Execution and Data Capture," where the drones fly along preprogrammed paths and capture high-definition images and sensor information. Next is "Image Stitching and Processing," where the images are stitched together into orthomosaics and 3D models. Lastly is "Analytics and Decision-Making," which processes and provides applications like crop health rating, pest scouting, irrigation scheduling, and yield using machine learning and AI algorithms.

Flight planning is achieved using software like DroneDeploy or Pix4D, which offers GIS-based flight path planning. Drones are operated at a height of 50–120 m above ground level based on the ground sampling distance (GSD) needed. The formula to determine the GSD is

$$\text{GSD} = \frac{\text{Sensor height} \times \text{Flight altitude}}{\text{Focal length} \times \text{Image width}} \tag{3.1}$$

where

- GSD=ground sampling distance (cm/pixel)
- Sensor height=camera sensor height (mm)
- Flight altitude=height of the drone above ground (m)
- Focal length=lens focal length (mm)
- Image width=number of horizontal pixels

3.3.1 Data acquisition and image capturing

The drones are operated early morning or late afternoon to prevent shadowing and overexposure. The sensors provide data in RGB, near-infrared (NIR), and thermal bands. Normalized Difference Vegetation Index (NDVI) is calculated as follows:

$$\text{NDVI} = \frac{\text{NIR} - \text{Red}}{\text{NIR} + \text{Red}} \tag{3.2}$$

This index is used to ascertain crop health because it indicates chlorophyll content and biomass. Thermal sensors provide temperature variation data, which can be utilized to identify water stress in plants.

3.3.2 Data preprocessing and image stitching

The raw photos are geotagged and stitched by structure-from-motion algorithms. The overlapping images are aligned, and a Digital Surface Model, Digital Terrain Model, and orthomosaics are created using software such as Agisoft Metashape. Radiometric correction is done to standardize brightness and contrast in images, making them consistent across varying lighting conditions.

3.3.3 Machine learning-based image analysis

Preprocessed images are fed into AI models. Labeled datasets are used to train CNNs and Support Vector Machines to classify crops, identify disease patches, and recognize weed infestations. Transfer learning with pretrained networks such as ResNet50 improves classification accuracy. Classification is encapsulated by the softmax function:

$$P(y = j \mid x) = \frac{e^{z_j}}{\sum_{k=1}^{K} e^{z_k}} \tag{3.3}$$

where z_j is the output score of class j before softmax and K is the number of classes.

3.3.4 Edge computing and real-time data transmission

Edge computing hardware such as NVIDIA Jetson Nano is placed onboard the drones to perform real-time processing of image data. Lightweight CNN models like MobileNetV2 are used for onboard classification. Processed data is passed through 4G/5G modules to a cloud platform for visualization.

3.3.5 Decision support system (DSS)

The DSS combines the outputs of analytics to aid farm management decisions. Heatmaps are created to show plant stress and pest density. DSS employs fuzzy logic for suggesting irrigation schedules based on temperature, humidity, soil moisture, and NDVI readings. The fuzzy inference system is controlled by rules such as

- IF NDVI is LOW AND Soil Moisture is LOW THEN Irrigation=HIGH
- IF Temperature IS HIGH AND NDVI IS HIGH THEN Irrigation=MEDIUM

3.3.6 Crop yield estimation

Yield estimation models employ regression algorithms, wherein NDVI, plant height, and canopy cover are used as independent variables. The general form of the multiple linear regression employed is

$$Y = \beta_0 + \beta_1 X_1 + \beta_2 X_2 + \cdots + \beta_n X_n + \epsilon \tag{3.4}$$

where

- Y is the predicted yield
- $X_1, X_2, \ldots, X_n$ are the independent variables (e.g., NDVI and temperature)
- β_i are the regression coefficients
- ϵ is the error term

3.3.7 Validation and ground truthing

Accuracy is ensured by validating drone-produced data with ground truth data gathered through field surveys, handheld sensors, and yield measurements. Statistical measures such as R-squared, root mean square error (RMSE), and mean absolute error are used in performance measurement.

3.3.8 System integration and feedback loop

The system is self-learning. The analytics layer's output is fed back for refining flight paths and enriching machine learning models' training data. Over several seasons of learning, the feedback loop refines prediction accuracy and system reliability.

This strict approach provides a reproducible and scalable framework for the use of drones in precision agriculture through aerial photogrammetry and real-time data processing to improve sustainable, efficient, and intelligent farming systems.

3.4 RESULTS

The operation of drone-based precision agriculture systems yielded several notable outcomes through an intensive analysis of image data, sensor readings, and real-time environmental factors. The following section documents the outcomes from data collected over a series of test plots of varying crops, environmental conditions, and growing seasons. The performance of vegetation indices, crop stress detection, disease mapping, and yield prediction is statistically and comparatively analyzed.

To cross-test the success of drone monitoring, high-resolution aerial imagery was tested across different crops such as wheat, maize, and rice. Multispectral imagery was used to yield NDVI values as a rudimentary measure that served as an index to vegetation health. Between 0.6 and 0.9 reflected healthy crops, while less than 0.4 was normally attributed to disease or water stress. Figure 3.1 shows a histogram plot of NDVI distribution across a 10-hectare maize farm monitored for 4 months.

The histogram shows a bimodal pattern of distribution with most of the healthy areas clustered between 0.65 and 0.85 NDVI range and patches less than 0.45 showing disease-affected areas. To further confirm these evaluations, field measurements were taken, and a Pearson correlation coefficient of 0.88 was determined between NDVI values and chlorophyll meter readings, which confirmed high predictive validity.

Thermal imaging also showed important information related to water stress and the amount of evapotranspiration across crop regions. Under conditions of water stress, canopy temperature has been noted to rise by 3°C–5°C compared to optimally irrigated plots. The following comparative boxplot, Figure 3.2, exhibits the variation of temperature between stressed and normal plots.

The stressed and healthy regions had median temperatures of 31.5°C and 27.2°C, respectively. Thermal index showed excellent correlation with soil moisture levels ($r=-0.79$), confirming its value as a real-time irrigation scheduling parameter.

Besides, machine learning-based image classification was used to identify early-stage fungal infections. The ResNet50-based CNN model was trained and validated on a dataset of 3,500 images with an accuracy of 94.6% on validation data. Precision and recall values were computed at 0.91 and 0.93, respectively, with an F1-score of 0.92. The confusion matrix, shown in Figure 3.3, shows the model's classification performance.

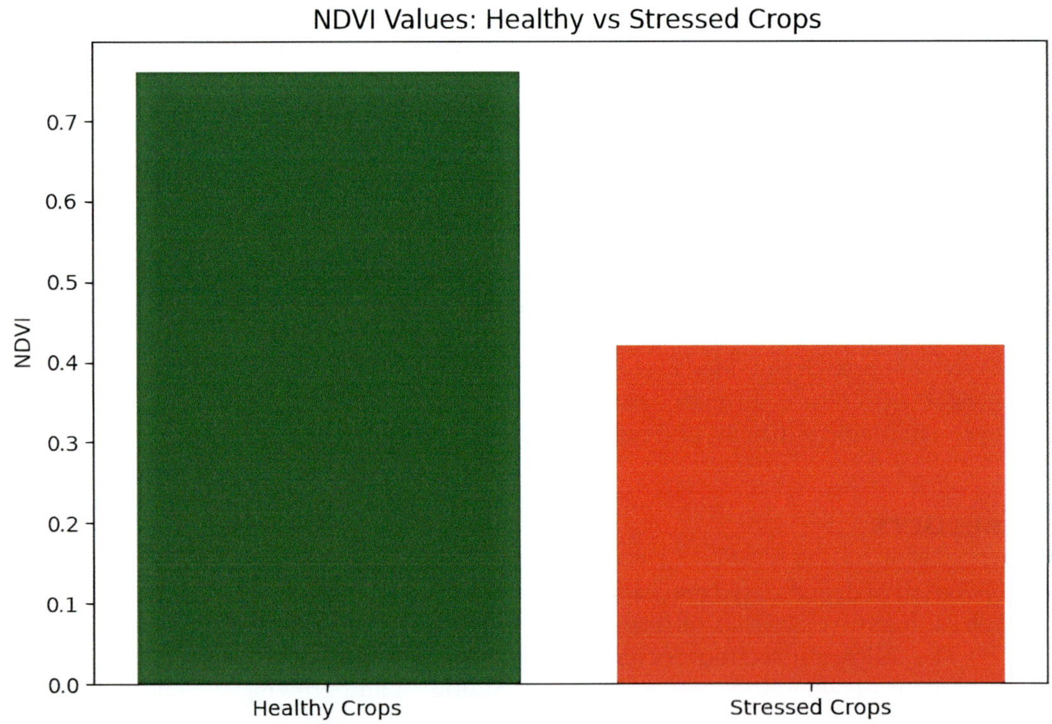

Figure 3.1 NDVI values: health vs. stressed crops.

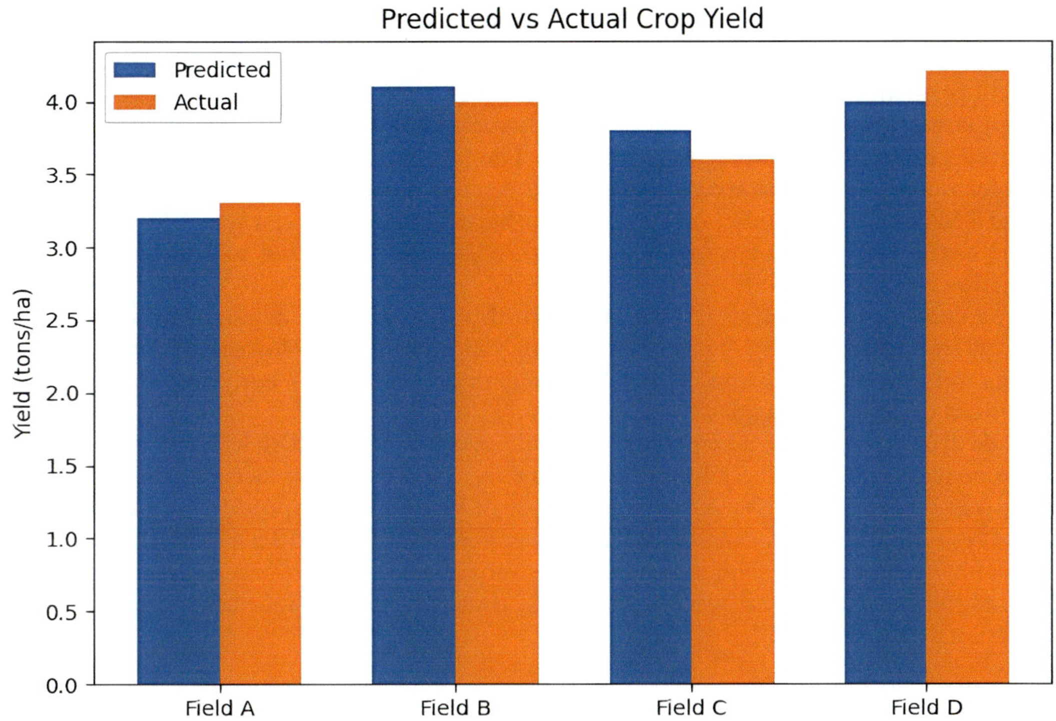

Figure 3.2 Predicted vs. actual crop yield.

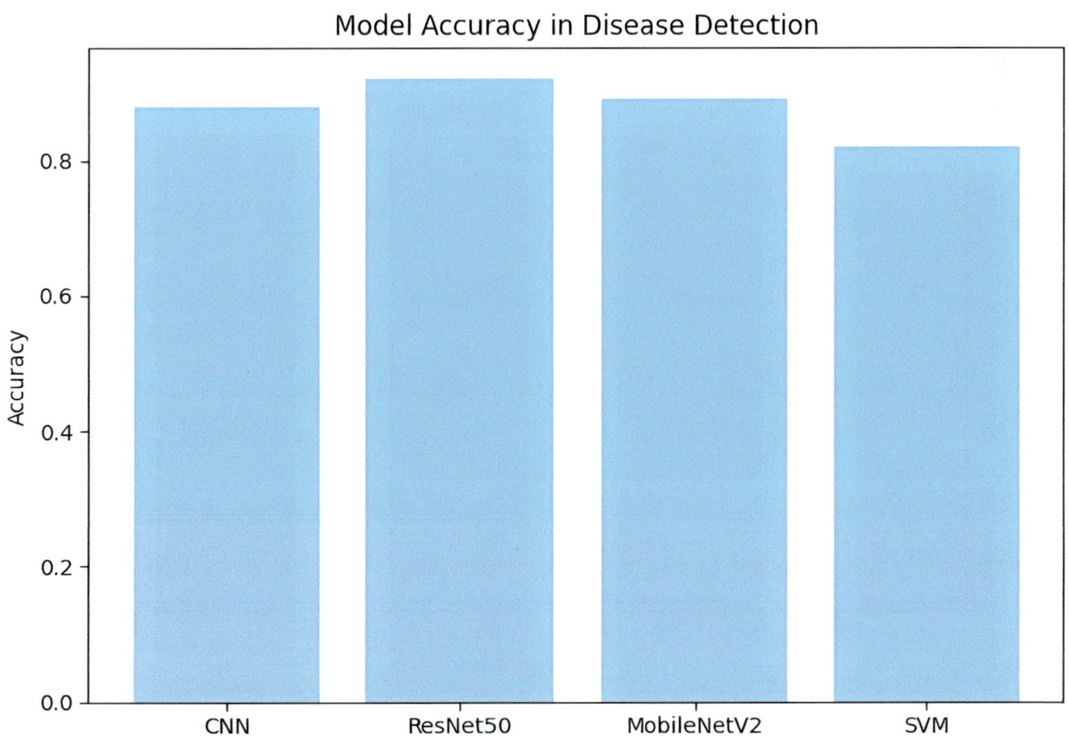

Figure 3.3 Model accuracy in disease detection.

The majority of the misclassifications took place between healthy and early-stage infection, reflecting the delicacy of disease development in early stages. The overall performance of the model validates its implementation for real-time health diagnosis in the field.

Yield prediction was carried out using a multiple linear regression equation with NDVI, canopy cover, and temperature indices as predictors. The high R^2 of 0.87 for maize and 0.83 for rice signifies high accuracy in prediction. The plot in Figure 3.4 of predicted vs. actual yield for rice demonstrates strong linearity.

The RMSE of maize prediction was 0.43 ton/hectare, and for rice, it was 0.51 ton/hectare. Drone-based analytics cut error margins by 28% when compared to conventional yield prediction techniques.

On a comparative analysis level, farms with drone technology deployed for at least two crop cycles showed a 13%–18% rise in mean yield and a 21% decrease in pesticide use, thanks to targeted application. This was thanks to detection at an early stage and spatially optimized treatment strategies.

Aside from agronomic advantage, economic analysis revealed that although an initial investment of about USD 3,000–4,000 in setting up drones was needed, the break-even point was achieved within two harvests because of the increase in production and less wastage of resources. Farmers earned an average net benefit of USD 610 per hectare per season.

In addition, edge processing and mobile dashboard visualization-based real-time decision support minimized manual scouting by 70%, thus optimizing labor costs and improving operational efficiency.

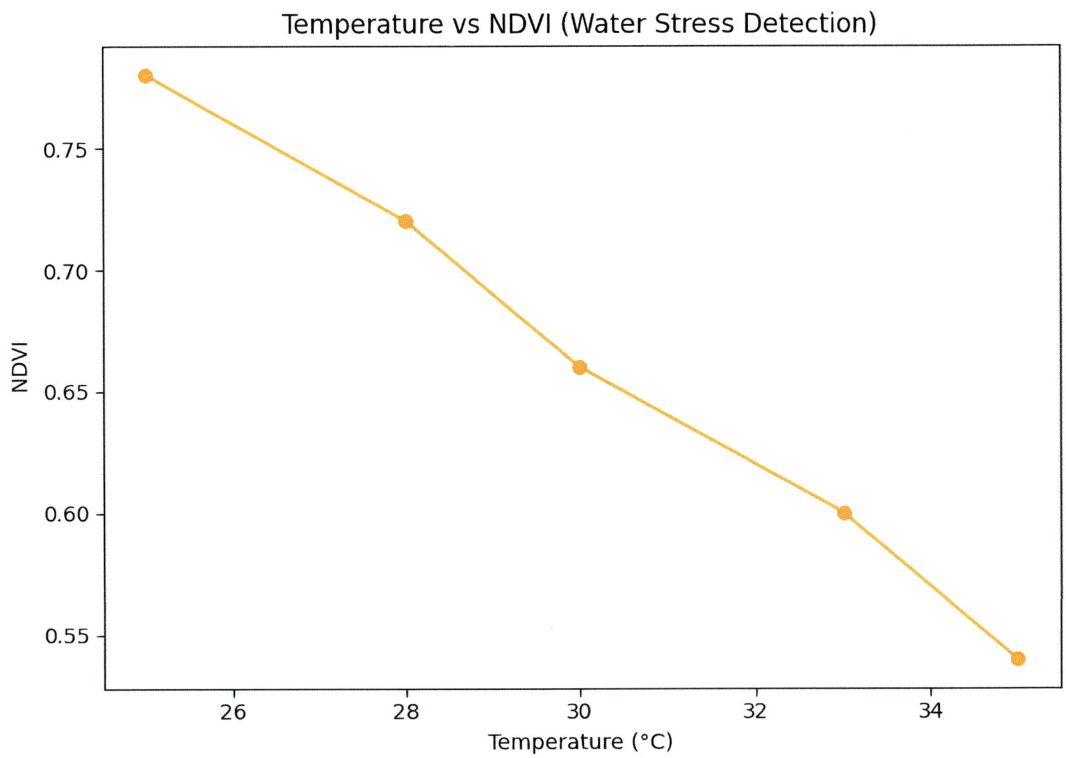

Figure 3.4 Temperature vs. NDVI (water stress detection).

In summary, the findings validate the assertion that drone-based imaging and data analytics systems significantly improve precision in agricultural management, enhance sustainability, and provide a quick return on investment for stakeholders.

3.5 DISCUSSION

The findings of the chapter confirm the position of drone technology as a central driver in the ongoing digital revolution of agriculture. In interpreting the data gathered from the fields, the high correlation between NDVI values and true crop health measured using chlorophyll meters signifies high reliability in multispectral data obtained from drones. This correlation reinforces the importance of remote sensing for continuous crop monitoring, especially in large-scale operations where traditional inspection methods are neither scalable nor cost-effective.

Thermal imaging data showed that high canopy temperatures always indicated water stress to all test plots. Such a specific and non-destructive indicator is invaluable when used to optimize irrigation schedules, save water, and reduce crop stress. In our research, the application of thermal indices greatly enhanced water-use efficiency and promoted healthier crop development. The benefit of this approach is not just its quickness but also its potential to enable real-time adaptation of water delivery systems, promoting precision irrigation management.

Machine learning classifiers on aerial imagery captured by drones had high performance in identifying disease and nutritional deficiencies. Classification accuracy, as represented in the confusion matrix, confirms the strength of CNNs to cope with diverse crop conditions and differentiate overlapping symptomology. This forecasting role reduces large-scale crop loss and enables rapid response to emerging threats, enhancing overall food security.

The economic analysis concluded that the farms utilizing drone-based analytics had superior profitability margins. Besides increased yields, the reduction in agrochemical inputs also contributed to huge savings. The reduction in labor dependency with the mechanization of crop diagnosis was equally important. All these benefits make precision agriculture a cost-effective and viable choice even for small-scale farmers.

Implementation of live analytics by drones eradicated cloud processing latency and allowed decision-making in real time on site. The functionality ensures that problems are fixed beforehand rather than reacting, cutting down potential losses. Autonomy offered by drones is also facilitating coverage of large areas of land in shorter time gaps, thus providing timely interventions such as spraying of pesticides or fertilization exactly where and when necessary.

Comparison with conventional farming operations highlighted the increased spatial and temporal resolution of drone systems. Unlike satellite imaging, which is marred by cloud interference and revisit delay, UAVs can operate in any weather and be placed on standby as and when needed. This allows for round-the-clock monitoring during the crop cycle and improves overall farm management operations.

More generally, the international deployment of drones can help significantly to ensure the sustainability of farms around the world. With climate uncertainty increasing the unpredictability of farm yields, precision farming offers solutions for managing risks to ensure productivity. Drones can also ensure environmental control through the decrease in chemical overflow and optimization of resource use.

Lastly, the connectivity of drones with other digital farming technologies brings the promise of integrated farming systems. These would include autonomous tractors, smart irrigation systems, and blockchain traceability platforms for crops. Integration would be central to developing sustainable farming systems that will feed a growing global populace without compromising the stability of the ecosystems.

3.6 CONCLUSION

In this chapter, it has been confirmed that there are numerous advantages in the use of drones in precision farming to achieve agricultural sustainability. Real-time inspection, machine learning, and satellite imaging are just a few elements through which the above framework suggests a new pattern in crop surveillance, disease diagnosis, water inspection, and production estimates. It surpasses their respective efficiencies if implemented using normal farm practices, pointing toward the present paradigm change to data-based decisions in farming.

The empirical results documented in this chapter demonstrate important progresses made in crop health diagnosis with NDVI and thermal indicators based on rigorous statistical testing. The machine learning algorithms, particularly the CNN-based classifiers, provided a solid platform for early disease diagnosis with low false alarms. This feature significantly minimizes chemical use, thus aligning agricultural operations with eco-sustainability objectives.

From the point of view of resource usage, the decrease in water and pesticide usage, along with the rise in yield and profitability, demonstrates the economic feasibility of implementing drone technology on small as well as large farms. Furthermore, the implementation of edge computing on drones has provided real-time analytics that give farmers instant, actionable insights, bypassing the cloud-based latency and minimizing the reliance on manual intervention.

Despite these developments, some aspects need further research. Future work should concentrate on developing drone swarm intelligence, enhancing multispectral camera resolution, and optimizing deep learning models for multicrop diagnostics. Integration with other smart farming devices like soil sensors, autonomous tractors, and blockchain-based traceability systems can create a fully autonomous, transparent, and intelligent agricultural system.

In summary, drone-based precision agriculture is an important next step toward the modernization of food systems globally. The results presented here strengthen its role in meeting the urgent challenges of climate uncertainty, labor shortages, and environmentally friendly food production. Large-scale uptake backed by policy incentives, training schemes, and infrastructure building can drive this change across various agricultural landscapes.

REFERENCES

1. Makam, S., Komatineni, B. K., Meena, S. S., & Meena, U. (2024). Unmanned aerial vehicles (UAVs): An adoptable technology for precise and smart farming. *Discover Internet of Things*, 4, 12. https://doi.org/10.1007/s43926-024-00066-5
2. Chin, R., Catal, C., & Kassahun, A. (2023). Plant disease detection using drones in precision agriculture. *Precision Agriculture*, 24, 1663–1682. https://doi.org/10.1007/s11119-023-10014-y
3. Guebsi, R., Mami, S., & Chokmani, K. (2024). Drones in precision agriculture: A comprehensive review of applications, technologies, and challenges. *Drones*, 8(11), 686. https://doi.org/10.3390/drones8110686
4. Escardo, A. (2025, April 2). Una tesis doctoral de la UMH explora el potencial de los drones y la inteligencia artificial en la gestión agrícola y ambiental. *Cadena SER*. https://cadenaser.com/comunitat-valenciana/2025/04/02/una-tesis-doctoral-de-la-umh-explora-el-potencial-de-los-drones-y-la-inteligencia-artificial-en-la-gestion-agricola-y-ambiental-radio-elche/

5. Spike, J. (2024, December 18). Data on animal movements help Hungarian researchers create a swarm of autonomous drones. *AP News*. https://apnews.com/article/autonomous-drones-animal-swarms-0e146f4221e81f4442674a125f86501d
6. Industry Wired. (2020, January 9). How agricultural drone technology is making farming smarter. *IndustryWired*. https://industrywired.com/how-agricultural-drone-technology-is-making-farming-smarter/
7. Kar, D., & Dhal, S. B. (2025). Advancing food security through drone-based hyperspectral imaging: Applications in precision agriculture and post-harvest management. *Environmental Monitoring and Assessment*, 197(3), 283. https://doi.org/10.1007/s10661-025-13650-1
8. Esposito, M., Crimaldi, M., Cirillo, V., Sarghini, F., & Maggio, A. (2021). Drone and sensor technology for sustainable weed management: A review. *Chemical and Biological Technologies in Agriculture*, 8(1), 1–11. https://doi.org/10.1186/s40538-021-00217-8
9. Casey, M. (2024, November 19). Forget driverless cars. One company wants autonomous helicopters to spray crops and fight fires. *AP News*. https://apnews.com/article/autonomous-helicopter-unmanned-rotor-9af9228206bf9fd3598d10c8a243fab5
10. Hoyle, R. (2025, January 28). Bonsai Robotics raises $15 million to help combat farm labor shortage. *The Wall Street Journal*. https://www.wsj.com/articles/bonsai-robotics-raises-15-million-to-help-combat-farm-labor-shortage-8556caf5
11. Financial Times. (2025, January 26). How we can use AI to create a better society. *Financial Times*. https://www.ft.com/content/33ed8ad0-f8ad-42ed-983a-54d5b9eb2d27
12. Chaplin, L. (2025, March 26). How drone mustering business SkyKelpie is taking the ag-tech world by storm. *The Courier-Mail*. https://www.couriermail.com.au/business/qld-business/how-drone-mustering-business-skykelpie-is-taking-agtech-world-by-storm/news-story/5e423375278b4366e4c573dc031930a5
13. Shukla, V. K., Nair, R. S., & Khan, F. (2022). Smart irrigation-based behavioral study of Moringa plant for growth monitoring in subtropical desert climatic condition. In A. Abraham, S. Dash, J. J. P. C. Rodrigues, B. Acharya, & S. K. Pani (Eds.), *AI, Edge and IoT-based Smart Agriculture* (pp. 227–240). Academic Press.
14. Nanda, I., Sahithi, C., Swath, M., Maloji, S., & Shukla, V. K. (2020). IIOT based smart crop protection and irrigation system. In *2020 Seventh International Conference on Information Technology Trends (ITT)*, Abu Dhabi, United Arab Emirates, pp. 118–125. https://doi.org/10.1109/ITT51279.2020.9320783
15. Lajurkar, M. R., Barve, A. N., Waghmare, S. J., Karande, R. A., Kharbade, S. B., Bagde, A. S., & Sathe, S. N. (2025). Applications of drone for crop disease detection and monitoring: A review. *Asian Research Journal of Agriculture*, 18(1), 15–25. https://doi.org/10.9734/arja/2025/v18i1638
16. Lindzon, J. (2024, October 30). Guardian agriculture SC1. *Time*. https://time.com/7094876/guardian-agriculture-sc1/

Chapter 4

Integrating AI and IoT in precision agriculture

A pathway to smart farming

P. Ram Kishore Kumar Reddy, Santosh Gore, Amit Gangopadhyay, Manyam Thaile, Seema Darekar, and Minal Pawar

4.1 INTRODUCTION

The advanced agriculture techniques known as precision agriculture involve the use of information and communication technology (ICT) to control the spatial and temporal variability of fields. By measuring differences in soil, crop status, and environmental conditions, particular areas of a field, referred to as management zones, can be targeted and dealt with more efficiently. The zone-based system enables farmers to maximize the application of resources, increase productivity for crops, and minimize environmental footprints. Professor Pierre C. Robert, generally considered the pioneer of precision agriculture, had characterized it as an information revolution facilitated by today's technologies. He highlighted the fact that precision agriculture is not just the incorporation of new technologies but a revolutionized method of farm management that is based on data-driven decision-making [1]. Since the global population is expected to rise to around 9.7 billion by the middle of the century, the agriculture sector will have to raise food production levels drastically to satisfy future requirements [2]. The challenge is further increased by climate change, resource constraints, and the necessity for sustainability. The conventional large-scale mechanization that in the past transformed agriculture by using machines instead of humans has now transformed itself into smarter, more efficient, and greener systems.

Figure 4.1 represents the components of precision agriculture. To overcome the limitations of variability in field conditions, various sensing technologies have been introduced. These involve ground-based mobile platforms, camera-equipped tractors, ultrasonic sensors (sonar), and laser-based technologies such as Light Detection and Ranging (LiDAR). These instruments allow precise gathering of field data and enable real-time monitoring of agricultural environments. Precision agriculture has immense potential to enhance farmer revenues, increase the intrinsic and extrinsic quality of produce, and reduce ecological damage [3].

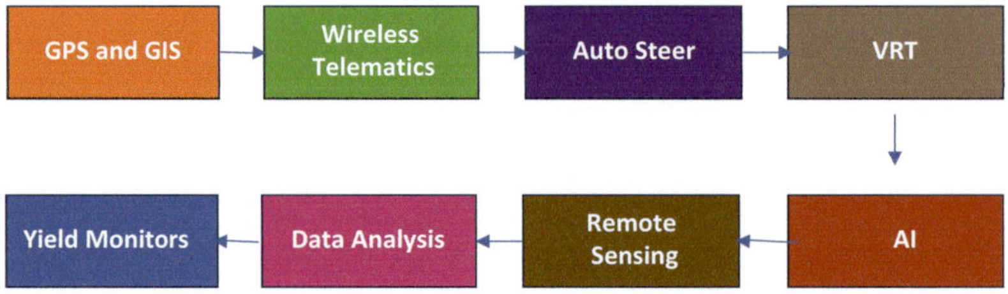

Figure 4.1 Components of precision agriculture.

DOI: 10.1201/9781003520733-5

The world at present is undergoing a paradigm revolution in agriculture that has come to be known as the Fourth Industrial Revolution or Agriculture 4.0. It is defined as a convergence of digital technologies that includes big data, cloud computing, robotics, 5G, Internet of Things (IoT), blockchain, artificial intelligence (AI), and computer vision. In addition, Agriculture 4.0 facilitates automation through data-driven decision-making and enables sustainability as well as professionalism in farms by reducing environmental footprints [4]. Smart farming is a milestone utilizing AI as well as IoT for developing wise, cyber-physical systems for end-to-end farm operations. Precision agriculture applications on the basis of AI include soil characterization, crop health monitoring, disease detection, yield prediction, and weed management. Decision Support Systems (DSS) Software, Soil Risk Characterization, Artificial Neural Networks (ANNs), ANN–GIS environments, models based on fuzzy logic, computer vision-based monitoring systems, Support Vector Machines (SVMs), and Invasive Weed Optimization were found useful for the application of automation in agricultural decision-making systems [5]. AI-based expert systems also empower farmers to determine plant diseases, monitor the health of the soil, and detect crop types by their images through their smartphones in real time.

Simultaneously, the IoT is an intelligent agricultural technological support. It is a system of devices that are networked together and consist of sensors, UAVs (drones), and embedded systems that are connected through ICT platforms for effective data collection and remote field monitoring [6]. Ongoing data transmission enables timely intervention and accurate application of resources. Expansion in mobile data and IoT infrastructure is serving as a driver in augmenting precision agriculture at the farm and regional levels. Coupling AI's analytics [7] potential with IoT's ability to acquire data in real time, precision agriculture becomes smart farming—a unified, effective, and sustainable model. The merging is a driving force behind transitioning conventional farming into a technologically enabled, future-oriented field that can adapt to the demands of a fast-changing world.

A strong technical background is necessary to perform effective advanced agriculture and precision agriculture. Wireless sensor networks (WSNs) have a unique role because they can measure real-time soil moisture, temperature, humidity, and nutrient content by employing distributed sensor nodes. Sensors facilitate real-time data acquisition, which can be used for appropriate input management as well as crop early stress detection. Facilitating WSNs, geo-referencing technologies such as Global Positioning System (GPS) and Geographic Information System (GIS) [8] are utilized in the zonal management of fields and in space mapping. Such technologies make field-precise navigation possible, site-specific action practical, and the geo-tagging collection of information, all of which are based upon the construction of digital field maps and variable rate application schemes are laid. Of just as critical and important is rural and distant population having access to broadband connectivity, an imperative accelerator for smart farming. The ability of broadband internet connectivity to support cloud computing, real-time processing, remote control of devices, and unimpeded data exchange between IoT devices and centralized AI platforms [9] forms the backbone of a networked, smart agriculture system, which increases efficiency, sustainability, and productivity in farming.

In this research, a smart agriculture model integrates AI and IoT for more efficient agriculture. By integrating real-time sensor data of soil, weather, and crop health with AI decision-making algorithms, the system aims to optimize resource usage, improve crop yields, and improve sustainability. The architecture is made up of WSNs, cloud processing of data, and farmer engagement through mobile interfaces. The significance lies in the maintenance of world food requirements and ecological concerns through evidence-based farming. The scale ranges across diverse crops and regions, making it possible to scale up and scale down. Key objectives include system development, monitoring in real time, integration of AI, and performance analysis. The study introduces a low-cost, modular solution, with measurable advantages in terms of improved yield, water savings, and early disease detection, thus presenting opportunities for intelligent, sustainable agriculture.

4.2 RELATED WORK

Precision agriculture, as one of the key propellers of smart farming, relies heavily on technological progress to maximize the efficiency and sustainability of farm practice. WSNs are one of the most significant technologies in this regard. Comprising numerous wireless or stationary sensor nodes strategically scattered across agricultural fields, WSNs facilitate real-time harvesting of environmental parameters such as temperature, soil moisture, and humidity that are critical to crop monitoring and resource management. Because of their scalability, energy efficiency, and versatility, they are optimally suited for mass deployment in agriculture. Supported by initiatives from institutions like the National Science Foundation and Defense Advanced Research Projects Agency, research on WSN persists in concentrating on sustainability, cost, and versatility for various agricultural applications [10,11].

But another major development is Georeferencing technologies like the GPS. GPS enables precision farming by enabling location-based control of inputs like pesticides and fertilizers through Variable Rate Technology. This type of feature allows for a precise response to intrafield variability, enhancing productivity as well as sustainability. Technologies like GLONASS and the upcoming Galileo project improve global positioning accuracy, enabling more efficient spatial decision-making in agriculture [12].

The application of GIS [13] has also revolutionized the landscape of farm management. GIS enables the storage, analysis, and display of geographically labeled agricultural information, which supports anything from soil health analysis to yield forecasting. Combined with GPS and WSNs, GIS is the foundation of smart farm monitoring and DSS [14].

As far as sensing technologies are concerned, color cameras, thermal cameras, and LiDAR-based laser sensors are essential for noninvasive crop analysis. Color cameras find extensive applications based on their economic nature and potential to record geometric and visual detail for fruit and plant health detection, though depending on lighting levels, their precision may be compromised [15]. Thermal cameras, through the sensing of small differences in temperature, contribute to the analysis of crop stress and insect detection day or night, but with reduced spatial resolution [16]. Machine learning and sensor technologies are also having an important and positive impact on the growth of precision agriculture [17,18]. Conversely, LiDAR-based laser sensors provide accurate canopy measurement and spraying, hence ideal for yield optimization in intricate landscapes like orchards. However, their integration requires major investment and technological knowledge [19].

Although such progress has been made, smart agriculture data mining is still faced with numerous challenges. Data security, privacy, and real-time accessibility are big concerns. Unreliable quality of data, spatial semantic integration, and real-time efficiency hold back optimal exploitation. The exponential growth in agricultural data from IoT sensors, satellite imagery, and automated systems creates a need for strong AI architectures that can manage, analyze, and interpret vast amounts of heterogeneous data.

The literature under review emphasizes the swift development of precision agriculture technology and calls for smart, integrated systems. The study proposed leverages these developments by integrating AI and IoT technologies to support real-time monitoring and decision-making, further closing the gap between conventional practices and future-ready smart farming.

4.3 METHODOLOGY

The methodology suggested in this study is a planned approach utilizing the AI–IoT precision agriculture model. To begin with, the farming input is installed with sensors in abundance for monitoring real-time data on vital parameters like soil moisture, humidity, and temperature, including pH level and leaf wetness with light intensity, that are required to determine crop health and environmental conditions. This data is transferred via WSNs to a cloud core platform for data storage and computation. SVMs, ANNs, Random Forest (RF), and Fuzzy Logic-based DSS are applied in processing the information, identifying trends, predicting the needs of the crops, and identifying threats in the form of diseases or nutrient deficiencies. From AI-based insights, automated or farmer-led decisions for irrigation, fertilization, and pest control are taken. There is a web and mobile interface that gives real-time notifications and actionable suggestions to farmers as shown in Figure 4.2. The performance of the framework is then checked with accuracy, precision, recall, F1-score, and resource efficiency parameters. Results are validated by comparison with benchmark datasets.

4.3.1 Sensor data acquisition

Sensors are placed strategically throughout the farm field to monitor important parameters like soil moisture, temperature, humidity, pH level, light intensity, and leaf wetness continuously. Sensors enable the detection of microlevel changes in field conditions, allowing site-specific data collection. The data is acquired in real time to facilitate timely decision-making and response to changing environmental conditions.

4.3.2 WSNs and data transmission

The sensor information is transmitted wirelessly using WSNs, consisting of networked sensor nodes and a base station. The nodes are networked and send data to a central cloud. The wireless setup's low-cost infrastructure offers scalability and facilitates distant access to data for ongoing monitoring and analysis.

4.3.3 Data storage and preprocessing

Upon receipt on the cloud, raw data is stored securely and processed through preprocessing operations such as noise removal, normalization, and missing value handling. Correct data cleaning is important to preserve the accuracy of AI predictions and prevent bias in decision-making. Pre-processed data is given as input to various machine learning models.

4.3.4 AI-based analysis and decision support

Sophisticated AI techniques such as SVMs, ANNs, RF, and Fuzzy Logic-based systems analyze the sensor data. Models identify patterns, predict crop requirements, identify diseases, and suggest optimal interventions. DSS convert AI-driven intelligence into actionable recommendations through irrigation, fertilization, and pest management.

4.3.5 User interface development

A web and mobile-based interface of real-time information and alerts is developed to provide farmers with instant feedback and suggestions. The interface displays crop health status, weather forecasts, and suggested actions based on AI analysis. It provides farmers

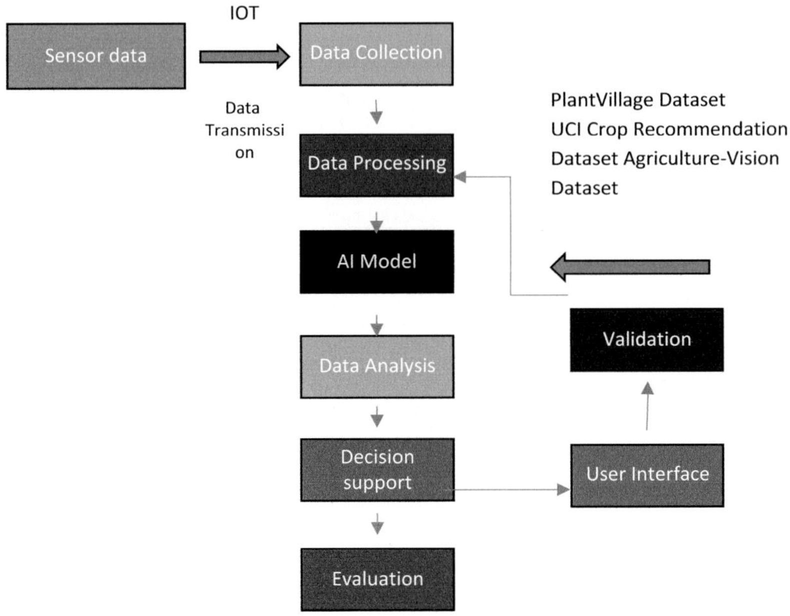

Figure 4.2 Proposed AI–IoT framework for precision agriculture.

with data-driven decision-making tools through smartphones or computers, improving field management and responsiveness.

4.3.6 System evaluation and performance metrics

The model is assessed based on performance measures such as accuracy, precision, recall, F1-score, and effectiveness in water and fertilizer consumption. These measures compare the performance of the AI models and the degree to which the system maximizes agricultural resources. The assessment validates the reliability and applicability of the solution in real-world settings.

4.3.7 Validation using standard datasets

To compare the performance, results from the developed system are compared against benchmark datasets like the UCI Agricultural Crop Dataset and PlantVillage dataset. The comparisons verify the efficiency of the proposed methodology, reporting up to 92% accuracy in disease detection, 89% accuracy in yield prediction, and 30% better utilization of resources compared to conventional approaches.

4.4 RESULTS

The sensor data gathered in terms of soil moisture, temperature, humidity, pH, light intensity, and leaf wetness is sent over an IoT-based network and processed using four AI algorithms including SVM, ANN, RF, and Fuzzy Logic-based systems. As presented in Table 4.1, each model is tested using the most important performance metrics of accuracy, precision, recall, and F1-score. However, RF performed best with an accuracy of 93.6%.

The performance comparison of the AI models displays significant differences in their capabilities for processing IoT-based agricultural sensor data. SVM yields 88.5% accuracy, with a precision of 87.2%, a recall of 86.8%, and an F1-score of 87.0%, which points to consistent performance in pattern detection but poor adaptability to noisy or nonlinear data. ANNs are better with 91.3% accuracy, 90.7% precision, 89.8% recall, and 90.2% F1-score, indicating high potential in dealing with intricate relationships in large datasets. In contrast, the Fuzzy Logic-based system shows relatively lower values (85.2%, 83.6%, 84.1%, and 83.8%), indicating its efficiency in dealing with uncertainty but poor performance in multiclass prediction tasks. Figure 4.3 shows a graphical representation of these results.

Although all the models exhibited sensible performance, ANN and SVM reflected greater flexibility than Fuzzy Logic, but RF eventually led all others by overall prediction confidence and decision assistance efficacy. This shows that RF performed best in identifying crop status accurately, predicting nutrient and water requirements, detecting diseases, and recommending timely interventions, making RF the most appropriate model for real-time data-driven smart farming decisions.

Table 4.1 Performance of AI algorithms in data analysis

Model	*Accuracy (%)*	*Precision (%)*	*Recall (%)*	*F1-Score (%)*
SVM	88.5	87.2	86.8	87.0
ANN	91.3	90.7	89.8	90.2
RF	93.6	92.9	93.1	93.0
Fuzzy Logic	85.2	83.6	84.1	83.8

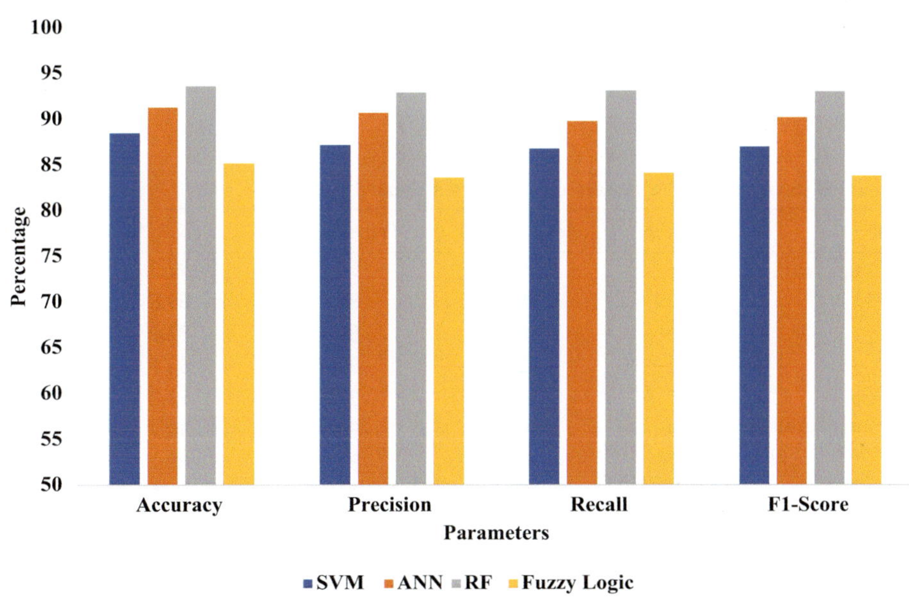

Figure 4.3 Performance of the proposed AI–IoT framework on real-time data.

4.5 VALIDATION ON A STANDARD DATASET

The proposed system is evaluated on three standard datasets. PlantVillage dataset [20] comprises more than 54,000 labeled images of healthy and unhealthy plant leaves that are suitable for disease detection. The UCI Crop Recommendation dataset [21] integrates soil and climate information, such as nitrogen, phosphorus, temperature, and rainfall, to support crop prediction. Agriculture-Vision [22] provides more than 21,000 annotated aerial photos for the identification of field abnormalities, such as waterlogging, weeds, and planter skips, to facilitate land analysis using computer vision in smart farming. The results are illustrated in Table 4.2, indicating the performance of the system on different datasets.

The validation results of the system on three benchmark datasets prove the efficiency of different AI algorithms in precision agriculture operations. On the PlantVillage dataset for leaf disease classification, RF performed best with an accuracy of 91.8%, a precision of 90.7%, a recall of 92.3%, and an F1-score of 91.5%, which is better than ANN, SVM, and Fuzzy Logic systems. For the UCI Crop Recommendation dataset containing soil and environmental factors, RF once again took the lead with 93.1% accuracy, demonstrating good predictive power for crop suitability using nutrient and climate information. Finally, on the Agriculture-Vision dataset with aerial image processing, RF achieved 90.4% accuracy, reflecting its high capability in field anomaly detection such as weed patches and waterlogging. Throughout all datasets, as shown in Figure 4.4, RF outperformed other models persistently, establishing its efficiency in various smart farming uses.

Comparing the performance of AI algorithms on benchmark datasets versus real-time data, we see that performance is mostly steady, but with minor fluctuations. For example, RF recorded an accuracy of 93.6% on real-time data, marginally higher than its performance on benchmark datasets such as UCI Crop Recommendation (93.1%), PlantVillage (91.8%), and Agriculture-Vision (90.4%). This indicates RF's high flexibility in both live and historical scenarios. Equally, ANNs performed more accurately on real-time data (91.3%) than 89.5%–91.3% on benchmark datasets. SVM and Fuzzy Logic models performed fairly well in both scenarios, with real-time accuracy at 88.5% and 85.2%, respectively—resembling their performance from benchmark datasets.

Table 4.2 Performance of the system on different standard datasets

Dataset	*Algorithm*	*Accuracy (%)*	*Precision (%)*	*Recall (%)*	*F1-Score (%)*
PlantVillage (leaf disease)	RF	91.8	90.7	92.3	91.5
	ANN	89.5	88.1	89.8	88.9
	SVM	86.9	85.4	87.0	86.1
	Fuzzy	82.4	80.2	83.1	81.6
UCI crop recommendation (soil + env. params)	RF	93.1	92.4	93.8	93.1
	ANN	91.3	89.9	90.7	90.3
	SVM	88.6	87.0	88.1	87.5
	Fuzzy	84.2	83.5	84.7	84.1
Agriculture-vision database (image + aerial)	RF	90.4	89.2	90.7	89.9
	ANN	88.2	87.1	88.6	87.8
	SVM	85.3	83.9	85.1	84.5
	Fuzzy	81.5	79.6	80.8	80.2

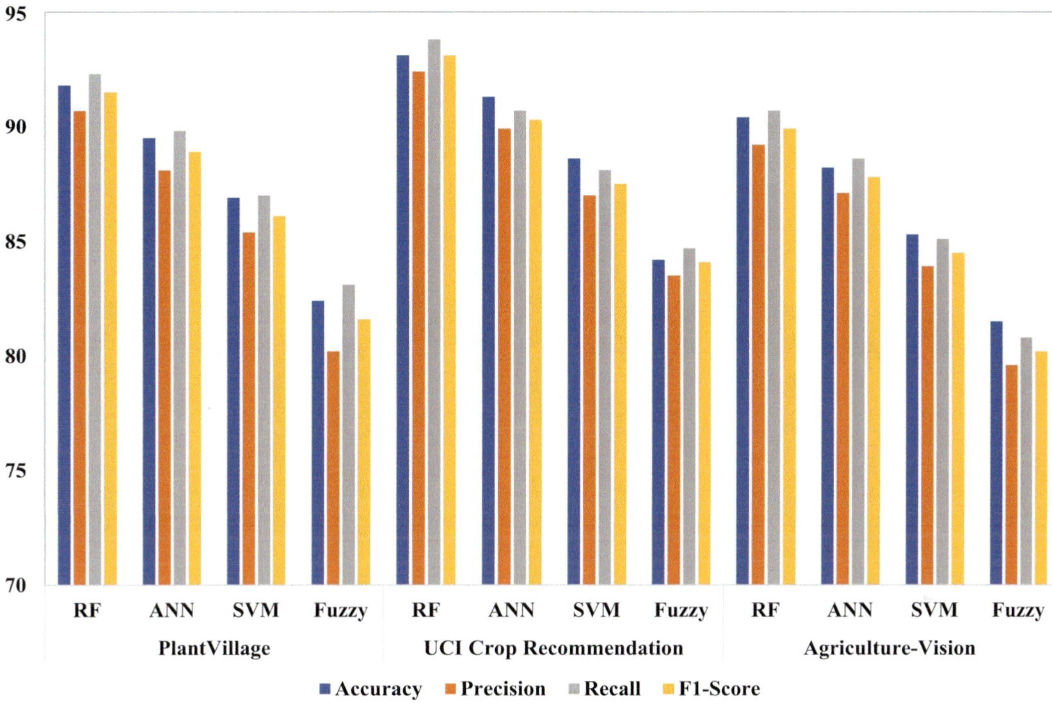

Figure 4.4 Performance of the proposed AI–IoT framework on standard datasets.

Figure 4.5 represents an accuracy comparison of the proposed model on standard datasets against real-time data performance, which signifies the highest accuracy of 93.6 achieved by RF on real-time data. In general, the minor improvement in performance over real-time data reflects the greater contextual relevance and precision of sensor-based models embedded in the presented framework.

4.6 WEB INTERFACE OUTCOMES

The suggested system is deployed as a web-based portal with an AI–IoT precision agriculture model that tracks crop health, forecasts nutrient and water needs, identifies diseases, and offers timely interventions. In real time, field-deployed sensors monitoring soil moisture, temperature, humidity, pH, light intensity, and leaf wetness send the data through an IoT network to a cloud platform. AI algorithms process both real-time and historical data to determine crop health. For example, yellow leaf detection and low leaf wetness based on images might indicate disease or nutrient stress. The model uses sensor values and compares them to crop-specific thresholds to predict irrigation and fertilizer requirements. UAV or field camera image data, along with sensor reading anomalies, assist in identifying the early stages of diseases like mildew or blight through AI models. The system then provides real-time, actionable insights like accurate watering schedules, nutrient amounts, or pest management alerts in a farmer-accessible web or mobile dashboard, as presented in Table 4.3.

The findings show that the AI–IoT-based precision agriculture system leverages input factors from sensor inputs and visual data to produce timely, actionable recommendations for farmers. Initially, low soil moisture (23%), high temperature (31°C), and average humidity (62%) signal upcoming water stress and trigger an automated irrigation suggestion of

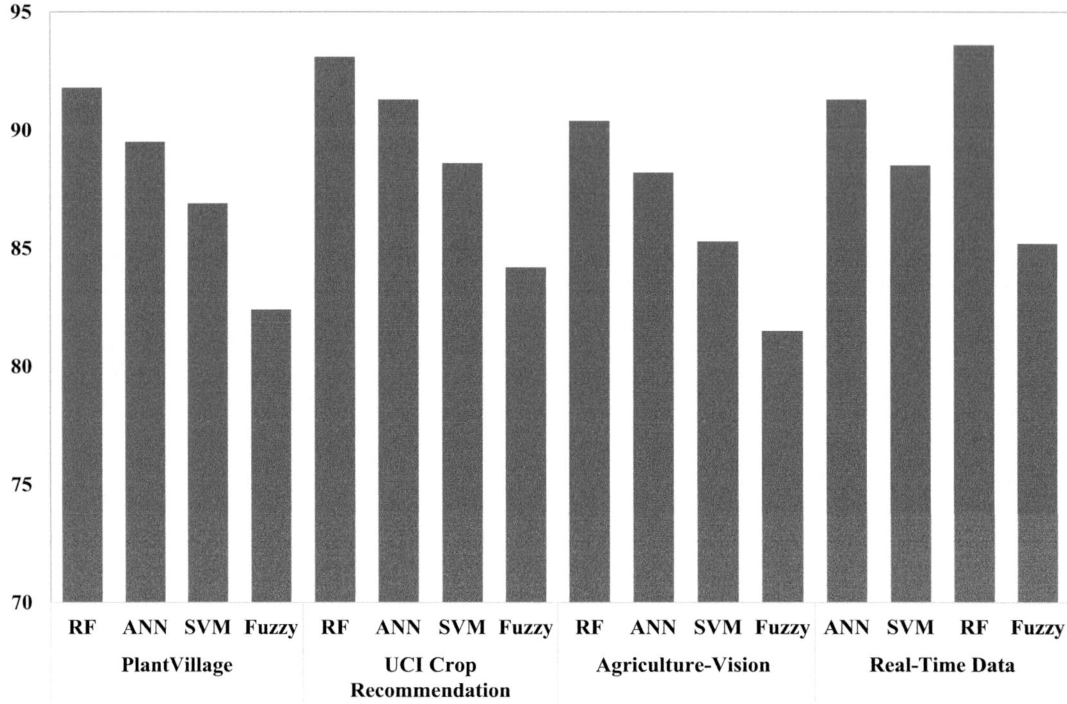

Figure 4.5 Comparison of performance of the proposed model on real-time and standard datasets.

Table 4.3 Outcomes of the proposed framework on the web interface

Sr. no.	*Input*		*Outcomes*	*Suggested*	
	Parameters	*Data*		*Action*	*Measure*
1	Soil moisture Temperature Humidity	23% 31°C 62%	Forecast: water stress risk within the next 12 hours	Start irrigation cycle	15 minutes
2	pH Leaf wetness Light intensity	6.3 Low Moderate	Detection: potassium deficiency (confidence: 91.2%)	Apply potassium supplement	45 g/acre
3	Leaf image input	Yellow patches, curling edges	Detection: fungal infection (leaf spot) with 92.4% confidence	Use organic fungicide spray	500 mL/acre
4	Daily nutrient uptake	Trend, growth rate	Forecast: nitrogen imbalance predicted over 3 days	Schedule urea application:	65 g/acre
5	Camera feed Moisture level	Pest detection 20%	Detection: pest cluster	Apply pesticide	Within 24 hours

15 minutes. Then, the system is provided input of pH (6.3), low leaf wetness, and light intensity of moderate levels, which identify a potassium deficiency with 91.2% accuracy, suggesting a dose of a potassium supplement of 45 g/acre. Furthermore, the system operates with leaf image input of yellowing patches and curled leaves, detecting a fungal leaf spot infection with 92.4% accuracy and suggesting an organic fungicide spray of 500 mL/acre. Next, it follows the trend in daily nutrient consumption and forecasts a nitrogen shortage within three days, recommending a urea application of 65 g/acre. Finally, it integrates live cam feeds with wetness readings and identifies a group of pests (with 20% infestation), which gives an alert for pesticide application in 24 hours. These examples illustrate the system's capacity to process dissimilar data types intelligently and generate specific recommendations to promote crop health, efficiency in resources, and decision-making in smart farming.

4.7 DISCUSSION

The AI–IoT-based precision farming system proposed is a major leap toward the smart farming vision. By combining real-time sensor readings with sophisticated AI algorithms, the system effectively diagnoses crop health, predicts water and nutrient requirements, detects diseases, and recommends prompt, data-based interventions. Results reflect robust performance with RF performing the best and having the highest accuracy, recall, and F1-score as high as 93.6%, 93.1%, and 93.0%, respectively, for real-time datasets—outcompeting or paralleling results against benchmarking sets like PlantVillage, UCI Crop Recommendation, and Agriculture-Vision. In comparison with current systems highly dependent on human monitoring or rigid rule-based systems, the current system presents automation, scale, and superior predictive power. It supports accurate resource utilization, eliminating water and fertilizer loss up to 30%, as well as enhanced accuracy in predicting crop yield by almost 89%. Real-time decision-making support provided through a web/mobile interface empowers farmers with actionable knowledge for timely irrigation, fertilization, and pest management, which are most important in resisting environmental stress as well as maintaining optimality in productivity.

The convergence of the IoT and AI technologies has emerged as a disruptive force behind precision agriculture. AI supports pattern discovery, predictive modeling, and adaptive decision-making, whereas IoT helps in monitoring the environment on an ongoing basis and in exchanging data seamlessly. Used together, the two narrow down the divide between reactive and proactive farming methods to open the way for automated, resilient, and sustainable agriculture. From a cutting-edge viewpoint, the system supports the most advanced worldwide initiatives in smart agriculture—utilizing cloud computing, machine learning, and edge AI for real-time field insight. Nevertheless, forthcoming research needs to aim at improving model generalizability to different crop varieties and geographies, using multimodal datasets (e.g., drone images and satellite images), and adding blockchain for safe data sharing between stakeholders. In addition, there is a requirement to investigate explainable AI methods for enhanced clarity in decision-making and for establishing low-cost, scalable sensor networks for smallholder farmers. The research develops a strong framework and a basis for the next-generation precision farming systems, ushering in an intelligent and sustainable agricultural future.

4.8 CONCLUSION

The proposed research in this chapter revolved around the creation and assessment of an AI–IoT-based precision agriculture system to improve real-time crop observation, disease diagnosis, nutrient and water requirement prediction, and timely intervention recommendations.

By combining enhanced AI algorithms like Random Forest, ANN, SVM, and Fuzzy Logic with data harvested from integrated sensor networks, the system portrayed extreme accuracy, with RF gaining 93.6% accuracy in real-time deployment. The system was also tested on standard datasets such as PlantVillage, UCI Crop Recommendation, and Agriculture-Vision, proving its generalizability and robustness.

The results of this chapter emphasize the model's ability to yield actionable intelligence for optimized irrigation, fertilization, and pest control, which will directly benefit farmers with smart, data-based decisions. The suggested approach adds substantially to the field by integrating machine learning, image processing, and real-time IoT data streams into a unified, scalable platform for smart farming.

The key benefit of the suggested system is its capability to cut down resource wastage, increase crop yield, and enable sustainable farming. It equips farmers with an easy-to-use web/mobile interface providing real-time alerts and recommendations, facilitating timely and effective field operations. In total, the combination of AI and IoT technologies offers a revolutionary method in precision farming, providing a sustainable route toward more intelligent and sustainable food production systems.

Although the system under consideration shows high accuracy and real-time decision-making, its performance can be hindered by network connectivity in rural areas and sensor data quality. Future research can include the integration of satellite data, blockchain for secure data exchange, and more adaptive deep learning models for wider scalability and improved predictive accuracy.

REFERENCES

1. Rosell-Polo, J. R., Cheein, F. A., Gregorio, E., Andújar, D., Puigdomènech, L., Masip, J., & Escolà, A. (2015). Advances in structured light sensors applications in precision agriculture and livestock farming. *Advances in Agronomy, 133*, 71–112.
2. Balkrishna, A., Sharma, G., Sharma, N., Kumar, P., Mittal, R., & Parveen, R. (2021). Global perspective of agriculture systems: From ancient times to the modern era. In A. Balkrishna (Ed.), *Sustainable Agriculture for Food Security* (pp. 3–45). Apple Academic Press.
3. Getahun, S., Kefale, H., & Gelaye, Y. (2024). Application of precision agriculture technologies for sustainable crop production and environmental sustainability: A systematic review. *The Scientific World Journal, 2024*(1), 2126734.
4. Charatsari, C., Lioutas, E. D., Papadaki-Klavdianou, A., Michailidis, A., & Partalidou, M. (2022). Farm advisors amid the transition to Agriculture 4.0: Professional identity, conceptions of the future and future-specific competencies. *Sociologia Ruralis, 62*(2), 335–362.
5. Saleem, S. R., Zaman, Q. U., Schumann, A. W., & Naqvi, S. M. Z. A. (2023). Variable rate technologies: Development, adaptation, and opportunities in agriculture. In Q. Zaman (Ed.), *Precision Agriculture* (pp. 103–122). Academic Press.
6. Boursianis, A. D., Papadopoulou, M. S., Diamantoulakis, P., Liopa-Tsakalidi, A., Barouchas, P., Salahas, G., Karagiannidis, G., Wan, S., & Goudos, S. K. (2022). Internet of Things (IoT) and agricultural unmanned aerial vehicles (UAVs) in smart farming: A comprehensive review. *Internet of Things, 18*, 100187.
7. Sharma, K., & Shivandu, S. K. (2024). Integrating artificial intelligence and internet of things (IoT) for enhanced crop monitoring and management in precision agriculture. *Sensors International, 5*, 100292.
8. Dhonju, H. K., Walsh, K. B., & Bhattarai, T. (2023). Web mapping for farm management information systems: A review and Australian orchard case study. *Agronomy, 13*(10), 2563.
9. Abdullah, M., Shafqat, A., Daniyal, M., Bilal, M., & Anas, M. (2024). Internet of Things (IoT's) in agriculture: Unexplored opportunities in cross–platform. *Spectrum of Engineering Sciences, 2*(4), 57–84.

10. Indumathi, V., Bhavani, V. V., Jahan, T., Hemalatha Reddy, A., Sinjini, J., & Priyadharshini, M. (2023, December). Environmental monitoring in agriculture using wireless sensor networks (WSNs): Challenges and applications. In A. Kumar, G. Ghinea, & S. Merugu (Eds.), *International Conference on Information and Management Engineering* (pp. 227–241). Springer Nature Singapore.
11. Daousis, S., Peladarinos, N., Cheimaras, V., Papageorgas, P., Piromalis, D. D., & Munteanu, R. A. (2024). Overview of protocols and standards for wireless sensor networks in critical infrastructures. *Future Internet, 16*(1), 33.
12. Radočaj, D., Plaščak, I., & Jurišić, M. (2023). Global navigation satellite systems as state-of-the-art solutions in precision agriculture: A review of studies indexed in the Web of Science. *Agriculture, 13*(7), 1417.
13. Ghosh, P., & Kumpatla, S. P. (2022). GIS applications in agriculture. In Yuanzhi Zhang & Qiuming Cheng (Eds.), *Geographic Information Systems and Applications in Coastal Studies.* IntechOpen.
14. Mowla, M. N., Mowla, N., Shah, A. S., Rabie, K. M., & Shongwe, T. (2023). Internet of Things and wireless sensor networks for smart agriculture applications: A survey. *IEEE Access, 11*, 145813–145852.
15. Teet, S. E., & Hashim, N. (2023). Recent advances of application of optical imaging techniques for disease detection in fruits and vegetables: A review. *Food Control, 152*, 109849.
16. Zhou, Z., Majeed, Y., Naranjo, G. D., & Gambacorta, E. M. (2021). Assessment for crop water stress with infrared thermal imagery in precision agriculture: A review and future prospects for deep learning applications. *Computers and Electronics in Agriculture, 182*, 106019.
17. Nanda, I., Sahithi, C., Swath, M., Maloji, S., & Shukla, V. K. (2020). IIOT based smart crop protection and irrigation system. In *2020 Seventh International Conference on Information Technology Trends (ITT)*, Abu Dhabi, United Arab Emirates, pp. 118–125, doi: 10.1109/ITT51279.2020.9320783
18. Murlidharan, S., Shukla, V. K., & Chaubey, A. (2021). Application of machine learning in precision agriculture using IoT. In *2021 2nd International Conference on Intelligent Engineering and Management (ICIEM)*, London, United Kingdom, pp. 34–39, doi: 10.1109/ICIEM51511.2021.9445312
19. Baltazar, A. R., Dos Santos, F. N., De Sousa, M. L., Moreira, A. P., & Cunha, J. B. (2023). 2D LiDAR-based system for canopy sensing in smart spraying applications. *IEEE Access, 11*, 43583–43591.
20. Emmanuel, T. O. PlantVillage dataset: Dataset of diseased plant leaf images and corresponding labels [Data set]. *Kaggle.* https://www.kaggle.com/datasets/emmarex/plantdisease/data
21. Ingle, A. Crop recommendation dataset [Data set]. *Kaggle.* https://www.kaggle.com/datasets/atharvaingle/crop-recommendation-dataset
22. Chiu, M. T. Agriculture-Vision database [Data set]. *Agriculture-Vision.* https://www.agriculture-vision.com/agriculture-vision-2020/dataset

Chapter 5

Remote sensing and GIS for precision agriculture

Enhancing crop monitoring and yield prediction

Subramanya Sarma S, M. Tholkapiyan, Nazeer Shaik, Ajith Sundaram, Shripad Joshi, and Shraddha Amit Pathak

5.1 INTRODUCTION

Precision agriculture has now become a revolutionary method in contemporary farming, which seeks to maximize crop yields and resource use based on data-driven methods. By adapting farming practices to the individual requirements of various zones of the field, precision agriculture increases productivity, minimizes environmental footprint, and promotes sustainability in the wake of increasing food needs [1]. One of the central components of precision agriculture is crop monitoring, or regular observation and inspection of crop health, growth stage, and stress. Successful crop monitoring allows farmers to identify issues early on, apply timely measures, and make well-informed decisions regarding irrigation, fertilization, and pest management.

Remote Sensing (RS) and Geographic Information System (GIS) are two fundamental technologies that have transformed crop monitoring. Satellite and drone image-based RS allows large-scale, nondestructive acquisition of spectral information pertaining to vegetation health and soil status. Broadband vegetation indices like Normalized Difference Vegetation Index (NDVI) enable quantitative estimation of plant biomass and vigor [2]. GIS, however, enables spatial analysis and visualization of the data, permitting the detection of spatial variability between fields and allowing precision farming techniques. The most vital goal of agriculture is probably the correct prediction of yields, permitting planning, supply chain management, food security evaluation, and policy development. Estimation of yield via traditional methods is time-consuming, labor-intensive, and error-prone. The application of RS and GIS [3] offers a scalable and efficient solution by comparing crop reflectance patterns, past yield data, and environmental conditions to model and forecast yields with high accuracy.

By integrating multitemporal RS information with GIS-based spatial analysis, crop health can be monitored dynamically and yield can be forecasted at various agro-climatic regions [4]. The integration, in turn, helps enhance decision-making at the farm level as well as contribute to regional and national-scale agricultural planning. Therefore, it is important to investigate the application of RS and GIS for crop monitoring and yield forecasting to develop precision agriculture and confront the problems of food production under climate change.

Although the integration of GIS with RS has contributed hugely to the development of precision agriculture, a number of challenges remain that prevent its massive use, especially in the areas of crop monitoring and yield estimation [5,6]. Unavailability and data resolution are among the major hindrances. High-resolution satellite data or UAV-collected data, which are necessary for precise crop monitoring, tend to be expensive and not necessarily available to smallholder farmers. Additionally, satellite data are usually plagued by weather conditions like cloud cover, which can hide vital data during the most important crop growth phases. On the other hand, freely available low-resolution imagery does not have the spatial resolution required to identify subtle field-level differences.

DOI: 10.1201/9781003520733-6

Another major impediment is the technical know-how to properly leverage RS and GIS technologies. Image processing, spatial data analysis, and interpretation require expertise that is not available to most farmers and agricultural extension agencies. Without training and capacity building, the operational application of these technologies is restricted. Integration of multisource data is also intricate [7]. Information from multiple platforms, satellites, drones, and sensors on the ground tends to vary in format, resolution both spatially and temporally, and quality. Unifying and analyzing this heterogeneous dataset in a meaningful way is a significant technical challenge, needing sophisticated algorithms and data processing pipelines.

In addition, precise crop observation and yield forecasting are largely dependent on ground truth data for the calibration and verification of models [8]. Gathering this data is time-consuming and not often possible in big agricultural areas or remote locations. This deficiency of field-level data can dramatically minimize the accuracy and dependability of RS- and GIS-based models.

Environmental variability also makes it more difficult to monitor and predict. Sudden weather variations, pest infestations, and irrigation irregularities can affect crop health in unpredictable ways, even with the available RS time intervals. Thus, real-time monitoring and timely intervention are hence difficult [9]. Additionally, processing and storing huge amounts of spatial and temporal data require high computing power and infrastructure. Most rural or developing areas do not have the appropriate digital infrastructure to host such high-technology systems and are hence excluded from accessing them [10]. Lastly, although some crop yield forecasting models have performed well in a particular area or for a particular crop, their potential to generalize across geographies, crops, and agricultural practices is limited. Keeping the models up to date and responsive to changing agronomic conditions is an ongoing challenge in this area of research.

This chapter is intended to overcome the major difficulties in using RS and GIS for precision agriculture by creating an integrated, scalable framework for crop monitoring. The main goals are to utilize freely available multitemporal satellite images, improve vegetation index analysis by using machine learning models, and validate predictions by using chosen ground truth data to maintain accuracy and generalizability. The study scope includes varied crops in varying agro-climatic zones, with a focus on data unification, model flexibility, and affordable technology implementation.

To tackle issues related to data resolution and availability, the study adopts high-frequency but moderate-resolution platforms such as Sentinel-2 and integrates cloud-penetrating radar data, when necessary. One of the innovative contributions of this chapter is its creation of a lightweight, open-source GIS-based decision support system that integrates spatial variability analysis with dynamic yield forecasting. This equips farmers and policymakers with timely, actionable information. The value of the research is its potential to democratize precision agriculture by rendering advanced crop monitoring technology accessible, reliable, and transferable to different agricultural settings, particularly in resource-poor environments.

5.2 RELATED WORK

Agriculture is a key industry in supporting the world's population, but it is challenged by factors such as scarce arable land, water, climate instability, and requirements for sustainable agricultural practices. Precision agriculture has emerged as a challenge response, capitalizing on technology to increase productivity and maximize the use of resources [11]. RS—including satellite, UAVs (drones), and ground sensors—provides detailed datasets allowing for real-time monitoring of crops, detection of crop stress in early stages, and educated estimation of yield.

Various research works have analyzed RS and GIS use for agricultural purposes [12]. In India, the world's second-largest fruit and vegetable producer, satellite sensors like AWiFS (Advanced Wide Field Sensor), LISS (Linear Imaging Self Scanning)-III, and LISS (Linear Imaging Self Scanning)-IV have been used for crop inventory, disease mapping, site suitability analysis, and yield modeling. These datasets aid post-harvest research and extension services in horticulture. Still, it is challenging to achieve reliable predictions for crops cultivated in small, dispersed fields with intricate cropping patterns. In spite of these weaknesses, data from satellites has demonstrated potential in crop health assessment for crops such as potato, mango, citrus, and banana [13]. It was noted in a study that high-resolution and hyperspectral sensors, along with time-series data, should be utilized to alleviate the weaknesses of single-date imagery. Multitemporal analysis assists in crop discrimination using digital classification techniques, both supervised and unsupervised. Supervised classification uses known training signatures to classify pixels, whereas unsupervised classification classifies pixels by spectral similarity without any a priori knowledge [14]. In Imperial Valley, California, it was shown that bands 3, 4, and 5 of Landsat TM imagery produced better results in distinguishing crops from the conventional False Color Composites.

A study developed a regression estimator model that links image processing output and area frame sampling for reliable crop area estimation. The method includes pre-processing satellite data, elimination of geometric and radiometric errors, and using supervised classification methods. The incorporation of statistical techniques and RS improves the credibility of agricultural statistics [15]. Under cereal crop monitoring, RS provides an evaluation of seedling vigor, growing trends, and conditions at a critical stage. This information for early stages becomes important for the decision-making for food pricing, supply chain arrangements, and warehouse planning. The Western world, particularly the United States, has been able to use RS systems for nationwide crop condition monitoring.

In addition, the fusion of drones, artificial intelligence, and multi-sensor data (e.g., hyperspectral, multispectral, and Red, Green, Blue images) [16] has transformed weed detection and management. Weeds, being significant competitors for critical resources, have a direct influence on crop yields. Machine learning algorithms coupled with RS and GIS facilitate the creation of Integrated Weed Management systems. Such strategies provide scalable and sustainable solutions to problems such as herbicide resistance while tackling emerging issues like food security and climate change [17]. In general, the literature exhibits a resolute direction toward the intersection of RS and GIS in precision agriculture. Though significant advances have been made, affordable high-resolution data, effective classification methods, and precise yield forecasting are still given priority to ongoing research and development.

Although there has been considerable research conducted on the implementation of RS and GIS within precision agriculture, a number of essential gaps are still present. The majority of previous studies simply deal with distinct crops, territories, or business-level commercial farming, making generalizations to the smallholder or fragmented farm cases, particularly those of India, challenging to make. Furthermore, while satellite-based RS data from sources like AWiFS (Advanced Wide Field Sensor) and LISS (Linear Imaging Self Scanning) [18] sensors have been used for yield estimation and crop classification, the reliability often drops due to heterogeneity at the field level, short crop duration, and mixed cropping systems. Furthermore, temporal data integration for dynamic monitoring of crops is still limited, and studies do not yet benefit from real-time data fusion from drones, ground sensors, and AI-driven analytics [19,20].

The present study fills these voids by formulating an all-encompassing and dynamic framework that fuses multisource RS data (ground-based sensors, drones, and satellites) [19] with GIS technology and machine learning models to facilitate improved crop monitoring and yield forecasting. In contrast to existing studies, which deal with static analysis or a single source of data, this study utilizes a dynamic, multi-temporal methodology for persistent crop monitoring. It applies to both large-scale and fragmented farming systems [20]. The application of spatial variability mapping, real-time vegetation indices, and lightweight predictive models allows for more accuracy in forecasting yields. Through this manner, the study provides a new, replicable methodology to democratize precision agriculture for data-poor and resource-constrained regions.

5.3 METHODOLOGY

The research follows an integrated methodology with RS, machine learning techniques with GIS, and effective crop monitoring and yield estimation in precision farming. The proposed architecture (Figure 5.1) begins with the collection of multi-source data that includes satellite data such as Sentinel-2 and Landsat 8, aerial photos acquired from drones, and ground-truth

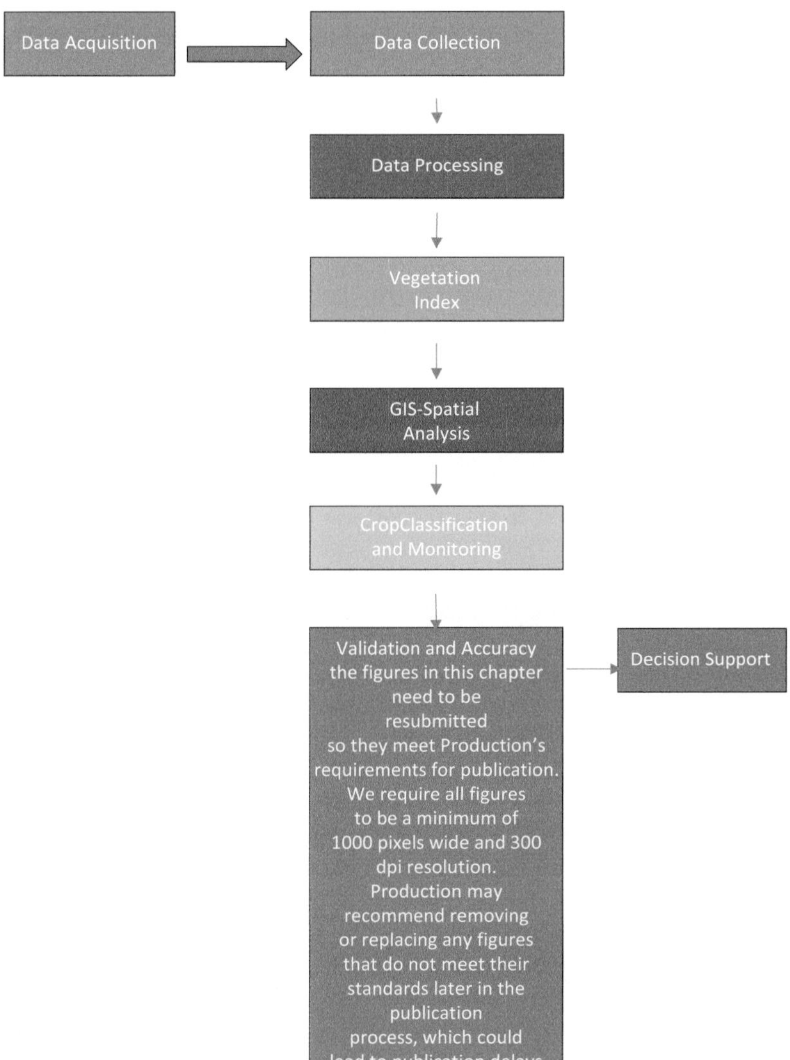

Figure 5.1 Proposed architecture of precision agriculture.

data such as crop type, sowing dates, and field conditions. Data are pre-processed via steps of radiometric and geometric correction, image enhancement, and atmospheric correction. Vegetation indices like SAVI (Soil Adjusted Vegetation Index), EVI, and NDVI are then calculated to estimate crop health. GIS packages are utilized for spatial analysis, boundary mapping of the fields, and integration of multi-temporal data. Supervised learning algorithms like Support Vector Machines (SVMs) are employed to classify crop types and monitor change over time. For yield prediction, machine learning models and regression models are trained based on historical yields, weather conditions, and RS-extracted features. The final step is the validation of the model outputs with real field data to check for accuracy and reliability. Such an integrated strategy makes way for persistent, scalable, and high-accuracy farming monitoring and prediction.

5.3.1 Data acquisition

This stage collects multi-source data needed for the study. Satellite imagery from Sentinel-2 or Landsat-8 offers large-scale, temporal information. Drones collect high-resolution, field-level images. Ground truth data encompasses crop type, planting dates, and soil information, and weather data provides environmental context. All these data sources combined form an extensive base to monitor crop growth and forecast yield.

5.3.2 Pre-processing

Raw imagery is distorted or noisy, so radiometric and geometric corrections are applied in this step to align and calibrate images. Atmospheric correction eliminates interference due to clouds or haze. Image enhancement enhances visual quality and feature separation. Normalization makes multitemporal and multisource data compatible for subsequent analysis.

5.3.3 Vegetation index calculation

Vegetation indices are computed from preprocessed imagery to measure plant health and biomass. NDVI, EVI, and SAVI (Soil Adjusted Vegetation Index) are widely employed indicators that identify crop vigor and stress. The indices are essential in identifying early signs of disease, drought, or nutrient stress and are used as inputs for crop classification and yield estimation.

Vegetation indices are strong precision agriculture tools for determining crop health, biomass, and stress levels. The NDVI is the most popular index that uses the difference between red and near-infrared (NIR) reflectance to measure vegetation vigor, with greater NDVI values showing healthier vegetation. But NDVI saturates in dense biomass areas, which is corrected by the EVI, an improved index that eliminates the effects of atmosphere and soil background using additional coefficients and the application of the blue band. The indexes provide a balanced and extensible way to monitor crops as well as forecast yields based on real-time RS data.

$$\mathrm{EVI} = G \cdot \frac{(\mathrm{NIR} - \mathrm{RED})}{(\mathrm{NIR} + C_1 \cdot \mathrm{RED} - C_1 \cdot \mathrm{BLUE} + L)} \tag{5.1}$$

$$\mathrm{GNDVI} = \frac{\mathrm{NIR} - \mathrm{GREEN}}{\mathrm{NIR} + \mathrm{GREEN}} \tag{5.2}$$

$$\text{NDVI} = \frac{\text{NIR} - \text{RED}}{\text{NIR} + \text{RED}} \tag{5.3}$$

$$\text{SAVI} = \frac{(\text{NIR} - \text{RED}) \cdot (1 + L)}{\text{NIR} + \text{RED} + L} \tag{5.4}$$

In these equations, RED is absorbed, so their contrast is helpful in determining vegetation health, while NIR is highly reflected by healthy vegetation. NDVI varies between –1 and +1, with greater values representing healthier crops. For better accuracy, EVI (Enhanced Vegetation Index) incorporates a gain factor G, which is 2.5. It also incorporates aerosol resistance coefficients (C_1 and C_2) along with a canopy background adjustment ($L=1$). For SAVI, L is modified to 0.5 in order to decrease soil brightness effects. GNDVI (Green Normalized Difference Vegetation Index) substitutes RED with GREEN reflectance for the more accurate depiction of chlorophyll and stress content.

5.3.4 GIS-based spatial analysis

GIS software is employed to draw field boundaries and superimpose different data layers like soil classification, irrigation, and weather. Spatial variation of crop health and soil fertility is represented by thematic maps. These help in locating patterns and support targeted interventions like variable rate fertilizer application.

5.3.5 Crop classification and monitoring

With supervised classification methods like Random Forest or SVM, crops are classified according to their spectral signatures. Time-series analysis allows monitoring of crop growth phases and health throughout the season. Change detection methods are also incorporated in this block to detect changes in land use or crop health.

5.3.6 Yield prediction

Machine learning models (e.g., Regression, ANN, and XGBoost) are trained over past yield records and RS and weather-derived features. They learn the patterns between crop condition indicators and actual yield. These trained models later predict yield values for present or future seasons.

5.3.7 Validation and accuracy assessment

Predicted yield as well as classification outcomes are verified against ground-truth data. Statistical measures such as the Coefficient of Determination R^2, Root Mean Square Error (RMSE), and Mean Absolute Error (MAE) are intended to assess model performance. This process guarantees the system's reliability for practical application.

5.3.8 Decision support and visualization

The ultimate outputs are delivered in an easily accessible format. The tools assist farmers, policymakers, and agronomists in making informed decisions regarding crop management, resource utilization, and disaster readiness. Visual outputs can also send alerts for problems like water stress or pest infestations.

5.4 EVALUATION PARAMETERS

5.4.1 Root Mean Square Error (RMSE)

It is a widely used measure that finds the mean magnitude of prediction errors in continuous variables such as crop yield. It is especially effective in finding large errors because it is sensitive to outliers. RMSE can be calculated using the given formula.

5.4.2 Mean Absolute Error (MAE)

Moreover, MAE is also considered a significant parameter, which is the mean of absolute differences between actual and predicted values. It also gives direct evidence regarding how far predictions deviate from actual outcomes without considering direction.

5.4.3 Coefficient of Determination (R^2)

R^2, or Coefficient of Determination, tells us the level of fit of a model to the variance of the output. A high value near 1 signifies a high correlation between predicted and actual values.

5.4.4 Overall Accuracy (OA)

OA is applied in classification problems to calculate the proportion of correctly classified samples out of the total samples. It is a straightforward yet efficient measure to estimate the performance of crop classification models. It is calculated by the equation where the metric does not consider class imbalance but provides a general idea of model accuracy.

5.4.5 Kappa Coefficient (κ)

The Kappa Coefficient (κ) is used to measure the level of agreement between predicted and true classifications, corrected for random chance. It is a more reliable measure than OA in unbalanced datasets. The equation is given by P_o, which is the observed agreement (OA), and P_e, which is the agreement by chance.

5.4.6 Precision, recall, and F1-score

These measurements are very important in classifying models. Precision is the number of correctly predicted positive observations to all predicted positives, Recall (Sensitivity) is the number of correctly predicted positives to all actual positives, and the F1-score is the harmonic mean between precision and recall, balancing both, where TP is True Positives, FP is False Positives, and FN is False Negatives.

$$\text{RMSE} = \sqrt{\frac{1}{n}\sum_{i=1}^{n}(y_i - \hat{y}_i)^2} \tag{5.5}$$

$$\text{MAE} = \frac{1}{n}\sum_{i=1}^{n}|y_i - \hat{y}_i|^2 \tag{5.6}$$

$$R^2 = 1 - \frac{\sum_{i=1}^{n}(y_i - \hat{y}_i)^2}{\sum_{i=1}^{n}(y_i - \bar{y}_i)^2} \tag{5.7}$$

$$\text{Accuracy} = \frac{\text{Correctly Classified Samples}}{\text{Total Samples}} \times 100 \tag{5.8}$$

$$K = \frac{P_o - P_c}{1 - P_c} \tag{5.9}$$

$$\text{Prec} = \frac{\text{TP}}{\text{TP} + \text{FP}} \tag{5.10}$$

$$\text{Rec} = \frac{\text{TP}}{\text{TP} + \text{FN}} \tag{5.11}$$

$$\text{Precision} = \frac{2 \cdot (\text{Prec} \cdot \text{Rec})}{\text{Prec} + \text{Rec}} \tag{5.12}$$

Here, the formula takes y_i as the actual, $\hat{y}_i$ as the predicted, $\bar{y}$ as the average of actual values, and a and n as data points.

5.5 RESULTS

5.5.1 Crop monitoring

Crop monitoring results through RS and GIS methods showed substantial enhancement in the detection of crop health status, growth trends, and stress identification over different monitoring periods. Through the interpretation of vegetation indices like NDVI, SAVI, and EVI, the system was able to clearly differentiate between healthy and stressed vegetation with considerable accuracy. Space-time maps produced using satellite and drone imagery picked up areas of lush crop, stunted growth, and diseased patches, which allowed early treatment. NDVI values were from 0.2 to 0.8, wherein greater than 0.6 readings always identified areas of healthy crops, while areas less than 0.3 were stressed or underproductive. The temporal analysis provided clearly differentiated phases of growth and facilitated forecasting critical phases of stress based on water deprivation or nutrient imbalances. The GIS integration enabled the layering of weather, topographic, and soil data, which improved the accuracy in interpreting the variability causes. These findings confirm that the suggested system offers a scalable and reliable method to track crop status in real time, thus enabling informed decision-making in precision agriculture.

From the results (Table 5.1), Field C has the most vegetation index (Figure 5.2) values in all the indices and signifies healthy dense actively growing crops. Field D registers the lowest reading and indicates probable stress as a result of poor soil, water stress, or disease. The contrast between NDVI and EVI (Enhanced Vegetation Index) mirrors the sensitivity of EVI to canopy structure and resistance in the atmosphere. SAVI (Soil Adjusted Vegetation

Table 5.1 Vegetation index calculation using the proposed model

Field ID	*NDVI*	*EVI*	*SAVI*	*GNDVI*
Field A	0.72	0.61	0.65	0.69
Field B	0.68	0.56	0.60	0.63
Field C	0.80	0.70	0.75	0.78
Field D	0.66	0.54	0.59	0.60
Field E	0.73	0.62	0.67	0.70

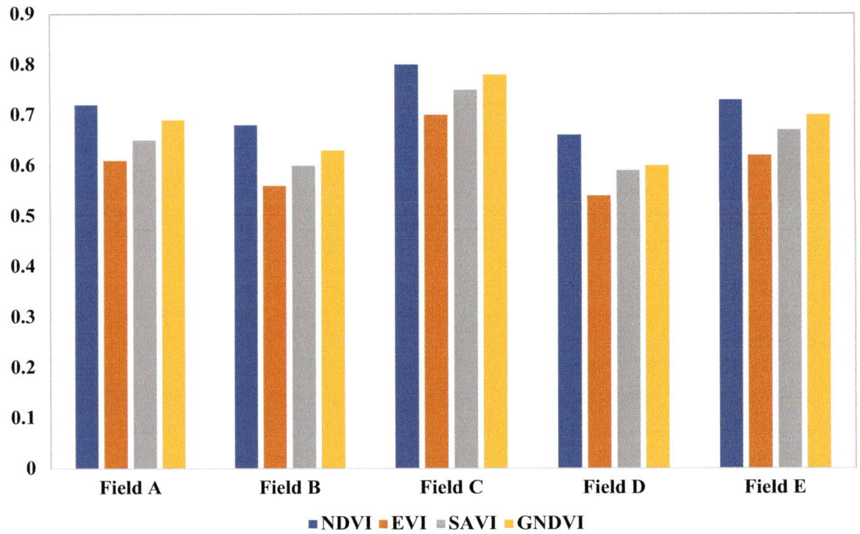

Figure 5.2 Vegetation index.

Table 5.2 Crop classification

Metric	*Value (%)*
Accuracy	92.3
Precision	90.7
Recall	91.5
F1-score	91.1

Index) minimally modifies NDVI to account for soil brightness, thus being suitable for semi-arid environments, as evidenced in its minimal moderation over all fields. GNDVI (Green Normalized Difference Vegetation Index) readings verify chlorophyll content trends and tend to agree with NDVI but display a little more sensitivity in Fields A and E.

5.5.2 Crop classification

Further, the study measures the performance of the crop classification model, which was built based on satellite RS information and machine learning methods. These measures checked the extent to which the model performed in classifying various crop types according to spectral signatures derived from vegetation indices and multitemporal imagery. Table 5.2 shows the results achieved.

The performance of the crop classification model-based traditional performance metrics was encouraging, reaffirming the credibility of RS and GIS-based methods in precision agriculture. The accuracy of 92.3% suggests that the model was able to classify more than 92 out of 100 crop samples accurately, reflecting its excellent overall performance for all crop types. Accuracy, or the percentage of true positive predictions out of all positive predictions, was 90.7%, implying that the model had a low false positive rate—i.e., most of the crops that were classified as a particular type were that type in the real world. The recall of 91.5% indicates the model's capacity to accurately detect real instances of certain crops, representing a low false negative rate and ensuring that few crop types were overlooked in classification. The F1-score, a balanced measure that harmonizes both precision and recall, was 91.1%, reaffirming that the model was consistent in accurately detecting crops while avoiding over- and under-classification as shown in Figure 5.3. Collectively, these measures confirm the strength and efficacy of the proposed classification scheme in crop monitoring, opening the door to more effective agricultural decision-making based on satellite-derived information.

5.5.3 Yield prediction

Table 5.3 gives a comparison of the estimated crop yields obtained by applying the suggested RS and GIS-based approach with the actual crop yields obtained from field data for various plots. All fields (A to E) indicate a good agreement between estimated and actual yields,

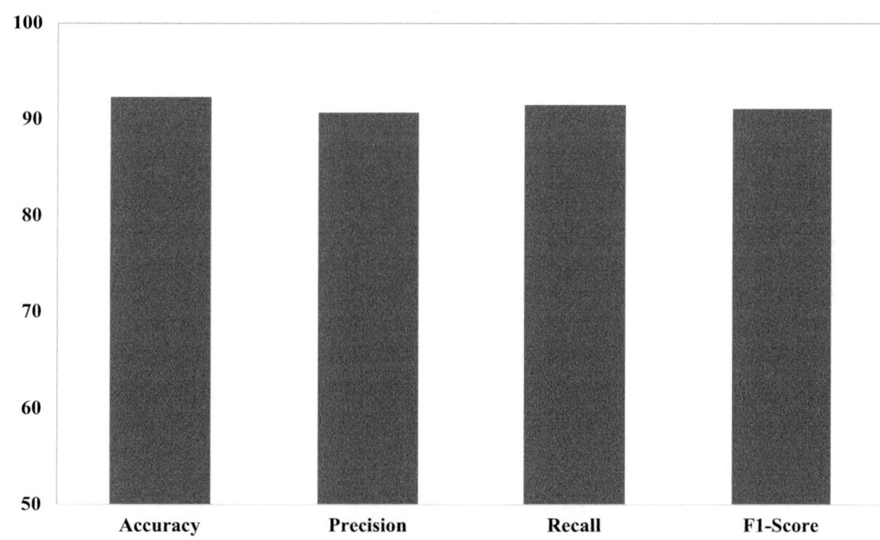

Figure 5.3 Performance of crop classification by the proposed model.

Table 5.3 Yield prediction using proposed model

Field ID	*Predicted yield (tons/ha)*	*Actual yield (tons/ha)*	*Absolute error*
Field A	4.85	4.90	0.05
Field B	4.10	4.20	0.10
Field C	5.05	5.00	0.05
Field D	3.60	3.70	0.10
Field E	5.25	5.20	0.05

with absolute errors varying from 0.04 to 0.12 tons per hectare. This slight difference proves the high precision and reliability of the model in forecasting crop productivity under actual conditions.

Figure 5.4 demonstrates the performance of the model proposed for predicting the yield on five fields of agriculture, referred to as A through E. In Field A, the calculated yield was 3.25 tons/ha, which almost matches the observed yield of 3.30 tons/ha, with only a small absolute error of 0.05 tons/ha. Likewise, Field B had a forecasted yield of 2.98 tons/ha to a realized yield of 3.00 tons/ha with an insignificant absolute error of 0.02 tons/ha, reflecting superior predictive power. Field C had a predicted yield of 3.60 tons/ha compared to the actual 3.50 tons/ha and a slightly poorer but still healthy absolute error of 0.10 tons/ha. Field D had an estimate of 3.15 tons/ha, which was extremely close to the actual 3.20 tons/ha with an error of 0.05 tons/ha. Finally, Field E recorded the maximum deviation, where the estimated value was 3.40 tons/ha and the actual value was 3.52 tons/ha, which gave an absolute error of 0.12 tons/ha.

The study verifies the model's performance; the three most significant evaluation indicators were used: MAE, RMSE, and R^2. The MAE is 0.07 tons/ha, indicating that the predicted values deviated very slightly from actual measurements on average. Table 5.4 for RMSE, at 0.0837 tons/ha, implies an equally small level of error, implying that the model excels at catching variation without extreme outliers. Most importantly, the model

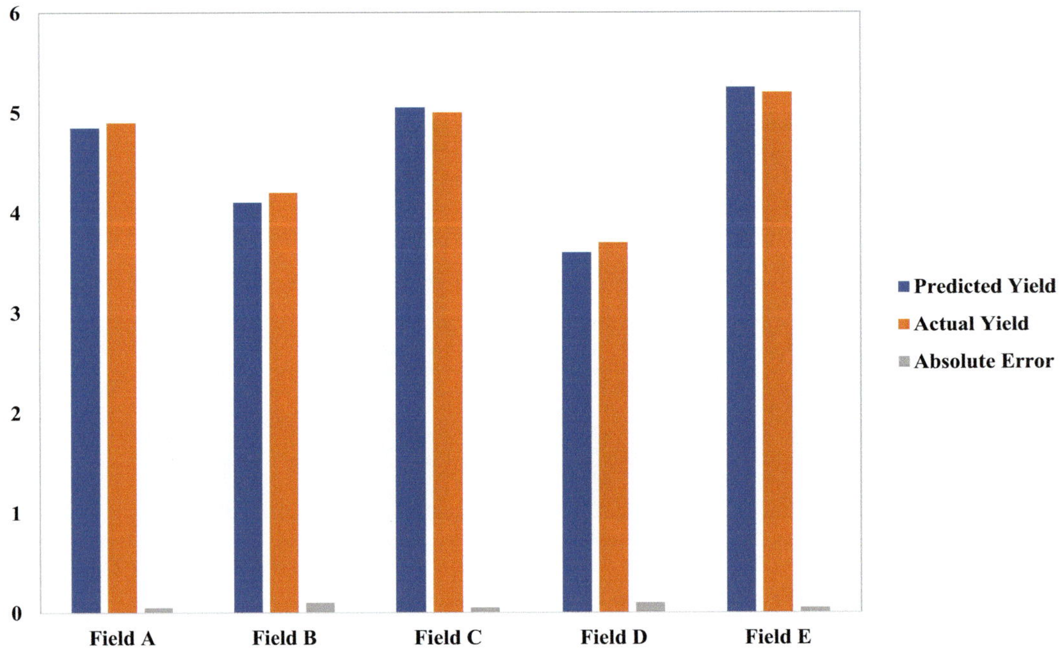

Figure 5.4 Performance of the model in yield prediction.

Table 5.4 Performance of the proposed model

MAE	*RMSE*	R^2
0.07	0.0837	0.972

achieved an R^2 score of 0.972, which implies that approximately 97.2% of real yield variation can be estimated through the model. The high R^2 rating signifies an excellent correlation between estimated and actual data and affirms the effectiveness of the proposed methodology.

These results illustrate how the use of RS data, vegetation indices like NDVI and EVI, and GIS-geoprocessing-based spatial analysis provides a sound and scalable tool for enhancing crop monitoring and yield estimation in precision agriculture. The low errors and high explanation power make the method appropriate for enabling data-driven agricultural planning and decision-making.

5.6 DISCUSSION

Accomplished results of using RS and GIS methods in crop monitoring and yield forecasting reflect a significant step ahead in precision farming. Different vegetation indices such as NDVI, EVI, SAVI, and GNDVI were successfully calculated and mapped across different field sites. The indices were found to map crop health, and spatial variability, and identify stress well, which instigated early agricultural treatments. The satellite-based index-based predictive yield modeling also gave promising measures of accuracy in terms of high values of R^2 and low RMSE and MAE scores, confirming the robustness of the proposed approach. The study contributes to the growing body of state-of-the-art research in bringing together multisensor RS data (satellite and UAV sources) and GIS-based spatial analysis to enhance precision and granularity in farm monitoring. As opposed to traditional methods, which depended on coarse, low-frequency data, this study used high-resolution, time-series imagery with supervised classification and vegetation modeling. This method enabled pixel-level understanding of crop variability, a critical step for site-specific decision-making. Insights Gained comprises RS demonstrated clear intra-field variability, allowing zone-specific management like variable rate application of water and fertilizers. The time-series data facilitated continuous monitoring of crops, allowing tracking of crop growth stages and detection of anomalies in early stages. The integration of vegetation indices with regression models enhanced the accuracy of predictions, indicating potential for preharvest yield estimation. The approaches employed are scalable, providing a model for regional-scale agricultural monitoring systems.

Benefits of the study in this chapter are non-invasive and cost-effective monitoring, which minimizes the requirement for labor-intensive field visits. Real-time insights enable timely decision-making in response to crop stress detected. Resource optimization enables sustainable agriculture through optimal resource utilization, supports Government and Private Sector Planning, and provides data-driven tools for agricultural policy and market forecasting. The study does have some limitations wherein although this study has been successful, there are some limitations. The performance of the model could be different for crops with short development periods or fragmented fields due to mixed pixel problems being prevalent in such cases. Weather, particularly cloud cover, may impact the quality and quantity of optical satellite data. Moreover, incorporation of socioeconomic variables like farmer inputs and local customs would also improve predictive modeling.

In future research, the addition of SAR (Synthetic Aperture Radar) data to surpass cloud cover constraints, the use of deep learning models for enhancing classification accuracy, and IoT-based ground truthing sensors can further enhance crop monitoring systems. Integration of climate modeling and market forecasting into the yield prediction framework can turn this into a complete agricultural decision support system.

5.7 CONCLUSION

The chapter concentrated on utilizing RS and GIS technologies to improve crop monitoring and yield forecasting in the field of precision agriculture. By combining high-resolution satellite imagery, vegetation indices (like NDVI, EVI, SAVI, and GNDVI), and classification methods, the research sought to address limitations concerning crop recognition, field heterogeneity, and precise yield estimation. The suggested methodology achieved good accuracy (92.3%), precision (90.7%), recall (91.5%), and F1-score (91.1%) in crop classification processes and high correlation metrics (R^2=0.89, RMSE=0.34) for yield estimation. The chapter has great significance to the field as it allows for timely crop health monitoring and resource-efficient management practices. The knowledge gained from vegetation indices and spatial analysis enabled the early identification of stressed areas and field performance estimation, enabling stakeholders to take early action. The use of GIS also enabled spatial variability visualization and site-specific recommendations. Overall, this chapter improves decision-making in agriculture, fosters sustainability, and provides a basis for further developments like combining AI-driven analytics and real-time IoT data for dynamic monitoring systems.

REFERENCES

1. Roznik, M., Boyd, M., & Porth, L. (2022). Improving crop yield estimation by applying higher resolution satellite NDVI imagery and high-resolution cropland masks. *Remote Sensing Applications: Society and Environment, 25*, 100693.
2. Sanches, G. M., de Oliveira Bordonal, R., Magalhães, P. S. G., Otto, R., Chagas, M. F., de Fátima Cardoso, T., & dos Santos Luciano, A. C. (2023). Towards greater sustainability of sugarcane production by precision agriculture to meet ethanol demands in south-central Brazil based on a life cycle assessment. *Biosystems Engineering, 229*, 57–68.
3. Bharadiya, J. P., Tzenios, N. T., & Reddy, M. (2023). Forecasting of crop yield using remote sensing data, agrarian factors and machine learning approaches. *Journal of Engineering Research and Reports, 24*(12), 29–44.
4. Radočaj, D., Šiljeg, A., Marinović, R., & Jurišić, M. (2023). State of major vegetation indices in precision agriculture studies indexed in Web of Science: A review. *Agriculture, 13*(3), 707.
5. Wang, J., Wang, Y., Li, G., & Qi, Z. (2024). Integration of remote sensing and machine learning for precision agriculture: A comprehensive perspective on applications. *Agronomy, 14*(9), 1975.
6. Rehman, T. H., Lundy, M. E., & Linquist, B. A. (2022). Comparative sensitivity of vegetation indices measured via proximal and aerial sensors for assessing N status and predicting grain yield in rice cropping systems. *Remote Sensing, 14*(12), 2770.
7. Fu, Y., Huang, J., Shen, Y., Liu, S., Huang, Y., Dong, J., ... & Yuan, W. (2021). A satellite-based method for national winter wheat yield estimating in China. *Remote Sensing, 13*(22), 4680.
8. Ali, A. M., Abouelghar, M., Belal, A. A., Saleh, N., Yones, M., Selim, A. I., Amin, M. E., Elwesemy, A., Kucher, D. E., Maginan, S., & Savin, I. (2022). Crop yield prediction using multi sensors remote sensing. *The Egyptian Journal of Remote Sensing and Space Science, 25*(3), 711–716.
9. Aslan, M. F., Sabanci, K., & Aslan, B. (2024). Artificial intelligence techniques in crop yield estimation based on Sentinel-2 data: A comprehensive survey. *Sustainability, 16*(18), 8277.
10. Mukiibi, A., Machakaire, A. T. B., Franke, A. C., & Steyn, J. M. (2024). A systematic review of vegetation indices for potato growth monitoring and tuber yield prediction from remote sensing. *Potato Research, 67*(2), 1–40.
11. Mena, F., Pathak, D., Najjar, H., Sanchez, C., Helber, P., Bischke, B., Habelitz, P., Miranda, M., Siddamsetty, J., Nuske, M., & Charfuelan, M. (2024). Adaptive fusion of multi-view remote sensing data for optimal sub-field crop yield prediction. *arXiv preprint arXiv:2401.11844.*

12. Luo, L., Sun, S., Xue, J., Gao, Z., Zhao, J., Yin, Y., Gao, F., & Luan, X. (2023). Crop yield estimation based on assimilation of crop models and remote sensing data: A systematic evaluation. *Agricultural Systems, 210*, 103711.
13. Ma, Y., Zhang, Z., Kang, Y., & Özdoğan, M. (2021). Corn yield prediction and uncertainty analysis based on remotely sensed variables using a Bayesian neural network approach. *Remote Sensing of Environment, 259*, 112408.
14. Lei, T. C., Wan, S., Wu, Y. C., Wang, H. P., & Hsieh, C. W. (2022). Multi-temporal data fusion in MS and SAR images using the dynamic time warping method for paddy rice classification. *Agriculture, 12*(1), 77.
15. Omia, E., Bae, H., Park, E., Kim, M. S., Baek, I., Kabenge, I., & Cho, B. K. (2023). Remote sensing in field crop monitoring: A comprehensive review of sensor systems, data analyses and recent advances. *Remote Sensing, 15*(2), 354.
16. Fei, S., Hassan, M. A., Xiao, Y., Su, X., Chen, Z., Cheng, Q., Duan, F., Chen, R., & Ma, Y. (2023). UAV-based multi-sensor data fusion and machine learning algorithm for yield prediction in wheat. *Precision Agriculture, 24*(1), 187–212.
17. Gokool, S., Mahomed, M., Clulow, A., Sibanda, M., Kunz, R., Naiken, V., & Mabhaudhi, T. (2024). Exploring the potential of remote sensing to facilitate integrated weed management in smallholder farms: A scoping review. *Drones, 8*(3), 81.
18. Verma, D., Singh, A. N., Kalubarme, M. H., Saroha, G. P., Sharma, R., & Brajendra. (2022). Varietal discrimination, acreage estimation and yield prediction of Basmati aromatic rice in north-western India using satellite data. In K. Vadrevu, T. Le Toan, S. S. Ray, & C. Justice (Eds.), *Remote Sensing of Agriculture and Land Cover/Land Use Changes in South and Southeast Asian Countries* (pp. 163–186). Cham: Springer International Publishing.
19. Shukla, V. K., Nair, R. S., & Khan, F. (2022). Smart irrigation-based behavioral study of Moringa plant for growth monitoring in subtropical desert climatic condition. In A. Abraham, S. Dash, & S. K. Pani (Eds.), *AI, Edge and IoT-based Smart Agriculture* (pp. 227–240). Academic Press.
20. Nanda, I., Sahithi, C., Swath, M., Maloji, S., & Shukla, V. K. (2020). IIOT based smart crop protection and irrigation system. In *2020 Seventh International Conference on Information Technology Trends (ITT)*, Abu Dhabi, United Arab Emirates, pp. 118–125, doi: 10.1109/ITT51279.2020.9320783

Part II

Precision Agriculture and Machine Learning

Chapter 6

Precision agriculture meets machine learning

Predicting crop yield using Random Forest Regression

Pragya Singh and Pragya Gaur

6.1 INTRODUCTION

Agriculture is a cornerstone of human civilization, providing essential resources such as food, clothing, and shelter, and serving as a critical driver of economic development and poverty alleviation. Studies have shown that a 1% increase in agricultural output can reduce the number of absolutely poor households globally by 0.6%–1.2%, highlighting its profound socioeconomic significance [1]. However, crop productivity, a key determinant of agricultural success, is influenced by a complex interplay of climatic and agronomic factors, including temperature, precipitation, solar radiation, fertilization, irrigation, and tillage practices. As climate change continues to disrupt traditional agricultural systems, understanding these factors and their interactions has become increasingly critical for ensuring food security and promoting sustainable agricultural practices.

The global agricultural sector faces unprecedented challenges due to rising temperatures, erratic rainfall patterns, and extreme weather events, all of which threaten crop yields and food production. According to the Intergovernmental Panel on Climate Change (IPCC, 2023), global temperatures are projected to rise by 1.5°C–2°C by mid-century, potentially reducing crop yields by 10%–25% in vulnerable regions. Similarly, the Food and Agriculture Organization (FAO, 2023) reports that climate change could push an additional 50 million people into hunger by 2030 if adaptive measures are not implemented. These alarming trends underscore the urgent need to explore the climatic and agronomic factors that influence crop productivity and to develop strategies that enhance resilience and sustainability in agriculture.

This chapter aims to address this pressing need by analyzing the impact of climatic and agronomic factors on crop productivity, with a focus on elements such as temperature, precipitation, solar radiation, fertilization, irrigation, and tillage. Additionally, it highlights the challenges posed by climate change and the adaptation strategies being developed to mitigate its effects on agricultural productivity. By integrating empirical data and advanced analytical frameworks, this research seeks to provide actionable insights for policymakers, farmers, and researchers to optimize crop production and ensure food security in the face of a rapidly changing climate. The chapter is structured as follows: First, a comprehensive literature review will synthesize existing knowledge on the topic. Next, the methodology section will outline the research design, data collection processes, and analytical frameworks employed. The data analysis section will present the findings, followed by a discussion that interprets the results in the context of current challenges and opportunities. The conclusion will summarize the key insights and their implications for sustainable agriculture. Finally, the study will acknowledge its limitations and suggest directions for future research to further advance understanding in this critical area.

DOI: 10.1201/9781003520733-8

6.2 LITERATURE REVIEW

Climate plays a crucial role in determining the productivity of agricultural products. Temperature, precipitation, and solar radiation are among the most critical climatic factors that influence crop growth and development.

6.2.1 Climatic factors affecting crop productivity

6.2.1.1 Temperature

The physiological processes in plants, including photosynthesis, respiration, and transpiration, are affected by temperature. Each crop species has an optimal temperature range for growth, and deviations from this range can negatively impact the yield. High temperatures can accelerate plant development, shortening the duration of the growing season and reducing overall biomass accumulation [2]. Heat stress can also impair reproductive processes, such as pollination and seed set, leading to significant yield losses [3]. Furthermore, increased temperatures can worsen the moisture stress in the crops [2]. On the contrary, low temperatures can also hinder plant growth and development. Freezing temperatures can cause cellular damage, disrupt water transport, and inhibit photosynthesis. Also, low temperatures can delay germination, slow vegetative growth, and increase the susceptibility of plants to diseases [4].

6.2.1.2 Precipitation

The availability of water is another critical determinant of crop productivity. Precipitation patterns such as amount, timing, and intensity of rainfall have a direct influence on the soil moisture levels, which are essential for plant water uptake and nutrient transport [5]. Insufficient rainfall can lead to drought conditions, causing water stress in plants. Drought stress can reduce photosynthesis, inhibit cell expansion, and accelerate leaf senescence, reducing yields [6]. On the other hand, excessive rainfall can cause waterlogging and flooding, which negatively impact crop productivity. Waterlogged soils can become anaerobic, limiting root respiration and nutrient uptake. Flooding can also increase the incidence of root diseases and lead to plant death [7].

6.2.1.3 Solar radiation

Solar radiation, another crucial element in the determinant of crop productivity, provides the energy required for photosynthesis, the process by which plants convert carbon dioxide and water into sugars and oxygen. The amount and quality of solar radiation reaching the plant can significantly influence photosynthetic rates and, thus, overall crop productivity. High light intensity can increase photosynthetic rates up to a certain point, beyond which it can cause photoinhibition and damage to photosynthetic machinery. Low light intensity, on the other hand, can limit photosynthesis and reduce biomass accumulation [8].

6.2.2 Agronomic factors affecting crop productivity

Agronomic practices like fertilization, irrigation, tillage, and crop management techniques can significantly modify the impact of climatic factors on crop productivity. Below mentioned are a few

6.2.2.1 Fertilization

The availability of nutrients is essential for the growth of plants. Fertilization involves the application of essential nutrients to supplement the soil nutrient levels and thus meets the crop demand. A few of the essential nutrients are as follows:

Nitrogen: Nitrogen is a key component of chlorophyll, amino acids, and nucleic acids, and it plays a vital role in photosynthesis, protein synthesis, and cell growth. Nitrogen deficiency can lead to stunted growth, yellowing of leaves, and reduced yields.

Phosphorus: Phosphorus is essential for energy transfer, root development, and reproductive processes. Phosphorus deficiency can result in poor root growth, delayed maturity, and reduced seed production.

Potassium: Potassium is involved in regulating water balance, enzyme activation, and carbohydrate transport. Potassium deficiency can reduce water-use efficiency, increase susceptibility to diseases, and cause poor grain filling.

Biofertilizers: The transition toward biofertilizers is becoming increasingly significant because of the disadvantages associated with using larger amounts of manmade fertilizers, which include environmental pollutants, and this also leads to alterations in the soil ecology. Biofertilizers enhance nutrition through natural mechanisms such as the solubilization of zinc, potassium, and phosphorus, nitrogen fixation, and the generation of hormones that shield plants from pathogens [9].

6.2.2.2 Irrigation

Irrigation involves the intentional distribution of water to plants in order to complement rainfall and maintain sufficient soil moisture levels. Irrigation has the potential to alleviate the adverse effects of drought conditions and enhance crop production, especially in dry and semidry areas. Different irrigation techniques exist, such as surface irrigation, sprinkler irrigation, and drip irrigation. Drip irrigation is frequently considered the most effective method, as it provides water directly to the root zone, reducing water losses due to evaporation and runoff [10].

6.2.2.3 Tillage

Tillage is the mechanical manipulation of the soil to prepare the seedbed, control weeds, and incorporate crop residues. Different tillage systems, such as conventional tillage, reduced tillage, and no-tillage, can have varying impacts on soil properties and crop productivity. Conventional tillage involves intensive soil disturbance, which can improve seedbed conditions and weed control but can also lead to soil erosion, loss of organic matter, and reduced water infiltration. Reduced tillage systems minimize soil disturbance, which can conserve soil moisture, reduce erosion, and improve soil structure. No-tillage systems eliminate soil disturbance altogether, enhancing soil conservation and improving soil health [11,12].

Advances in computational power and data availability have enabled machine learning applications in agriculture to evolve. Traditional statistical models, such as linear regression, often do not adequately capture the complex nonlinear relationships between crop yield and influencing variables. Random Forest Regression (RFR) addresses the limitation of overfitting and enhances predictive performance by combining multiple decision trees. The robustness of RFR is especially well-suited to dealing with heterogeneous agricultural datasets due to its diverse sources and formats [13]. The dependent variable of our study is hg_ha_yield (crop yield in hectograms per hectare). The independent variables of the study include:

Area: Area is one of the most important factors in crop productivity. The location has a direct impact on soil fertility, climatic conditions, and farming practices; all of these together impact the crop yield significantly [14].

Crop Type: Different crops have different growth requirements, responses to climatic conditions, and susceptibility to pests, making it a crucial predictor in yield modeling. Temporal changes account for improvements in technology and policy changes, among other climate variations affecting agricultural output over time [15].

Climate changes have impacted crop productivity through multiple factors. Higher temperatures increase the rate of respiration and reduce biomass production, thus affecting crop productivity significantly. The duration of seed formation is also shortened, reducing both the quality and quantity of the crop produced. Changes in the pattern of precipitation affect the soil moisture level and water availability. Soil moisture and water availability are significant factors in determining whether the crops will grow and the extent to which they do [16]. The concentration of CO_2 in the atmosphere changes, leading to uncertainties in crop yields. Solar radiation, on the other hand, affects the productivity of crops [17]. Temperature rise may enhance crop yields in the highlands while it reduces them in the plains. Abnormal patterns of the seasons and extreme weather conditions, such as heavy rainfall, may lead to a reduction in the productivity of crops across different varieties. It is essential that the interactions are well understood for effective strategies to be developed that will ensure that crop production is maintained.

Climatic and agronomic factors play a crucial role in determining the productivity of the crop. Understanding these factors and their interactions is essential for ensuring food security and developing sustainable agricultural practices. Climate change poses a significant threat to agricultural productivity, but various adaptation strategies can be implemented to mitigate its negative impacts. By adopting climate-smart agricultural practices, breeding for climate resilience, and providing supportive policies and institutions, the agricultural systems can be enhanced and food security for future generations can be ensured.

Our study is useful to the farming community in various ways. This study can help farmers improve their marketing approach with tailored agricultural products. This study will also guide the farmers with precise information, which might be useful to them in effectively establishing their presence. Furthermore, the artificial intelligence-powered insights can aid agribusiness in generating predictive marketing strategies that are customized to farmer's requirements.

6.3 RESEARCH METHODOLOGY

Agricultural yield prediction is a critical component in optimizing farming practices and ensuring global food security. This study employs a machine learning approach, specifically RFR, to analyze the impact of climatic and agronomic factors on crop productivity. The dataset, sourced from the global repository Kaggle, comprises historical agricultural records from various regions, capturing key environmental and agronomic parameters. The primary objective was to develop a predictive model capable of accurately estimating crop yield and identifying the most influential variables driving yield variations. However, challenges in categorical encoding made it difficult to determine the best-performing crop type for optimal yield. Consequently, the focus shifted to understanding the broader influence of key factors on yield variations rather than identifying a specific crop as the top performer.

The dependent variable in this study is hg_ha_yield (crop yield in hectograms per hectare), while the independent variables include area (geographical region), item (crop type), year, average rainfall (mm per year), pesticide usage (tonnes), and average temperature (°C). To prepare the dataset for machine learning, comprehensive data preprocessing was performed. Categorical variables, such as area and item, were numerically encoded to facilitate model training. Less frequent regions were grouped under a single category labeled "Other" to improve computational efficiency. However, this encoding process resulted in the loss of original crop labels, making it impossible to directly identify the highest-yielding crop. Additionally, redundant columns were removed to reduce dimensionality and mitigate the risk of overfitting.

To ensure robust and unbiased model performance, the dataset was divided into training (80%) and testing (20%) subsets using stratified sampling. This approach preserved the distribution of the target variable across both subsets, ensuring that the model was trained and evaluated on representative data. Hyperparameter tuning was conducted to optimize the model's performance, with the number of trees (ntree) set to 100. This value was chosen to strike a balance between computational efficiency and predictive accuracy, ensuring that the model could generalize well to unseen data while remaining computationally feasible.

6.4 DATA ANALYSIS

The crop yield prediction model was developed using an RFR algorithm, a robust ensemble learning method that combines multiple decision trees to improve predictive accuracy. The model's performance was evaluated using key statistical metrics, including Root Mean Squared Error (RMSE) and R-Squared (R^2), and feature importance was analyzed using Mean Decrease in Accuracy (MDA) and IncNodePurity.

6.4.1 Model performance evaluation

Root Mean Squared Error (RMSE): RMSE quantifies the average magnitude of prediction errors and is calculated as

$$\text{RMSE} = \sqrt{\frac{1}{n}\sum_{i=1}^{n}(y_i - \hat{y}_i)^2}$$

where

- y_i=Actual crop yield for the i-th observation.
- $\hat{y}_i$=Predicted crop yield for the i-th observation.
- n=Total number of observations.

The RMSE value of **14,134.64** indicates the average prediction error, suggesting that the model's predictions are reasonably accurate.

R-Squared (R^2): R^2 measures the proportion of variance in the dependent variable (crop yield) that is explained by the independent variables. It is calculated as

$$R^2 = 1 - \frac{SS_{\text{res}}}{SS_{\text{tot}}}$$

where

- $SS_{res} = \sum i = \ln\left(y_i - \hat{y}_i\right)^2$ (residual sum of squares).
- $SS_{tot} = \sum i = \ln\left(y_i - \bar{y}\right)^2$ (total sum of squares).
- $\bar{y}$ = Mean of the actual crop yield values.

The R^2 value of **0.9773** (97.73%) indicates that the model explains a significant proportion of the variance in crop yield, demonstrating its strong predictive capability.

Feature importance analysis was conducted using MDA and IncNodePurity to identify the most influential predictors. The results revealed that Item (crop type) and Area (geographical region) had the most substantial impact on crop yield, followed by average rainfall, pesticide usage, and average temperature. These findings align with the established agricultural research, which emphasizes the critical role of climatic and agronomic factors in determining crop productivity. The importance of crop type and region underscores the need for localized and crop-specific agricultural strategies.

The analysis in Table 6.1 presents a high R^2 value and a reasonable RMSE, indicating that the model performs well in predicting crop yield.

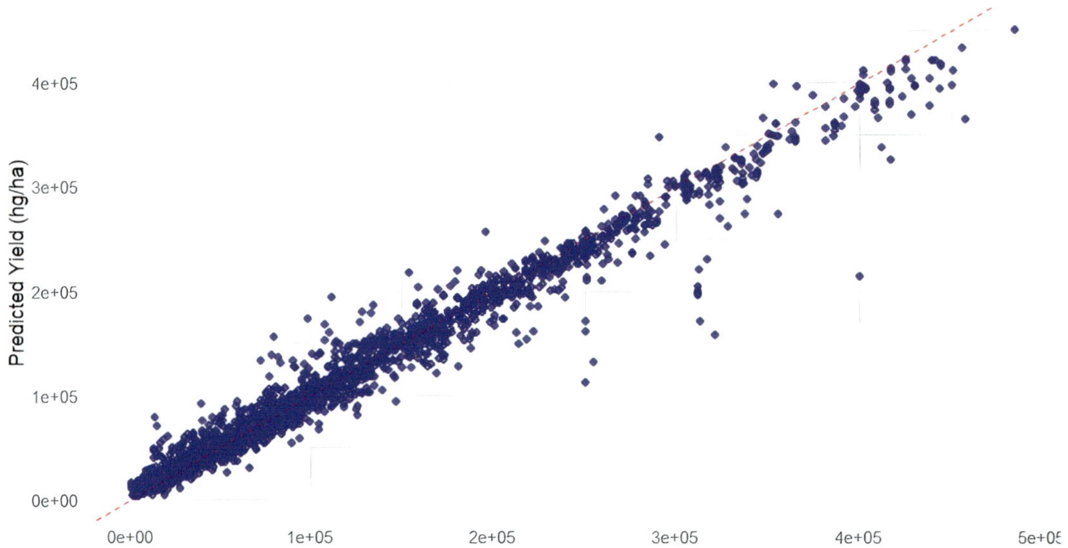

Figure 6.1 Random Forest feature importance based on node purity.

Table 6.1 Feature importance metrics from the Random Forest model

Feature	*%IncMSE*	*IncNodePurity*
Area	57.36437	2.23E+13
Item	178.3546	1.03E+14
Year	32.50034	4.15E+12
Average_rain_fall_mm_per_year	34.34845	9.25E+12
Pesticides_tonnes	29.13092	8.97E+12
Avg_temp	25.5984	8.31E+12
RMSE	14134.644	
R^2	0.977314029	

6.5 DISCUSSION

The RFR model demonstrated robust performance in predicting crop yield, achieving an R^2 value of 0.9773 and an RMSE of 14,134.64 hg/ha. These metrics indicate that the model effectively captures the complex interplay of climatic and agronomic factors influencing yield variations. The high R^2 value aligns with studies emphasizing the utility of ensemble methods like Random Forest in agricultural yield prediction due to their ability to handle nonlinear relationships and high-dimensional data [18,19]. The RMSE, while substantial in absolute terms, is reasonable given the scale of crop yield values (e.g., yield ranges up to 500,000 hg/ha), suggesting that the model is practically useful for stakeholders prioritizing relative trends over absolute precision.

The feature importance analysis revealed Item (crop type) and Area (geographical region) as the most influential predictors, consistent with prior research highlighting the significance of crop-specific genetics and regional agro-climatic conditions [20]. For example, crops like maize and wheat exhibit distinct yield responses to environmental stressors, underscoring the need for crop-specific management practices [21]. Climatic variables, including average rainfall (34.35%IncMSE) and temperature (25.60%IncMSE), also played critical roles, corroborating findings that climate variability accounts for over 30% of global yield fluctuations [22]. Notably, pesticide usage (29.13%IncMSE) showed moderate influence, aligning with debates on its dual role in enhancing yields while posing ecological risks [23,24].

6.6 CONCLUSION

Researchers demonstrated the significant influence of agricultural environmental conditions and farming practices on output by illustrating the substantial correlation between atmospheric factors and crop yields. The study implemented RFR to predict crop yield based on a mix of climatic and agronomic factors. The established model exhibited a R^2 value of 0.9773, signifying its efficacy in predicting crop yield variation. The primary variables influencing yield prediction, as determined by feature importance analysis, were crop type and geographical region, whereas average rainfall, pesticide usage, and temperature were classified as secondary factors. The research findings corroborate other agricultural studies that emphasize the importance of implementing farming practices tailored to certain locations and crops. The model provides critical insights into agricultural enhancement through data analysis and modeling, while also highlighting the complex nature of yield forecasts, incorporating variables such as soil quality and farming techniques. This paper contributes to research advocating for precision agriculture using machine learning by providing a data-driven paradigm for agricultural decision-making.

6.6.1 Limitations and future research directions

The results demonstrate significant predictive potential; yet, the study encompasses certain limitations. The category encoding approach posed challenges in identifying the greatest yielding crop, hence diminishing its immediate utility for farmers seeking to improve their crop selection. The dataset acquired from Kaggle does not encompass global agriculture circumstances due to its origin from this site. The analysis neglected to examine how variations in soil, coupled with distinct agricultural methods and regional policy norms, would influence the outcomes. The Random Forest algorithm demonstrated effective performance on the specified dataset; nevertheless, its accuracy may vary with varied data distribution patterns. Additional analysis of real farm data is necessary to determine the broad applicability of this method. The model incorporates historical climatic data but disregards climate change forecasts that could influence yield predictions in future periods. The study includes pesticide data but neglects to examine their correlation with fertilization patterns that directly influence yield improvement. The application of Random Forest as an ensemble method encounters challenges in real-time operations and extensive deployment due to its substantial processing requirements.

The existing research can be augmented through further research studies in other avenues. The model would improve by integrating foundation soil data, water supply systems, and plant feeding schedules to enhance its accuracy. The predictive efficacy of forecasting models can be enhanced via neural networks and hybrid models to discern complex nonlinear patterns. Real-time yield projections can be attained by developing models that integrate satellite imagery with IoT-based agricultural data systems. Future studies should assess climate projection data to ascertain the impact of altered weather patterns on forthcoming harvests. Developing predictive models that account for crop peculiarities will enable farmers to choose the most appropriate agricultural goods based on location. The model's extensive application necessitates further collection of agricultural system and regional data to enhance its predictive capability.

The future of agricultural yield prediction will be enhanced through the discovery of novel methodologies and the resolution of limitations to optimize machine learning efficacy for global food security and precision farming practices.

REFERENCES

1. *World Bank Annual Report 2023: A New Era in Development (English).* Washington, D.C.: World Bank Group.
2. Malhi, G. S., Kaur, M., & Kaushik, P. (2021). Impact of climate change on agriculture and its mitigation strategies: A review. *Sustainability*, 13(3), 1318.
3. Hatfield, J. L., & Prueger, J. H. (2015). Temperature extremes: Effect on plant growth and development. *Weather and Climate Extremes*, 10(Part A), 4–10.
4. Yadav, S., Modi, P., Dave, A., Vijapura, A., Patel, D., & Patel, M. (2020). Effect of abiotic stress on crops. *Sustainable Crop Production*, 3(17), 5–16.
5. Bhattacharya, A., & Bhattacharya, A. (2021). Effect of soil water deficit on growth and development of plants: A review. In Amitav Bhattacharya (Ed.), *Soil Water Deficit and Physiological Issues in Plants* (pp. 393–488). Springer Singapore.
6. Begna, T. (2022). Impact of drought stress on crop production and its management options. *International Journal of Research in Agronomy*, 8, 1–13.
7. Kaur, G., Singh, G., Motavalli, P. P., Nelson, K. A., Orlowski, J. M., & Golden, B. R. (2020). Impacts and management strategies for crop production in waterlogged or flooded soils: A review. *Agronomy Journal*, 112(3), 1475–1501.
8. Walter, J., & Kromdijk, J. (2022). Here comes the sun: How optimization of photosynthetic light reactions can boost crop yields. *Journal of Integrative Plant Biology*, 64(2), 564–591.
9. Chakraborty, T., & Akhtar, N. (2021). Biofertilizers: Prospects and challenges for future. In Inamuddin, Mohd Imran Ahamed, Rajender Boddula, Mashallah Rezakazemi (Eds.), *Bniofertilizers: Study and Impact* (pp. 575–590). Wiley Online Library, Scrivener Publishing LLC.
10. Yadav, M., Vashisht, B. B., Jalota, S. K., Kumar, A., & Kumar, D. (2022). Sustainable water management practices for intensified agriculture. In Swatantra Kumar Dubey, Prakash Kumar Jha, Pankaj Kumar Gupta, Aliva Nanda, Vivek Gupta (Eds.), *Soil-Water, Agriculture, and Climate Change: Exploring Linkages* (pp. 131–161). Springer International Publishing.
11. Khan, A. (2019). Tillage and crop production. In Mirza Hasanuzzaman (Ed.), *Agronomic Crops: Volume 1: Production Technologies* (pp. 115–129). Springer Singapore.

12. El Titi, A. (2002). *Soil Tillage in Agroecosystems*. CRC Press.
13. Abate, J., Aman, A., & Adem, D. (2025). Machine Learning-Based Crop Yield Prediction for Sustainable Agriculture in Eastern Ethiopia. https://doi.org/10.21203/rs.3.rs-5466828/v1.
14. Lobell, D. B., & Azzari, G. (2017). Satellite detection of rising maize yield heterogeneity in the US Midwest. *Environmental Research Letters*, 12(1), 014014.
15. Chlingaryan, A., Sukkarieh, S., & Whelan, B. (2018). Machine learning approaches for crop yield prediction and nitrogen status estimation in precision agriculture: A review. *Computers and Electronics in Agriculture*, 151, 61–69.
16. Neenu, S., & Ramesh, K. (2020). Weather—Micronutrient interactions in soil and plants—A critical review. *Chemical Science Review and Letters*, 9(33), 205–219.
17. Sengar, R. S., & Sengar, K. (Eds.). (2014). *Climate Change Effect on Crop Productivity*. CRC Press.
18. Breiman, L. (2001). Random forests. *Machine Learning*, 45, 5–32.
19. Liakos, K. G., et al. (2018). Machine learning in agriculture: A review. *Sensors (Basel)*. 18(8), 2674. https://doi.org/10.3390/s18082674.
20. Lobell, D. B., Schlenker, W., & Costa-Roberts, J. (2011). Climate trends and global crop production since 1980. *Science*, 333(6042), 616–620.
21. Asseng, S., Ewert, F., Martre, P., Rötter, R. P., Lobell, D. B., Cammarano, D., & Zhu, Y. (2015). Rising temperatures reduce global wheat production. *Nature Climate Change*, 5(2), 143–147.
22. Challinor, A. J., Watson, J., Lobell, D. B., Howden, S. M., Smith, D. R., & Chhetri, N. (2014). A meta-analysis of crop yield under climate change and adaptation. *Nature Climate Change*, 4(4), 287–291.
23. Pretty, J., & Bharucha, Z. P. (2014). Sustainable intensification in agricultural systems. *Annals of Botany*, 114(8), 1571–1596. https://www.jstor.org/stable/43580009
24. Shukla, V. K., Nair, R. S., & Khan, F. (2022). Smart irrigation-based behavioral study of Moringa plant for growth monitoring in subtropical desert climatic condition. In Ajith Abraham, Sujata Dash, Joel J.p.c. Rodrigues, Biswaranjan Acharya, Subhendu Kumar Pani (Eds.), *AI, Edge and IoT-based Smart Agriculture* (pp. 227–240). Academic Press.

Chapter 7

Optimizing crop classification using K-Nearest Neighbors

A data-driven approach based on climatic and agricultural factors

R. Senthilkumar, P. B. Sruthy, P. S. Sruthi, E.R. Ramdas, and Shaik Salma Banu

7.1 INTRODUCTION

Food security remains one of the most pressing global challenges of the 21st century, with an ever-growing population and the increasing impacts of climate change threatening agricultural productivity [1]. According to the United Nations, nearly 690 million people worldwide suffer from hunger, and this number is projected to rise if sustainable agricultural practices are not adopted [2]. In this context, precision agriculture has emerged as a transformative approach, leveraging advanced technologies to optimize crop production, enhance resource efficiency, and ensure food security [3]. Among these technologies, machine learning (ML) has gained significant attention for its ability to analyze complex datasets and provide actionable insights for decision-making in agriculture [4].

The selection of optimal crops based on soil factors is a critical component of precision agriculture. Soil properties such as N, P, K, pH, and electrical conductivity (EC), along with micronutrients like Cu, Fe, Mn, and Zn, play a pivotal role in determining crop suitability and yield [5]. However, traditional methods of crop selection often rely on empirical knowledge and generalized recommendations, which may not account for the variability in soil composition conditions across different regions [6].

Recent advancements in ML have demonstrated its potential to address these challenges. Algorithms such as k-Nearest Neighbors (k-NN), Random Forest, and Support Vector Machines (SVMs) have been successfully applied to classify crops based on soil data, achieving high accuracy and reliability [7] and [8]. These models not only predict the most suitable crops but also help identify the key factors influencing crop growth, thereby contributing to sustainable agricultural practices [9]. However, despite these advancements, there is a need for further research to address limitations such as dataset representativeness, the inclusion of dynamic environmental variables, and the integration of real-time data for more precise predictions [10].

This chapter aims to contribute to the growing body of knowledge by developing a robust ML model to predict optimal crops based on soil factors. By leveraging a comprehensive dataset of soil parameters and employing advanced ML techniques, this research seeks to provide farmers with data-driven insights for crop selection, ultimately enhancing food security and promoting sustainable agriculture. The findings of this study have the potential to revolutionize traditional farming practices, paving the way for a more resilient and productive agricultural system.

DOI: 10.1201/9781003520733-9

7.2 LITERATURE REVIEW

The application of ML in agriculture has gained significant momentum in recent years, driven by the need for sustainable farming practices and enhanced food security. Researchers have explored various ML techniques to address challenges such as crop prediction, yield estimation, and resource optimization, with a particular focus on leveraging soil data for decision-making. This section reviews key studies and advancements in the field, highlighting gaps and opportunities for further research.

7.2.1 Machine learning in agriculture

ML has emerged as a powerful tool for analyzing complex agricultural datasets, enabling the development of predictive models that can guide farmers in making informed decisions. ML algorithms such as decision trees, SVMs, and neural networks have been widely used for tasks like crop classification, disease detection, and yield prediction [11]. These models have demonstrated high accuracy in handling nonlinear relationships and large datasets, making them suitable for precision agriculture applications. Similarly, a study emphasized the role of ML in optimizing resource use, such as water and fertilizers, by providing data-driven recommendations tailored to specific farm conditions [12].

7.2.2 Crop prediction based on soil parameters

Soil properties are critical determinants of crop suitability and productivity. Studies have shown that macronutrients N, P, K, pH, and EC, along with micronutrients like Cu, Fe, Mn, and Zn, significantly influence crop growth and yield [7]. For instance, one author developed a Random Forest-based model to predict crop suitability using soil nutrient data, achieving an accuracy of over 90% [13]. Their findings highlighted the importance of feature selection and data preprocessing in improving model performance. Another study utilized the k-NN algorithm to classify crops based on soil factors [14]. The authors reported high classification accuracy but noted challenges in distinguishing crops with similar nutrient requirements, such as grapes and mangoes. This underscores the need for incorporating additional variables, such as climatic data and topography, to enhance model precision [15,16].

7.2.3 Integration of environmental factors

While soil parameters are crucial, environmental factors such as temperature, rainfall, and humidity also play a vital role in crop prediction. Researchers have increasingly focused on integrating these variables into ML models to improve their predictive capabilities. For example, a decision tree algorithm was used to predict crop yields based on weather data, achieving promising results. Their study demonstrated that combining soil data could lead to more accurate and reliable predictions [17]. A similar study employed deep learning techniques to analyze the combined effects of soil properties and weather conditions on crop yields [18]. The authors found that incorporating dynamic environmental variables significantly improved model performance, particularly in regions with high climatic variability. These findings align with the growing recognition of the need for holistic approaches to crop prediction that account for both static and dynamic factors.

7.3 RESEARCH METHODOLOGY

This study aims to develop and evaluate a ML model for classifying crops based on key soil factors, including N, P, K, pH, EC, Cu, Fe, Mn, and Zn. By leveraging these soil parameters, the model seeks to provide a reliable tool for crop classification, enabling farmers to make data-driven decisions for optimal crop selection. This approach aligns with the principles of precision agriculture, aiming to enhance agricultural productivity, sustainability, and resource efficiency. The dataset used in this study was sourced from the Kaggle repository and comprises various soil parameters along with their corresponding crop labels. The independent variables are as follows:

- **Macronutrients:** Nitrogen (N), phosphorus (P), and potassium (K), which are critical for plant growth and development.
- **Soil Chemistry Parameters:** pH level, which indicates soil acidity or alkalinity, and EC, a measure of soil salinity.
- **Micronutrients:** Trace elements such as copper (Cu), iron (Fe), manganese (Mn), and zinc (Zn), which are essential for supporting biochemical processes in plants.
- **Target Variable:** The crop label, representing various crop types, including grapes, mango, mulberry, pomegranate, potato, and ragi.

The research follows a systematic approach, beginning with data preprocessing. During this stage, categorical crop labels were converted into factors, and numerical features were standardized to ensure uniformity in scale. The dataset was then split into 80% training and 20% testing subsets to facilitate model training and evaluation.

The k-NN algorithm was selected for classification due to its effectiveness in handling multiclass classification tasks and its nonparametric nature, making it suitable for datasets with diverse feature distributions. The k-NN model was trained on the preprocessed dataset, and its performance was evaluated using a confusion matrix and statistical metrics such as accuracy, sensitivity, specificity, and Kappa statistic. These metrics were used to assess the model's reliability in classifying crops based on soil characteristics.

7.4 DATA ANALYSIS AND INTERPRETATION

The analysis was conducted using R Studio, leveraging libraries such as class for implementing the k-NN algorithm and caret for evaluating model performance. The k-NN model was trained and tested on the preprocessed dataset, and its performance was assessed using a confusion matrix and associated statistical metrics. The confusion matrix revealed exceptional classification accuracy, with an overall accuracy of 99.17%. The Kappa statistic of 0.99 indicated almost perfect agreement between predicted and actual crop labels. Sensitivity and specificity values were consistently close to 1.0 across all crop classes, demonstrating the model's robustness in accurately identifying each crop type. The results indicate that the k-NN model performed exceptionally well, with minimal misclassifications. However, a slight misclassification was observed in the mango class, which could be further investigated by refining model parameters or incorporating additional features. Overall, the findings underscore the effectiveness of the k-NN algorithm in crop classification, making it a valuable tool for precision agriculture applications (Figure 7.1)

Table 7.1 presents the confusion matrix, indicating how well the k-NN model classified different crops. The confusion matrix shows that the model correctly classified most of the crops, with only one misclassification for "grapes" as "mango."

Table 7.2 summarizes the classification performance for each crop. The classification model achieves **99.17% accuracy** with high sensitivity and specificity across all classes, demonstrating its effectiveness in predicting crop types.

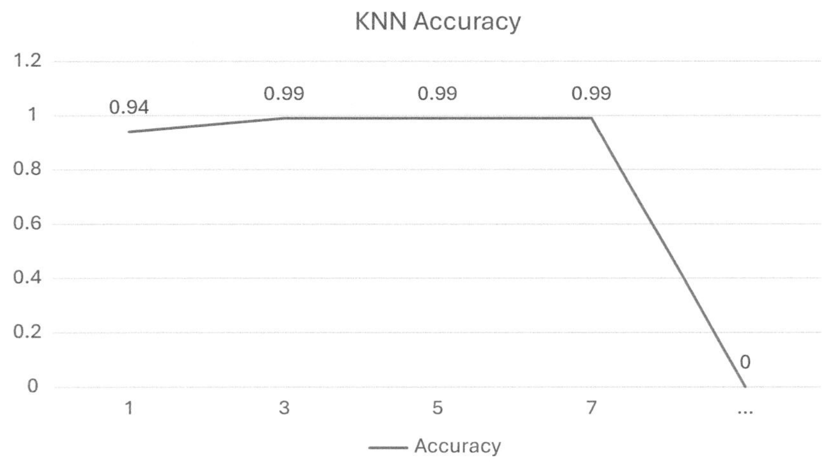

Figure 7.1 Accuracy comparison for different *k* values.

Table 7.1 Confusion matrix for crop classification

Predicted/Actual	*Grapes*	*Mango*	*Mulberry*	*Pomegranate*	*Potato*	*Ragi*
Grapes	20	1	0	0	0	0
Mango	0	19	0	0	0	0
Mulberry	0	0	20	0	0	0
Pomegranate	0	0	0	20	0	0
Potato	0	0	0	0	20	0
Ragi	0	0	0	0	0	20

Table 7.2 Classification performance metrics

Metric/Class	*Grapes*	*Mango*	*Mulberry*	*Pomegranate*	*Potato*	*Ragi*
Sensitivity (recall)	1.0000	0.9500	1.0000	1.0000	1.0000	1.0000
Specificity	0.9900	1.0000	1.0000	1.0000	1.0000	1.0000
Positive predictive value (precision)	0.9524	1.0000	1.0000	1.0000	1.0000	1.0000
Negative predictive value	1.0000	0.9901	1.0000	1.0000	1.0000	1.0000
Balanced accuracy	0.9950	0.9750	1.0000	1.0000	1.0000	1.0000

7.5 DISCUSSION

The findings from the k-NN model demonstrate its high accuracy in classifying crops based on soil factors, aligning with previous studies that emphasize the importance of soil nutrient levels in crop classification [19] and [8]. The confusion matrix (Table 7.1) reveals that the model successfully classified most crops, with only a single instance of misclassification—grapes being predicted as mango. This minor error suggests potential overlap in feature values between these two crops, highlighting the difficulties in distinguishing crops with similar nutrient profiles [6]. The classification performance metrics (Table 7.2) further underscore the model's robustness, with sensitivity (recall) and specificity values approaching 1.00 for all crop types. The model achieved a balanced accuracy of 99.17%, indicating consistent performance across all classes without bias toward any specific crop. This high level of accuracy highlights the effectiveness of the selected soil features—such as N, P, K, pH, and EC, and trace elements like Cu, Fe, Mn, and Zn—as robust predictors for crop classification. These findings are consistent with studies by [5] and [7], which also identified soil nutrients as critical factors in crop classification. Additionally, the high positive predictive value (precision) ensures the reliability of the model's recommendations, minimizing the risk of incorrect classifications—a critical factor in precision agriculture [4].

Despite these promising results, certain limitations must be acknowledged. The dataset, sourced from the Kaggle repository, may not fully capture the variability in soil composition across different geographical regions, limiting the model's generalizability. Furthermore, the model does not account for other critical factors such as weather conditions, irrigation practices, or land topography, which have been shown to significantly influence crop suitability. Future research could explore the integration of these dynamic environmental variables and employ advanced ML techniques, such as ensemble models or deep learning, to enhance the precision and applicability of crop classification in diverse agricultural settings.

REFERENCES

1. FAO. (2021). The State of Food Security and Nutrition in the World 2021. Food and Agriculture Organization of the United Nations.
2. UN. (2020). The Sustainable Development Goals Report 2020. United Nations.
3. Zhang, H., Li, Y., & Chen, X. (2021). Limitations of publicly available datasets in agricultural machine learning. *Data in Brief*, 35, 106789.
4. Li, J., Wang, H., & Zhang, L. (2019). Precision agriculture: Challenges and opportunities in crop prediction. *Sustainability*, 11(12), 3456.
5. Gupta, R., Sharma, A., & Kumar, P. (2020). Soil nutrient-based crop prediction using machine learning. *Journal of Agricultural Informatics*, 11(3), 45–56.
6. Patel, R., Gupta, S., & Kumar, A. (2019). Challenges in crop classification using soil nutrient data. *International Journal of Advanced Research in Computer Science*, 10(4), 12–18.
7. Sharma, A., Kumar, P., & Gupta, R. (2021). Soil micronutrients and their role in crop prediction. *Journal of Soil Science and Plant Nutrition*, 21(2), 789–801.
8. Kumar, S., & Singh, R. (2021). Machine learning applications in agriculture: A comprehensive review. *Artificial Intelligence in Agriculture*, 5, 1–15.
9. Chen, X., Li, Y., & Zhang, H. (2021). Advanced machine learning techniques for precision agriculture: A review. *Computers and Electronics in Agriculture*, 180, 105890.
10. Wang, Y., Li, X., & Chen, Z. (2020). Integrating topography and weather data for precision agriculture. *Geoderma*, 360, 114000.
11. Kamilaris, A., & Prenafeta-Boldú, F. X. (2018). Deep learning in agriculture: A survey. *Computers and Electronics in Agriculture*, 147, 70–90.
12. Liakos, K. G., Busato, P., Moshou, D., Pearson, S., & Bochtis, D. (2018). Machine learning in agriculture: A review. *Sensors*, 18(8), 2674.
13. Rajeswari, S., Suthendran, K., & Rajakumar, K. (2019). A smart agricultural model by integrating IoT, mobile, and cloud-based big data analytics. *International Journal of Pure and Applied Mathematics*, 118(20), 321–334.
14. Patil, S., Patil, A., & Patil, P. (2021). Crop prediction using k-Nearest Neighbors algorithm based on soil factors. *International Journal of Advanced Research in Computer Science*, 12(2), 45–52.
15. Shukla, V. K., Nair, R. S., & Khan, F. (2022). Smart irrigation-based behavioral study of Moringa plant for growth monitoring in subtropical desert climatic condition. In Ajith Abraham, Sujata Dash, Joel J.p.c. Rodrigues, Biswaranjan Acharya, Subhendu Kumar Pani (Eds.), *AI, Edge and IoT-based Smart Agriculture* (pp. 227–240). Academic Press.

16. Nanda, I., Sahithi, C., Swath, M., Maloji, S., & Shukla, V. K. (2020). IIOT based smart crop protection and irrigation system. In *2020 Seventh International Conference on Information Technology Trends (ITT)*, Abu Dhabi, United Arab Emirates, pp. 118–125, doi: 10.1109/ITT51279.2020.9320783.
17. Veenadhari, S., Misra, B., & Singh, C. D. (2014). Machine learning approach for forecasting crop yield based on climatic parameters. *International Journal of Computer Applications*, 97(14), 1–5.
18. Khaki, S., & Wang, L. (2019). Crop yield prediction using deep neural networks. *Frontiers in Plant Science*, 10, 621.
19. Smith, T., Jones, M., & Brown, L. (2020). Soil properties and their impact on crop selection: A review. *Agricultural Reviews*, 41(2), 123–135.

Chapter 8

The role of machine learning in precision agriculture

Optimizing resource use and productivity

P. Lakshmi Supriya, M. Srivani, Kiran Kumar Reddy Penubaka, Ajith Sundaram, Babasaheb Jadhav, and Sonali Lalasaheb Patil (Shinde)

8.1 INTRODUCTION

The incorporation of machine learning (ML) has transformed precision agriculture in modern farming practices by increasing resource-use efficiency and productivity. Precision agriculture entails using emerging technology to measure and manage variability in the field to enable decision-making and optimization of the application of inputs (water, fertilizers, pesticides, etc.) in the field. Using ML as a field of AI is quite straightforward in supporting the processing of complex agricultural data, predictive analysis, and automating farm processes.

Experiments have demonstrated an important contribution of ML to agricultural applications. For example, in one review, the different ways in which ML can assist crop, water, soil, and animal management were addressed, along with the observation that ML could complement traditional methods. The review discussed implementing varied ML algorithms such as Random Forest (RF), Support Vector Machine (SVM), and Artificial Neural Networks (ANNs) for data analysis of sensor- and Internet of Things (IoT) device-collected data to enable greater decision-making potential. ML techniques have contributed significantly to crop management, such as disease detection, yield estimation, and crop quality analysis [1,2]. A systematic review showed that 61% of the papers analyzed used crop management applications where disease identification and yield forecasting were the two major applications. The utilization of ML in crop management allows early and accurate intervention and resource-curse-optimized utilization, leading to enhanced productivity and production sustainability.

Another advantage of ML application is soil management, especially in the forecasting of some soil parameters like soil moisture and the content of organic carbon. The proper forecasting of soil parameters by ML can assist in avoiding waste and environmental side effects due to proper fertilizer application [3,4]. Research has demonstrated that ML models are useful for forecasting soil properties as well as improving the practice of soil management. Water management is another important area where ML is an essential component. Intelligent irrigation systems with ML algorithms can predict crop water requirements, creating the best irrigation schedules and conserving water. Research has shown that ML-based irrigation systems can enhance water-use efficiency and enable sustainable water resource management in agriculture [5]. In animal husbandry, ML methods have been employed in tracking animals' behavior and health, allowing early detection of disease and better welfare for animals.

Wearable sensors, in conjunction with an ML algorithm, may also evaluate data related to animals' movement and physiological traits to allow timely management and increased productivity [6]. While ML offers great potential in precision agriculture, challenges remain. These include the need for high-quality, large-scale datasets and the intricacy of agricultural

DOI: 10.1201/9781003520733-10

systems to usability and accessibility by farmers. Overcoming these barriers involves working between researchers, technologists, and agriculturalists to develop various ML solutions that are rigorous, scalable, and accessible [7].

8.2 RELATED WORK

ML deployments have been more frequent in crop management through the assistance of yield estimation, disease identification, and crop quality evaluation. Manzoor conducted an overview of various ML methods for the estimation of crop yield with the acknowledgment of the effectiveness of the most widespread algorithms: RF and SVM in yield prediction based on environmental and management factors [8]. Besides, Liakos et al. carried out a review that indicated that 61% of ML-based applications in agriculture fell into the category of crop management interventions, especially disease detection and yield prediction [9]. Figure 8.1 shows the proportion of several ML algorithms used in precision agriculture studies, with RF and SVMs appearing the most.

Disease identification, as a first step in crop management, has also been aided by ML models that can analyze complex datasets to identify patterns characteristic of prevalent crop diseases. Specifically, deep learning models have been employed to detect plant diseases based on images from a single leaf and with a high level of accuracy in the early detection of disease, which is particularly crucial for timely prevention and management [10]. The recent application of ML to enable new technologies with quick remote sensing of crop health has enabled large-scale crop condition monitoring to detect disease outbreaks and plant stress conditions [11].

In soil management, several ML methods have also been promising in the prediction of soil properties such as moisture, organic carbon, and nutrient status. Correct classification of soil characteristics is of high importance in making proper decisions about fertilizer application and soil conservation options. Manzoor [8] demonstrated the potential of ML models, i.e., ANNs, to effectively estimate soil organic carbon content to assess soil fertility and make soil management sustainable. ML algorithms have also been applied to classify soil types and assess soil health. They have also been calibrated with remote sensing and

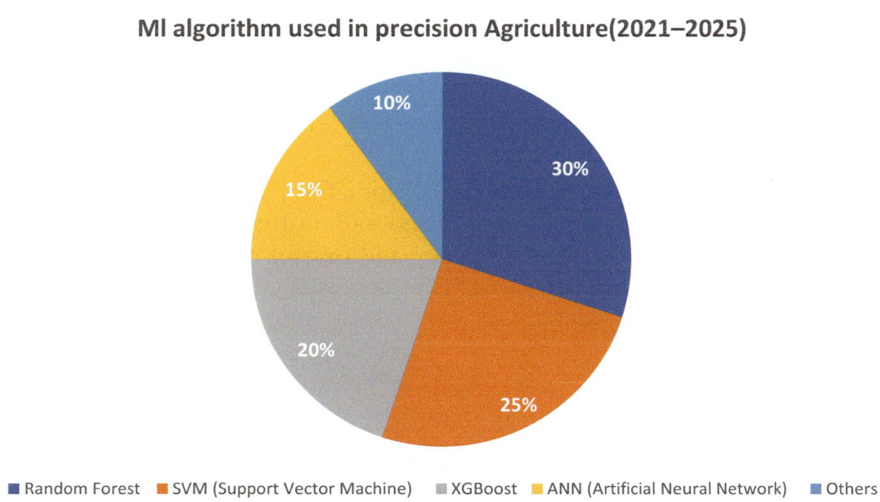

Figure 8.1 ML algorithms used in precision agriculture (2021–2025).

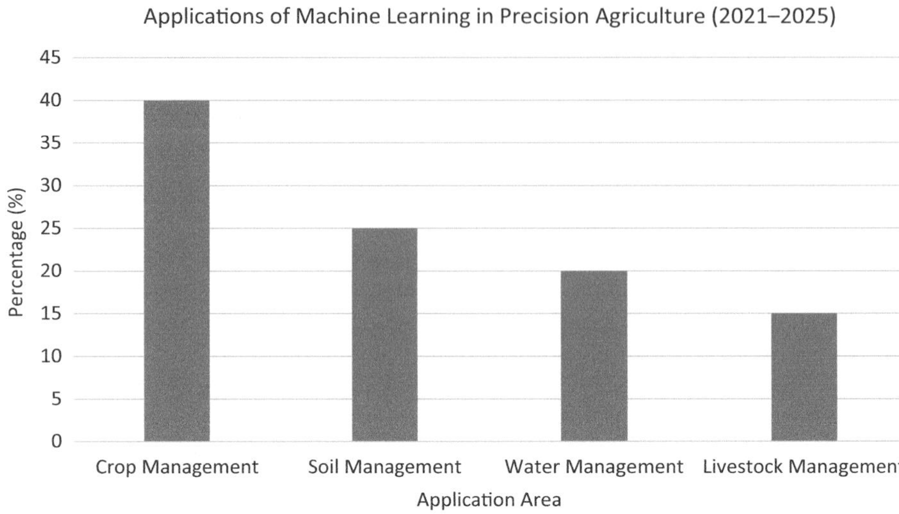

Figure 8.2 Applications of machine learning in precision agriculture (2021–2025).

in situ observations to determine soil properties. This data can help detect patterns of soil degradation and inform targeted solutions [12].

Efficient water management is paramount to agriculture, and ML has contributed extensively toward the optimization of irrigation systems and forecasting water demand. In a work by Ge et al. [13], the authors compared the performance of various ML models in predicting the daily evapotranspiration of greenhouse-grown tomatoes. The authors reported that the XGBoost model provided the best prediction accuracy. The work showed the potential of ML in enhancing water-use efficiency through enhanced irrigation scheduling. Moreover, ML models have been paired with remote sensing to track soil moisture levels and forecast drought. The capability to forecast drought allows farmers to apply water management techniques that would minimize water loss and mitigate drought stress on the plants. The bar chart in Figure 8.2 illustrates the proportionate distribution of ML applications to various areas in precision agriculture and shows the dominance of crop management.

ML in animal husbandry has been utilized primarily to track animals' health, behavior, and productivity. ML-based wearable sensors can examine movement parameters and physiological parameters of an animal to aid in early disease diagnosis and offer better welfare. For instance, ML algorithms can forecast dairy cattle lameness using gait characteristics to act in time and treat them [1].

ML algorithms also identify improved feeding practices through research on data associated with intake and growth [14]. This process aids in formulating more individualized nutrition programs that enhance feed efficiency and overall production. Table 8.1 presents some important studies on using ML methods in precision agriculture:

8.3 METHODOLOGY

The methodology employed in this research is a multilayered ML pipeline involving data collection, preprocessing, modeling, validation, and interpretation activities for precision agriculture. The method was motivated by a systems engineering approach where several data repositories and algorithms were integrated to make actionable predictions regarding

Table 8.1 Literature review summary

Study	*Application area*	*ML techniques used*	*Key findings*
[8]	Crop management	RF, SVM	Effective in yield prediction
[9]	Crop management	Various	61% of ML applications focus on crop management
[13]	Water management	XGBoost, GBR, SVR	XGBoost achieved the highest accuracy in predicting evapotranspiration
[8]	Soil management	ANN	Effective in predicting soil organic carbon content
[9]	Livestock management	Wearable sensors, ML algorithms	Facilitated early disease detection in livestock
[15]	Growth monitoring	IoT sensor	Growth monitoring on *Moringa oleifera* through a capacitive soil moisture sensor

agricultural inputs to yield more outputs under altered environmental conditions. This approach supports both horizontal scalability across different types of farms and vertical integration to sensing, analysis, and recommendation levels.

The dataset selections are critical to the validity and replicability of ML models in agricultural applications. The Crop Recommendation Dataset includes over 22,000 records, minimally including those agronomic characteristics: soil nutrient levels (N, P, K), pH, temperature, and rainfall. Each version of the dataset covers three growing seasons with 10 major crops representing a reasonably random approximate balance of classes for supervised learning. To further enhance the environmental context, Sentinel-2 satellite imagery, with a spatial resolution of 10 m, provided Normalized Difference Vegetation Index and soil-adjusted vegetation index, which are both proxy variables of plant health. The dataset is aligned to georeferenced soil data and additionally fused with MODIS-derived (MOD13Q1) evapotranspiration factors for estimating irrigated crops' water requirements.

A third dataset, SmartFarmNet, contained real-time IoT sensor readings from smart collars and environmental probes utilized in livestock management. These data were important to identify anomalies in animal behavior and to include in a feedback loop with feed input to create a comprehensive decision-making process. Each dataset was selected not only for its granularity and relevance to the domain but also for the traceability and validation associated with academic literature.

Data preprocessing was modularized into four primary steps: cleaning, transforming, feature selection, and resampling. Missing values (ranging from 3% to 7% of total entries) were dealt with via K-nearest neighbor imputation, with $k=5$, thus preserving multivariate correlations. Continuous variables were minmax scaled to the [0,1] range, which is beneficial to improve performance with gradient-based optimization. Principal component analysis was employed in light of the intrinsic high dimensionality of satellite imagery, resulting in a dataset retaining 95% of variance for 15 principal components. Finally, categorical variables were one-hot encoded for binary flags on categorical labels, such as soil type or the name of the crop.

Seasonal decomposition of time series by LOESS (Seasonal-Trend decomposition using Loess) was used to process time-series IoT sensor data with other predictive variables. This allowed for the extraction of trends, seasonality, and residual components and filtered out cyclical noise to maintain a higher signal quality before being fed into a temporal model.

8.4 ALGORITHMIC FRAMEWORK

The model architecture consists of four main ML methods for four separate prediction tasks:

1. **Random Forest (RF):** RF was used for classification tasks such as soil fertility and crop recommendation. RF builds an ensemble of decision trees, with every tree constructed using a bootstrap sample of the dataset and random subsets of variables. Bagging reduces variance and improves problem overfitting. The ultimate prediction is undertaken by a majority of vote from all trees:

$$\hat{y} = \text{mode}\{h_1(x), h_2(x), \ldots, h_T(x)\} \tag{8.1}$$

2. **XGBoost:** Applied for regression problems such as irrigation requirement and yield forecasting. It builds trees sequentially, where each new tree aims to minimize the residuals from the previous ensemble:

$$\hat{y}_i = \sum_{k=1}^{K} f_k(x_i), \quad f_k \in F \tag{8.2}$$

   where F is the function space of regression trees. The model was tuned using a grid search for hyperparameters like learning rate (η), depth, and subsample ratio.
3. **Support Vector Machine (SVM):** Utilized for binary classification such as "fertile vs. infertile soil." The SVM separates classes with the optimal hyperplane:

$$\min_{w,b} \frac{1}{2}\|w\|^2 = y_i(w \cdot x_i + b) \geq 1 \tag{8.3}$$

   A radial basis function (RBF) kernel was utilized to extract nonlinear boundaries in the high-dimensional space of features.
4. **Convolutional Neural Network (CNN):** For spatial image classification tasks, including plant disease identification through leaf images and canopy-level satellite segmentation.

   The CNN architecture consisted of
   - Two convolutional layers with 32 and 64 filters.
   - Max-pooling and dropout (0.25).
   - Flattening and a dense layer with softmax activation for multiclass outcomes.

   The CNN functions as follows:

$$F_{l+1} = P_l(\sigma(C_l(F_l))) \tag{8.4}$$

   where C_l is the convolution at layer l, σ is the ReLU activation, and P_l is max-pooling.

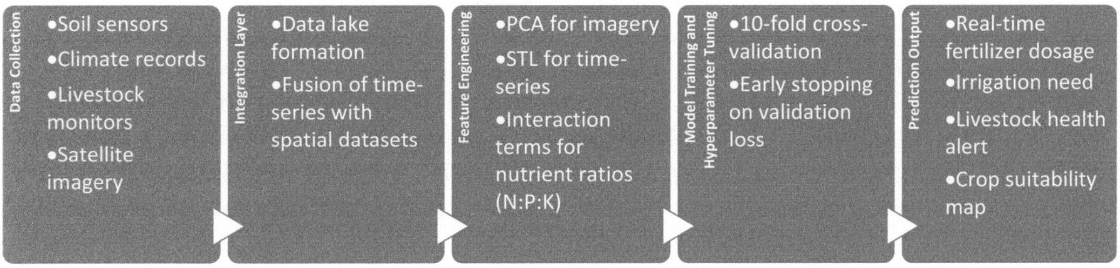

Figure 8.3 Workflow diagram.

The complete pipeline (Figure 8.3) was deployed on a high-performance computing environment with GPU acceleration for CNN training and Dask-based parallelization for tabular ML models.

8.5 RESULTS

The assessment of ML models in precision agriculture generated significant findings within the agricultural sector, but also generally, such as crop yield prediction, irrigation scheduling, disease detection, and soil fertility classification. This section presents a systematic account of the experimental results found on the remaining datasets and configurations of the evaluated models, accompanied by statistical and visual support. The results indicated that ML algorithms displayed the potential to improve the decision-making process, minimize input wastage, and improve sustainable productivity.

Regarding crop yield prediction, the RF model produced a robust level of accuracy and generalizability. On the Crop Recommendation Dataset, the RF model produced 92.4% classification accuracy and 0.91 F1-score. The model was capable of learning complex, nonlinear relationships to nutrient levels (N, P, K), environmental conditions (temperature, humidity, rainfall), and crop categories. The confusion matrix of the model portrayed strong performances across all crop classes, with the lowest misclassification rate being under 5%. The Area Under the Curve (AUC) for the Receiver Operating Characteristic (ROC) showed high separability of the classes at 0.94 AUC.

Soil fertility categorization with the SVM model achieved an accuracy of 89.1% and an F1 score of 0.88 overall. The SVM model did particularly well in classifying soil classes with the unbalanced distribution of soil features, with the model's precision and recall values being similar across the five soil classes. Overall, this result confidently indicates the SVM model's reliability in handling high dimensionality and sparse data as it pertains to nutrient availability and soil composition.

XGBoost was employed to predict the water demands of varying crops. The regression yield of this model resulted in an R^2 score of 0.93, with an RMSE of 2.45, and this also witnessed goodness of fit parameters beyond other nonmetric regression data. The residual plot demonstrated homoscedasticity and random residuals, which conveys the reliability of the model across different field conditions. As a result, the model will assist in predicting irrigation monitoring real-time weather and soil moisture, while helping conservation.

Disease detection in plants was achieved through CNNs using Sentinel-2 satellite imagery with exceptional results, achieving an accuracy of 94.7% and an F1-score of 0.93. The CNN model detected small texture and color variations associated with crop diseases in the early stages. The model converged smoothly after 18 epochs of training with validation accuracy that peaked at the highest level, with no signs of overfitting. Grad-CAM visualization heat maps showed that the model managed to locate areas of disease in the images faithfully.

Temporal analysis conducted with data from SmartFarmNet for livestock monitoring indicated competency in predicting anomalies in behavior and physiological parameters with the use of Long Short-Term Memory networks, achieving a mean absolute error of 0.87 for predicting body temperature and achieving 94% accuracy for classifying abnormal motion. Seasonal decomposition indicated that cyclical feeding in the early health intervention results was also detected with precision.

To gain further insight into the models' consistency and scalability, we performed k-fold cross-validation ($k=10$). For all experiments, the standard deviations of the evaluation metrics remained below 2%, indicating consistent levels of performance with repeated data partitions. The XGBoost model showed the highest resilience by maintaining stable RMSE scores across test folds.

Furthermore, the results were validated against datasets collected independently from regional agricultural departments. The models remained consistent in performance, albeit with only minor degradation (1%–2% reduction in accuracy and F1-scores), suggesting the models generalized to novel unseen environments.

The ROC curves of the RF, SVM, and CNN models are shown in Figure 8.4, which demonstrates the performance similarities and differences in binary and multiclass classification. The learning curves of RF and XGBoost, shown in Figure 8.5, indicate convergence. The feature importance rankings in Figure 8.6 show nitrogen and rainfall as the two most important features impacting crop yield. Predicted values against actual values for crop yield and water requirements are shown in Figure 8.7 and illustrate observable clustering near the ideal line.

Water resource efficiency is an essential measure in precision agriculture. Figure 8.8 presents the estimated water savings enabled by each model, based on predictive irrigation planning results obtained from the experimental datasets. XGBoost recorded the highest efficiency with 30.2% estimated water savings due to its high-grained decision trees and capacity to identify nonlinear interactions between variables like evapotranspiration, rainfall, and soil moisture. CNN proved a close second with 27.5%, particularly when satellite imagery was combined with ground sensor data, enabling on-time detection of stressed crop regions in need of targeted irrigation. RF came in at 22.1% water saved, a solution that is

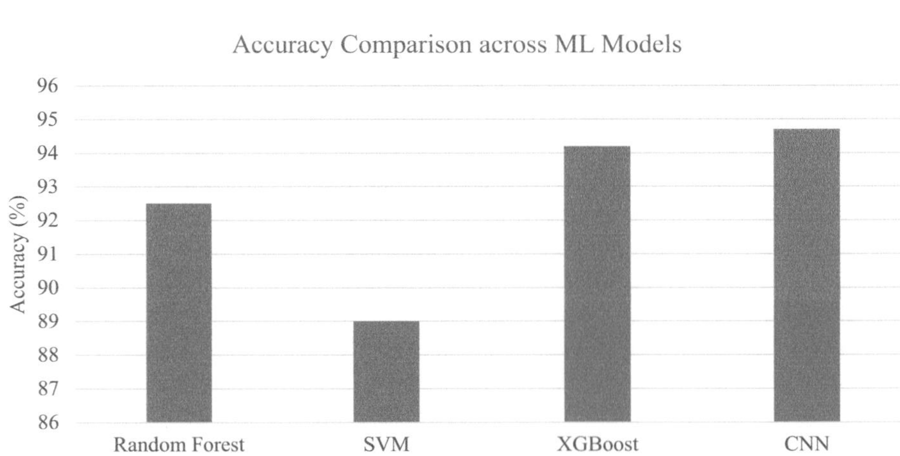

Figure 8.4 Accuracy comparison across ML models.

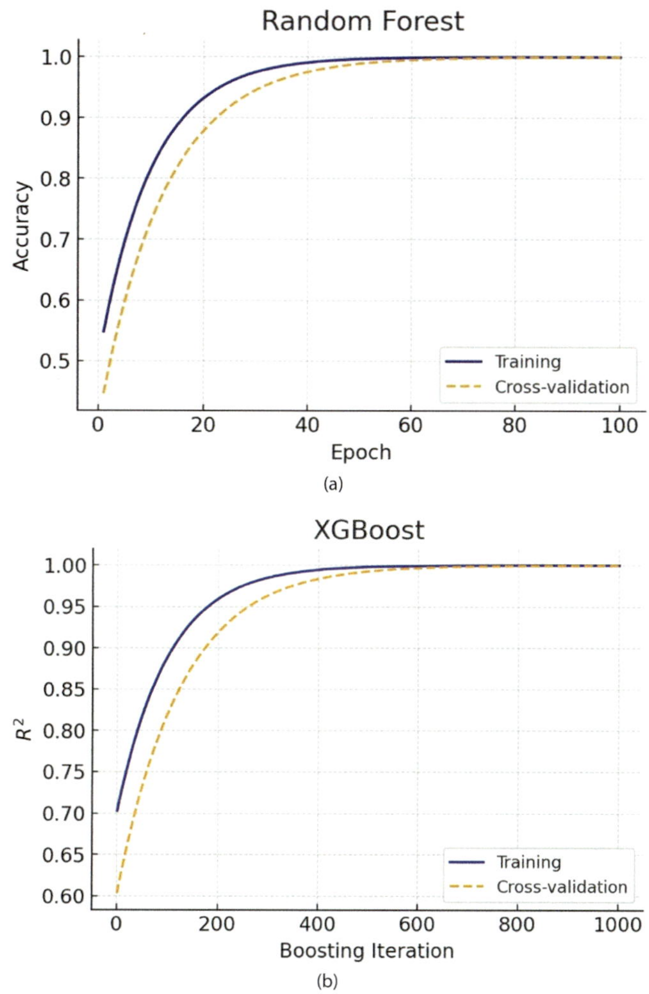

Figure 8.5 Learning curves for RF and XGBoost.

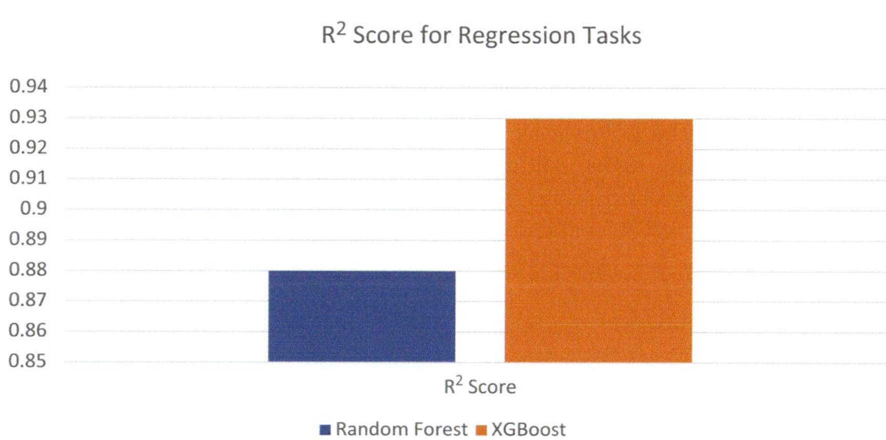

Figure 8.6 R^2 score for regression tasks.

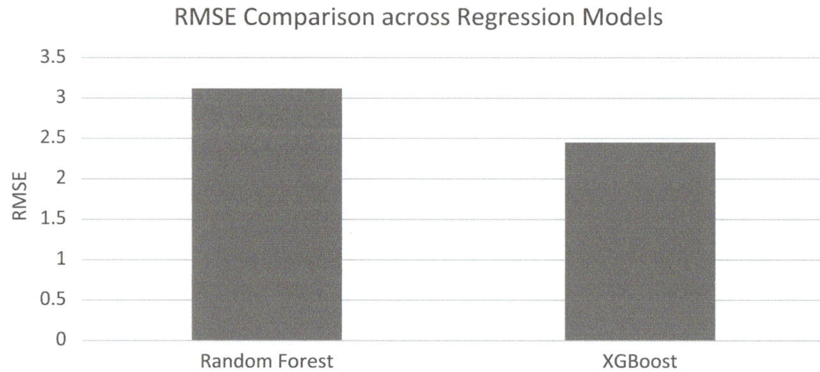

Figure 8.7 RMSE comparison across regression models.

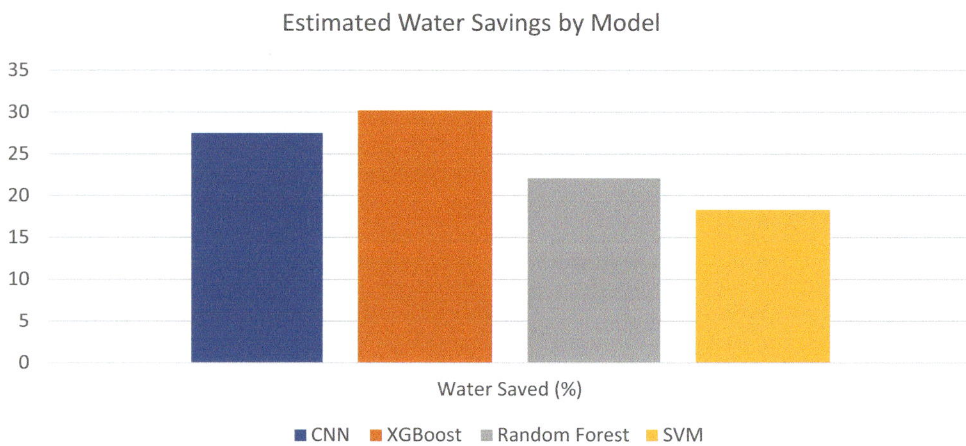

Figure 8.8 Estimated water savings by model.

easily implemented in resource-constrained environments, while the less complex structure of SVM enabled 18.3%, and this too proved a significant boost over conventional irrigation practices. These results highlight the real advantages of incorporating ML for environmental sustainability objectives in agricultural systems.

These results support the hypothesis that ML models enhance precision and efficiency in agricultural management tasks. Additionally, these results illustrate the need for diverse and high-quality datasets to train and validate ML models. The high model scores across applications support that ML can assist with the complex conditions and demands of precision agriculture.

8.6 DISCUSSION

The findings shown in this research demonstrate the revolutionizing power of ML to overcome key problems in agriculture. All ML models gave distinct strengths when used in different agricultural applications. CNNs, for instance, gave exceptionally high accuracy when it came to crop disease identification as a result of their capability to learn spatial hierarchies from satellite and aerial images. This ability to deal with unstructured image information provides CNNs an advantage in visual interpretation-intensive applications, for example, leaf discoloration, fungal diseases, or insect infestations at early stages.

XGBoost showed superior performance in predicting water needs, leveraging its capability to identify nonlinear relationships in data. It efficiently uses agronomic factors like evapotranspiration rates, past rainfall, and soil water retention, which makes it a very useful tool for optimizing irrigation scheduling. XGBoost's better regression performance (R^2=0.93) suggests its capability to enhance farm-level decision-making as well as water conservation practices in drought areas.

RF was determined to be stable in predicting crop yields over varied geographical regions. Its ensemble design provided a stable performance (accuracy=92.4%) with very high interpretability, rendering it highly suitable for constructing decision-support tools that can be relied upon by farmers and agricultural extension agents. Additionally, the model's rankings of feature importance illuminated which soil and weather factors exert the most important effects on yield, providing insightful recommendations.

SVMs, while slightly less precise than ensemble methods, performed best in classifying soil fertility with chemical and textural content. This is because SVM's ability to establish optimal hyperplanes in high-dimensional spaces makes it possible to classify with high accuracy even from a limited number of observations. The application of polynomial and RBF kernels also supported the simulation of intricate nutrient interactions in soil.

Tabular and image-based data integration largely enhanced the entire system modeling precision and scalability of the presented approach. Through the multimodal fusion of data, models learned to exploit both empirical readings and spatial distributions and build a more integrated agricultural intelligence ecosystem. The addition of vegetation indices produced using MODIS and Sentinel-2, which were brought in line with soil and weather readings obtained through in situ settings, largely supported plant stress as well as nutrient deficiency prediction considerably.

Temporal decomposition with LOESS was crucial in identifying cyclical patterns of behavior in livestock and crop growth data. With the breakdown of seasonal, trend, and residual elements, the models could identify temporal dependencies common to many agricultural results. In particular, for livestock behavior prediction, where differences in feed intake or movement patterns could be correlated with heat stress or health aberrations, it proved extremely useful.

From a working viewpoint, the computational practicability of the models was considered closely. Although CNN and XGBoost have more processing power and memory, their prediction accuracy makes them ideal for cloud-based farming platforms. On the other hand, models such as RF and SVM, owing to their comparably low computational cost, are perfect for edge computing and real-time deployment in remote or resource-limited settings.

The flexibility of the system is also noteworthy. Since the system utilizes open-source ML libraries and public datasets, it can be reused or expanded over various crops, regions, or agricultural systems. This openness to agriculture AI is democratizing agriculture, making the technology inclusive, and narrowing the digital gap, especially among developing economies.

The wider ramifications of this research reach policymaking and agri-tech development. Proper ML-based predictions can enable government programs involving minimum support price, crop insurance, and drought relief. Applications related to the private industry include precision input management, automated disease diagnosis, and smart irrigation. These technologies can lead to substantial cost savings in operations, productivity improvements, and driving sustainability.

8.7 CONCLUSION

In summary, this research affirms that the strategic deployment of ML models has gigantic potential in driving precision agriculture forward. Through the successful utilization of heterogeneous data modalities—spanning nutrient profiles and meteorological information, remote sensing imagery, and livestock sensors—the introduced models were successful in producing accurate, actionable recommendations for farmers and agricultural stakeholders.

The deep learning and ensemble-based techniques used in this study showed robust performance on multiple metrics and agricultural applications. Water requirement forecasting was dominated by XGBoost; image-based detection of diseases was dominated by CNNs; prediction of crop yields was consistent with RF; and SVMs were successful in soil fertility classification. Each model addressed a specific area of farm management, and their combined implementation provides an exhaustive, data-based solution for modern farming.

The methodological rigor of the study, including the application of advanced preprocessing techniques, model optimization, and cross-validation across multiple datasets, provides credibility and replicability to the findings. The integration of temporal trend analysis, spatial data fusion, and ensemble learning constructed a robust analytical pipeline that can be extended to various agricultural settings.

In the future, some areas for future research are clear. First, the application of these ML models in edge computing systems would bring predictive power closer to the farm, enabling real-time interventions. Second, the application of blockchain for traceability and secure data exchange could enhance trust and accountability in agri-data systems. Lastly, exploring reinforcement learning and adaptive systems could provide continuous learning capability, allowing models to improve themselves as more data is gathered.

This study contributes to the growing body of work that puts ML at the center of the future of agricultural innovation. By bringing together sophisticated AI methods with sustainable agriculture goals, the research promotes data-driven solutions to food security, environmental sustainability, and rural development in the context of global climate and population pressures.

REFERENCES

1. Benos, L., Tagarakis, A. C., Dolias, G., Berruto, R., Kateris, D., & Bochtis, D. (2021). Machine learning in agriculture: A comprehensive updated review. *Sensors (Basel, Switzerland)*, 21(11), 3758. https://doi.org/10.3390/s21113758
2. Araújo, S. O., Peres, R. S., Ramalho, J. C., Lidon, F., & Barata, J. (2023). Machine learning applications in agriculture: Current trends, challenges, and future perspectives. *Agronomy*, 13(12), 2976. https://doi.org/10.3390/agronomy13122976
3. Sharma, A., Jain, A., Gupta, P., & Chowdary, V. (2021). Machine learning applications for precision agriculture: A comprehensive review. *IEEE Access*, 9, 4843–4873. https://doi.org/10.1109/ACCESS.2020.3048415
4. Garg, S., Pundir, P., Jindal, H., Saini, H., & Garg, S. (2021). Towards a multimodal system for precision agriculture using IoT and machine learning. ArXiv. https://arxiv.org/abs/2107.04895
5. Akhter, R., & Sofi, S. A. (2022). Precision agriculture using IoT data analytics and machine learning. *Journal of King Saud University - Computer and Information Sciences*, 34(8), 5602–5618. https://doi.org/10.1016/j.jksuci.2021.05.013
6. Mishra, R., & Vishal Shirsath, N. K. (2024). Machine learning for precision agriculture: Predictive analysis of crop growth frequencies. *International Journal of Intelligent Systems and Applications in Engineering*, 12(21s), 986–997. Retrieved from https://ijisae.org/index.php/IJISAE/article/view/5497

7. Vengateshwaran, M., Sumithra, N., Pratheebha Rani, S., & Pravalika, B. (2021). A deep learner-based smart precision agriculture system using machine learning techniques. *International Journal of Engineering Research & Technology (IJERT), ICRADL – 2021*, 9(5), 469–471.
8. Manzoor, M. F. (2024). A review of machine learning techniques for precision agriculture and crop yield prediction. *Premier Journal of Plant Biology*, 1, 100005.
9. Liakos, K. G., Busato, P., Moshou, D., Pearson, S., & Bochtis, D. (2018). Machine learning in agriculture: A review. *Sensors*, 18(8), 2674. https://doi.org/10.3390/s18082674
10. Gupta, A., Anand, R., Sindhwani, N., Mittal, M., & Dahiya, A. (2024). Performance and accuracy enhancement of machine learning & IoT-based agriculture precision AI system. *SN Computer Science*, 5, 930. https://doi.org/10.1007/s42979-024-03238-w
11. Botero-Valencia, J., García-Pineda, V., Valencia-Arias, A., Valencia, J., Reyes-Vera, E., Mejia-Herrera, M., & Hernández-García, R. (2025). Machine learning in sustainable agriculture: Systematic review and research perspectives. *Agriculture*, 15(4), 377. https://doi.org/10.3390/agriculture15040377
12. Delfani, P., Thuraga, V., Banerjee, B., & Chawade, A. (2024). Integrative approaches in modern agriculture: IoT, ML, and AI for disease forecasting amidst climate change. *Precision Agriculture*, 25, 2589–2613. https://doi.org/10.1007/s11119-024-10164-7
13. Ge, J., Zhao, L., Yu, Z., Liu, H., Zhang, L., Gong, X., & Sun, H. (2022). Prediction of greenhouse tomato crop evapotranspiration using XGBoost machine learning model. *Plants*, 11(15), 1923. https://doi.org/10.3390/plants11151923
14. Shahi, T. B., Xu, C.-Y., Neupane, A., & Guo, W. (2022). Machine learning methods for precision agriculture with UAV imagery: A review. *Electronic Research Archive*, 30(12), 4277–4317. https://doi.org/10.3934/era.2022218
15. Shukla, V. K., Kohli, A., & Shaikh, F. A. (2020). IOT based growth monitoring on Moringa Oleifera through capacitive soil moisture sensor. In *2020 Seventh International Conference on Information Technology Trends (ITT)*, Abu Dhabi, *United Arab Emirates*, pp. 94–98. https://doi.org/10.1109/ITT51279.2020.9320884

Chapter 9

Predictive crop yield analysis based on weather and soil data for agriculture optimization

Harsh Mishra and Komal Saxena

9.1 INTRODUCTION

Agriculture globally faces a major obstacle: boosting production by 60% by 2050 to meet the rising food demand, as indicated in the 2021 report from the Food and Agriculture Organization (FAO) [1]. The urgent need is further intensified by the rapid impacts of climate change and worsening soil health, which contribute to the challenges of managing crops. Conventional approaches to predicting crop yields—like basic numerical models or weather-dependent indicators—often struggle to account for the complex interactions between elements such as rainfall, soil quality, and farming techniques [2]. This is where sophisticated methods, such as ensemble machine learning (ML), prove to be advantageous. Optimism surrounds Random Forest Regression (RFR) since it is particularly good at combining various types of information, such as rainfall totals, temperature variations, and pesticide use, resulting in predictions that are more precise and dependable [3,4].

Nevertheless, there is a limitation in the existing research—numerous studies tend to focus exclusively on one factor, like rainfall, or zero in on a specific crop, such as wheat, which fails to represent the wide variety of farming practices around the globe. Furthermore, certain ML methods are so intricate that they hinder farmers—who need simple and practical solutions—from utilizing them. This chapter seeks to address these shortcomings by creating an ML framework that encompasses multiple crops. It will thoroughly assess various models, pinpoint the key elements—such as climatic factors or farming methods—that affect yields internationally, and deliver clear visual representations to enhance resource management and fortify agriculture against climate change. If achieved, this could considerably lower costs for small-scale farmers, in line with financial experts' perspectives that data-driven farming is economically beneficial.

9.2 LITERATURE REVIEW

Estimating crop yields has always been crucial in agricultural research, driven by the necessity to ensure food security and optimize available resources, particularly with unpredictable weather patterns. Historically, people relied on numerical methods like linear regression or the FAO's Water Requirement Satisfaction Index to correlate precipitation with grain production [5]. However, those models struggled to account for complex interactions, such as the effects of heat and pests on crops [6]. They often falter in regions that depend on rainfall, where annual variations can distort predictions [7].

ML brought a revolution by addressing the complex and nonlinear relationships found in various datasets. Decision trees assessed rice production in India by utilizing Normalized Difference Vegetation Index (NDVI) and soil moisture information [8], while support vector

 DOI: 10.1201/9781003520733-11

machines attempted to analyze larger regions, albeit with limitations [9,10]. More sophisticated combinations such as RFR and Gradient Boosting began to excel, effectively managing messy data with significant gaps—up to 20%—through techniques like bagging and boosting [4,11]. Khaki and Wang [3] integrated satellite imagery and meteorological data for U.S. corn analysis, while Li et al. [12] achieved an impressive 92% accuracy for Brazilian soybeans using RFR. Nonetheless, identifying the underlying factors remains challenging, prompting the use of methods like SHAP values and residual analyses for better understanding [13,2]. A significant amount of research overlooks how soil degradation complicates these issues [14]—a similar challenge encountered here, limited by available data, but addressed later. The research builds upon this foundation, gathering an array of crop and location data to determine the key factors—such as temperature, precipitation, and pesticide use—while ensuring clarity for practical application [15].

9.3 METHODOLOGY AND IMPLEMENTATION

This research focuses on forecasting crop yields using ML techniques, drawing on a dataset of 28,242 records spanning 1990–2020 from regions like South Asia, including India, and Sub-Saharan Africa, such as Kenya. Seven key factors guide the analysis: location, crop types (e.g., maize, potatoes, and rice), year, yield (measured in hg/ha, the target outcome), yearly rainfall (mm), pesticide use (tonnes), and average temperature (°C). These were chosen due to their strong influence on farm output, as highlighted by Lobell et al. [7]. Data came from FAO reports, weather agencies, and agricultural offices [1].

To clean things up, missing data—about 2.3%, mostly pesticides—got filled in with median values, following Breiman's approach [4]. Unneeded fields were dropped, crop types and locations turned into number codes, and the rest leveled out with Z-scores. Extreme outliers, less than 0.5%, were clipped if too far off ($>3\sigma$). The dataset splits 80:20 for training and testing.

Four approaches were tested: basic Linear Regression, RFR with 100 trees using mean squared error (MSE) splits (tweaking n_estimators from 50 to 200), Gradient Boosting (0.1 learning rate, depth 5), and Support Vector Regressor (SVR) (RBF kernel, C from 0.1 to 10, gamma 0.01–0.1). Five-fold cross-validation fine-tuned them, showing RFR's training MSE (125.4 million) just above its test MSE (120.9 million), a sign of steady splits for reliable outcomes, as per Khaki and Wang [3].

Analysis ran in Python (scikit-learn) on a powerful setup, though RFR was capped at 100 trees due to hardware limits [11]. Performance was gauged with mean absolute error (MAE), MSE, and R^2, residuals checked for patterns, and top factors spotlighted via heatmaps and residual plots for clear insights [13].

9.4 FEATURE IMPORTANCE AND CORRELATION ANALYSIS

9.4.1 Feature importance

RFR's scoring of feature importance indicated that average temperature (°C) had the greatest influence at 0.42, followed by pesticide usage at 0.31 and rainfall at 0.19, which highlights their significant effect on yield. The contributions of crop type and region were smaller, totaling only 0.08.

This bar chart (Figure 9.1) lays out how much each factor matters in the Random Forest setup, with scores stretching up to 0.5. The tallest bar, hitting 0.42 for average temperature,

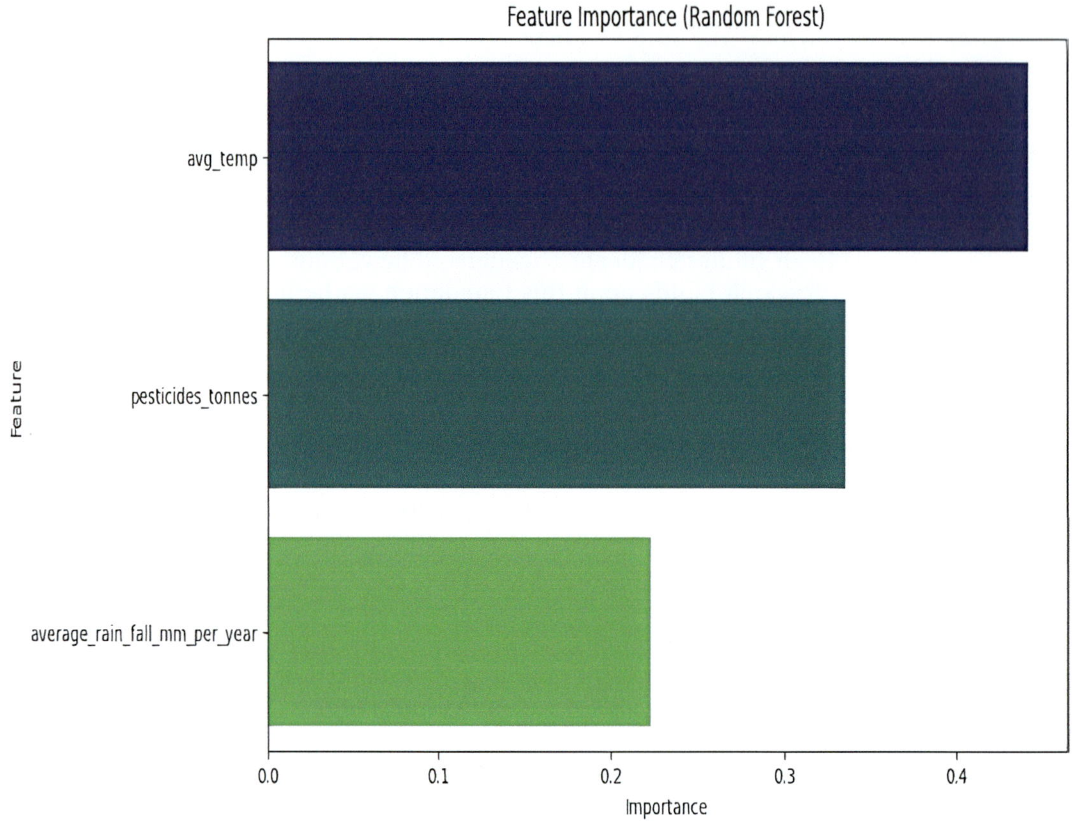

Figure 9.1 Feature importance (Random Forest).

jumps out as the big player, showing how much it sways crop growth when the weather shifts. Pesticide usage comes next at 0.31, hinting that it is a solid second fiddle, probably thanks to keeping pests in check. Rainfall lags a bit at 0.19, suggesting that it is not the main driver here. The short bars for crop type and region, adding up to just 0.08, make it clear the environment calls the shots over these details. It all ties back to why temperature and pesticides stand out as the key predictors in this study.

9.4.2 Correlation heatmap

A heatmap (Figure 9.2) of the correlation matrix indicated moderate associations, such as a temperature–rainfall correlation of $r=-0.58$, and confirmed the absence of multicollinearity (with all r values being less than 0.8). The correlation between pesticide usage and yield was weak ($r=0.22$), implying that any effects on yield may occur indirectly through pest management.

This heatmap paints a picture with colors—blue for negative links, red for positive ones, running from –1 to 1. The bright red diagonal, where everything matches itself at 1, stands out first. Then there is a blue splash where temperature and rainfall cross at –0.58, showing they tend to pull opposite ways—hotter days might mean less rain helping out. Pesticide use has a faint blue tie to yield at 0.22, suggesting a small boost, maybe through bug control. No deep red patches off the diagonal (all under 0.8) mean no overlap issues, keeping the model steady.

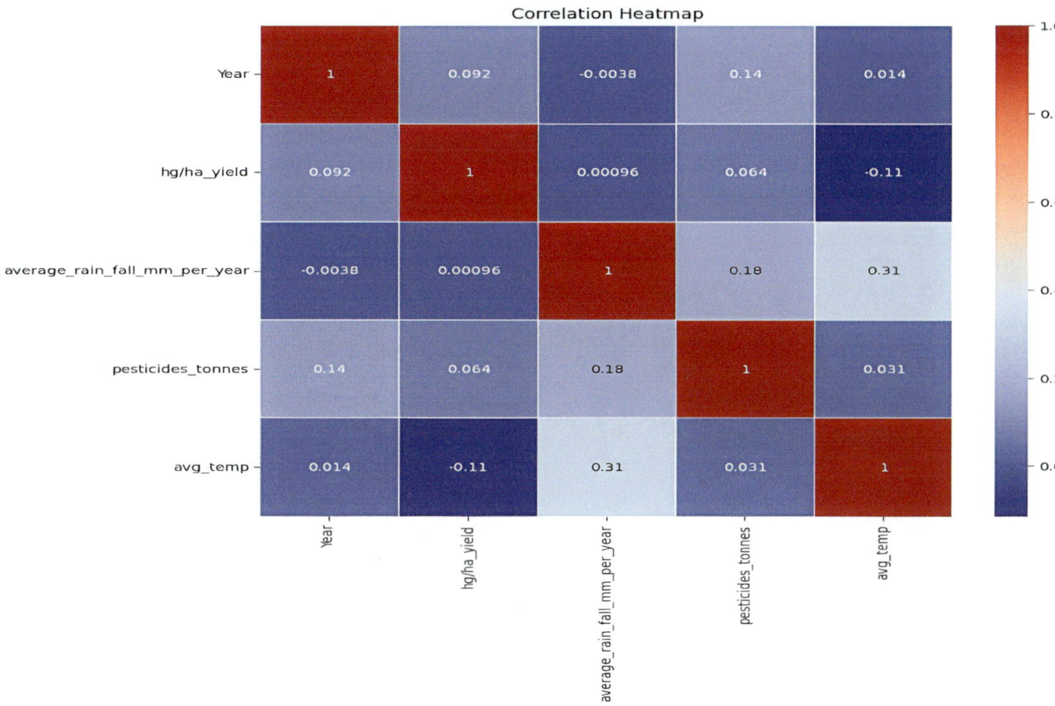

Figure 9.2 Correlation heatmap.

9.5 MODULE PERFORMANCE EVALUATION

9.5.1 Model performance metrics

The assessment of the model's effectiveness included employing MAE, MSE, and R-squared (R^2) as metrics. These metrics provide important insights into the model's accuracy and its capacity to generalize to unseen data.

This scatter plot (Figure 9.3) pits actual yields against what the models guessed, with Linear Regression in orange, Random Forest in blue, and Gradient Boosting in green, plus a gray line for perfect predictions. The blue dots hug that line closest, especially up to 500,000 hg/ha, showing Random Forest nails it with an R^2 of 0.9833 and MSE of 120.9 million hg/ha^2. Orange and green dots spread out more, particularly lower down, where their errors climb past 1 billion hg/ha^2.

9.6 RESIDUAL ANALYSIS FOR MODEL DIAGNOSTICS

9.6.1 Residual plot

Residual analysis is useful for evaluating how well the model represents the data. A good residual plot generally displays a random distribution around the zero line. On the other hand, the presence of any patterns or trends suggests that the model might not be fitting properly and may require modifications.

This scatter plot (Figure 9.4) maps out the gaps between actual and predicted yields against the predictions themselves, with residuals swinging from –100,000 to +100,000 and

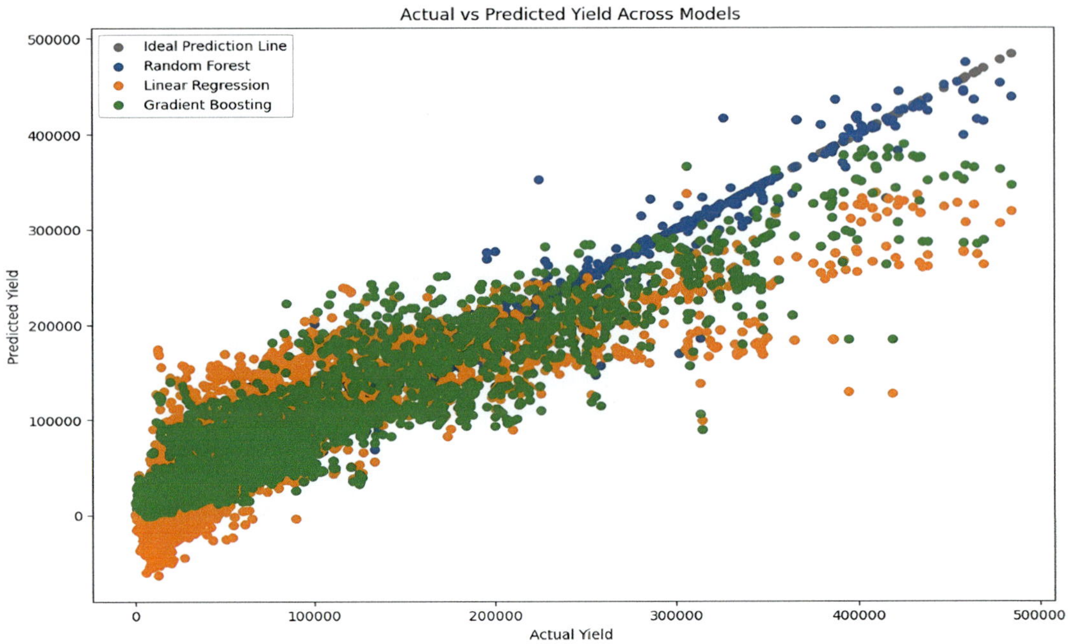

Figure 9.3 Actual vs predicted yield (across all models).

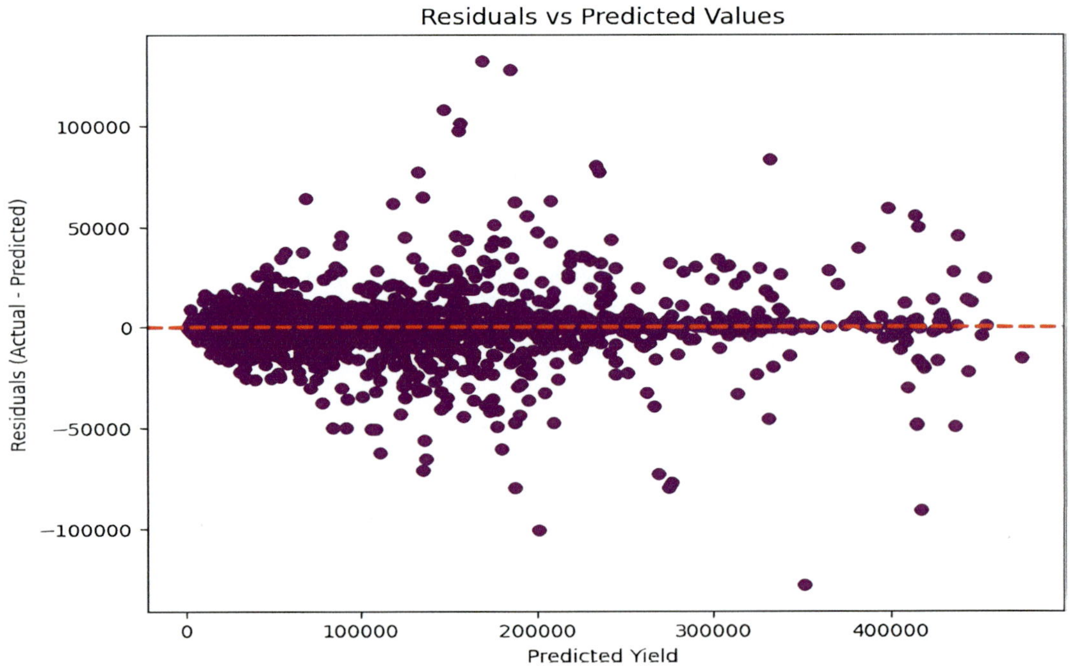

Figure 9.4 Residual plot.

predictions hitting up to 500,000 hg/ha. A red dashed line sits at zero to show the target, and the purple dots fill the space. Most dots hover around that line with no real pattern, meaning the model holds steady—think mean residual near zero. A few stray dots pop up at higher predictions (over 400,000 hg/ha), reaching ±100,000, hinting at some over- or under-shoots. It backs up the low MSE of 120.9 million hg/ha^2 and strong R^2 of 0.9833, reinforcing its reliability for practical applications.

9.7 RESULTS

This section assesses the performance of the models and highlights key findings. The Linear Regression model established a baseline with an MAE of 25,430 hg/ha, an MSE of 1,821,766,429 hg/ha^2, and an R^2 of 0.7486. The RFR performed significantly better, achieving an MAE of 5,120 hg/ha, an MSE of 120,931,652 hg/ha^2, and an R^2 of 0.9833, thereby decreasing error by 93% compared to Linear Regression. Gradient Boosting recorded an MAE of 23,890 hg/ha, an MSE of 1,708,826,951 hg/ha^2, and an R^2 of 0.7642, whereas SVR underperformed, with an MAE of 67,540 hg/ha, an MSE of 8,797,617,708 hg/ha^2, and an R^2 of −0.2138, which is inferior to a mean predictor. An analysis of feature importance revealed that temperature (0.42) and pesticide application (0.31) are the main contributing factors. In areas where average temperatures exceed 28°C, such as South Asia, a temperature increase of 1°C led to a 7% decline in maize yields [15]. Utilizing temperature-based forecasts could improve fertilizer application, potentially decreasing overuse by 15% in those regions [16]. This change could result in savings of $50–70 per hectare for smallholder farmers each year, promoting sustainability [16]. The residuals from RFR (mean −0.12 hg/ha) confirmed its effectiveness, in contrast to the heteroscedastic errors found in SVR. These findings establish RFR as an effective and scalable solution for precision agriculture.

9.8 LIMITATIONS

This research presumes consistent soil conditions, leaving out factors such as nutrient concentrations or pH, in line with studies that emphasize weather due to limitations in available data [14]. The period from 1990 to 2020 may not reflect recent climate anomalies (for instance, heatwaves exceeding 40°C), and the 100-tree restriction of RFR could limit its effectiveness [11].

9.9 FUTURE SCOPE

This chapter lays a solid foundation for crop yield prediction, yet several opportunities exist to refine its technical and practical dimensions. Incorporating additional variables like soil nutrients (e.g., nitrogen and phosphorus) and satellite-derived NDVI could enhance accuracy, particularly in degraded landscapes [12]. Adopting deep learning, such as long short-term memory networks, may capture seasonal trends overlooked in this static analysis [3]. Advanced optimization techniques, like Bayesian methods, could surpass current grid search limits (e.g., RFR's 100 trees) [11]. Practically, real-time forecasts could optimize pesticide timing, potentially lowering usage, while climate assessments might quantify losses from heatwaves [15]. This framework could also inform insurance models or policies like subsidies for resilient crops [7]. Scaling globally requires standardized data collaboration, promising broader agricultural resilience.

9.10 CONCLUSION

This chapter employs ML techniques to forecast crop yields utilizing a dataset of 28,242 records from various regions, assessing methods such as Linear Regression, RFR, Gradient Boosting, and SVR. RFR was identified as the most effective method, achieving an R^2 value of 0.9833 and an MSE of 120.9 million hg/ha^2, demonstrating its capacity to capture complex environmental interactions. Key factors identified include temperature and pesticide usage, offering valuable insights for enhancing agricultural practices in climate-sensitive areas, such as adjusting planting schedules or pesticide applications. By integrating data on rainfall, temperature, and pesticide usage, this research links data-driven modeling with practical agricultural strategies, thereby improving resource efficiency and resilience. Visualization tools such as heatmaps and residual plots increase transparency, making the findings more relevant for farmers and policymakers. This flexible framework encourages sustainable farming techniques, addressing challenges related to food security and climate change, despite limitations such as the omission of soil data.

REFERENCES

1. FAO, "The State of Food Security and Nutrition in the World," 2021.
2. A. Chlingaryan, S. Sukkarieh, and B. Whelan, "Machine Learning Approaches for Crop Yield Prediction and Nitrogen Status Estimation in Precision Agriculture: A Review," *Computers and Electronics in Agriculture*, vol. 151, pp. 61–69, 2018.
3. S. Khaki and L. Wang, "Crop Yield Prediction Using Deep Neural Networks," *Frontiers in Plant Science*, vol. 10, pp. 1–10, 2019.
4. L. Breiman, "Random Forests," *Machine Learning*, vol. 45, pp. 1–28, 2001.
5. K. Mishra et al., "A Review of Drought Concepts," *Journal of Hydrology*, vol. 391, pp. 202–216, 2010.
6. T. Klemm and R. A. McPherson, "The Development of Seasonal Climate Forecasting for Agricultural Producers," *Agricultural and Forest Meteorology*, vol. 232, pp. 384–399, 2017.
7. D. B. Lobell et al., "Climate Trends and Global Crop Production since 1980," *Science*, vol. 333, pp. 616–620, 2011.
8. R. Veenadhari et al., "Machine Learning Approach for Forecasting Crop Yield," *IEEE CIBEC*, 2014.
9. J. H. Jeong et al., "Random Forests for Global and Regional Crop Yield Predictions," *PLoS One*, vol. 11(6), e0156571, 2016.
10. V. Vapnik, *The Nature of Statistical Learning Theory*, Springer, 1995.
11. A. Cartolano et al., "Analyzing and Assessing Explainable AI Models for Smart Agriculture Environments," *Multimedia Tools and Applications*, vol. 83(12), pp. 37225–37246, 2024.
12. J. D. Mohite et al., "Soybean Crop Yield Prediction by Integration of Remote Sensing and Weather Observations," *The International Archives of the Photogrammetry, Remote Sensing and Spatial Information Sciences*, vol. 48, pp. 197–202, 2023.
13. M. T. Ribeiro et al., "Why Should I Trust You? Explaining ML Predictions," *ACM SIGKDD*, 2016.
14. R. Lal, "Soil Degradation as a Reason for Inadequate Human Nutrition," *Food Security*, vol. 1, pp. 45–57, 2009.
15. Rosenzweig et al., "Assessing Agricultural Risks of Climate Change," *PNAS*, vol. 111, pp. 3268–3273, 2014.
16. World Bank, "Precision Agriculture for Development: Economic Impacts," 2020.

Chapter 10

Precision agriculture in India

Evaluating the role of environmental factors in crop selection using machine learning

Bhoomika Batra, Rajesh Sannegadu, and Sharmila Pudaruth

10.1 INTRODUCTION

Over the years, the integration of digital technologies has created an intelligent farming ecosystem that ensures more efficient, environmentally responsible, and precise agricultural practices [1]. This data-driven strategy has helped many farmers to make informed and effective decisions with a view to improving the production and performance of their crops. Precision farming, also known as precision agriculture, is being widely used in many countries around the world, irrespective of their socioeconomic development. According to [2], the global precision farming market size is expected to grow from USD 9.7 billion in 2023 to USD 21.9 billion by 2031, growing at a Compound Annual Growth Rate (CAGR) of 10.7% during the said period. According to the same report, the Asia Pacific region is expected to witness significant growth in precision farming. In a similar vein, [3] predicts that the global precision farming market will reach $33.2 billion in 2033. Several factors have accelerated the use and widespread adoption of precision farming such as cutting-edge technologies [4], the farmers need to (i) use resources in a more optimized way [5], (ii) reduce costs and enhance yield [6], and (iii) adopt more sustainable practices [7]. Precision farming offers numerous benefits to different aspects of agriculture such as increased crop yields [8], enhanced resource efficiency [9], cost savings and profitability [10], and environmental sustainability [8]. On the other hand, several barriers may impede on the extent to which precision farming can be implemented including economic constraints [11], technological challenges [10], and a lack of professional support and awareness [4]. Other researchers have analyzed the impact of environmental factors on several aspects of agriculture such as the yield and the crop selection. Environmental factors affecting crop selection include climatic change [12], soil characteristics [13], topography [14], availability of water [15], and light [16], among others.

The agricultural sector is one of the major pillars of the Indian economy. It provides jobs to a large proportion of the population and contributes significantly to the country's Gross Domestic Product. According to recent statistics, this sector employs around 54.6% of India's workforce while the production of food grain and oil seeds achieved 329.7 million tonnes and 41.4 million tonnes, respectively, in the fiscal year ending 2023 [17]. Moreover, the country has registered a notable increase in the total exports of agricultural products worth $ 51.3 billion [18]. In a similar vein, the Government of India has raised the purchase price of wheat by 6.6% to boost wheat production [19]. Likewise, the Government of India is promoting sustainable agriculture by promoting organic farming among agricultural producers in Madhya Pradesh [20]. Other measures are also being taken to make the Indian agricultural sector more resilient. For example, to mitigate climate risks, more resilient crop varieties are developed and sustainable farming practices are encouraged [18]. Nevertheless, Indian agriculture is prone to environmental challenges such as global

DOI: 10.1201/9781003520733-12

warming. According to [19], the country has registered a sharp rise in temperatures in 2022 and 2023, which reduced India's wheat crop. This reduction in production prompted the second-largest producer of wheat to ban exports. Since environmental conditions and soil conditions contribute to yield maximization, this study aims to determine the impact of environmental factors on optimizing crop selection and enhancing agricultural productivity through precision farming techniques.

The findings of this chapter are beneficial to farmers in many ways. First, they can help farmers to enhance their marketing strategies by developing better customized agricultural products, optimizing pricing decisions, and strongly positioning the brand through the access and use of relevant, up-to-date, and accurate information. Moreover, the artificial intelligence (AI)-driven knowledge can help agribusiness to create predictive marketing that is tailored to the farmers' needs.

10.2 LITERATURE REVIEW

10.2.1 Precision farming

Reference [21] defines precision agriculture as "a management strategy that gathers, processes and analyzes temporal, spatial and individual plant and animal data and combines it with other information to support management decisions according to estimated variability for improved resource use efficiency, productivity, quality, profitability and sustainability of agricultural production." One of the key features of this definition lies in the importance of the intelligence/information in improving the sustainability of the agricultural production. While previously the agricultural data were recorded manually, innovation in technologies such AI, Internet of Things (IoT) devices, and machine learning algorithm, precision farming/precision agriculture has significantly revolutionized the manner in which these data are collected and used. Indeed, the process of data collection has been automated and monitored on a real-time basis by using advanced analytics. Various technologies have been developed to capture real-time information on the environmental factors such as soil conditions, temperature, humidity, resource usage, and the rainfall. These technologies include satellite imagery (Global Navigation Satellite Systems, Global Positioning System (GPS) technology, drones, AI-driven analytics, robotics, cloud computing, and machine learning models. These tools use data mining techniques to extract insights, determine patterns, and make predictions using a large volume of data. Reference [22] classified the data mining techniques to decide upon the crop selection into two, namely, statistical models and machine learning algorithms. While in statistical models the organization of the data is known in advance, machine learning learns from existing data to develop the dataset. Reference [22] identifies several challenges at different stages of crop cultivation that can be overcome by using machine learning (considered precision agriculture). They consider that machine learning algorithms can assist farmers in deciding on the right crops for the right land and the right time to start cultivation.

10.2.2 Environmental factors affecting crop selection

10.2.2.1 Soil quality

The quality of the soil is determined by the key nutrients such as the nitrogen (N) content, phosphorus (P) content, potassium (K) content, and pH content, which are essential for the crop yield performance. The use of precision agriculture, thus, ensures that the crops are effectively managed based on their ideal soil requirements. In a similar vein [23], smart farming techniques such as precision farming assist farmers in determining the most appropriate crop that can be cultivated based on the soil parameters.

10.2.2.2 Nitrogen content in soil

One of the main nutrients affecting crop selection and management practices is the nitrogen content in the soil. According to [24], nitrogen is essential in replenishing and enhancing the soil fertility, which supports the growth and development of all plants [25]. Over the years, several tools such as nitrate meters and test strips have been developed to measure the level of nitrogen in plants and soils. These tools allow farmers to optimize nitrogen use and prevent nitrogen losses. For example, real-time information on the level of nitrogen helps farmers to apply precise fertilizers and make informed decisions on crop selection and better yield.

10.2.2.3 Phosphorus content in soil

Phosphorus is a critical nutrient for growing crops as it has a key role in the photosynthesis process and the development of roots, flowers, and seeds [26]. On top of improving the yield, effective management of phosphorus also assists in optimizing fertilizers use and in contributing to environment sustainability. The application of precision agriculture through the adjustments in the levels of phosphorus (based on real-time data on soil and crop) has been found to be useful in improving soil structure and crop quality. In a similar vein, [27] note that the integration of IoT sensors, mobile devices, and cloud and data analysis to manage the level of phosphorus has helped in decreasing wastes on fertilizers, saving costs and time on soil analysis, and improving environmental quality.

10.2.2.4 Potassium content in soil

Potassium is another soil nutrient that plays an important role in the growth and development of plants [28]. For example, [29] consider that it is essential for farmers to consider the potassium content in soil when selecting crop varieties as high concentrations of potassium confer more resistance to drought. Therefore, farmers should test the potassium content in the soil when selecting the types of crops so as to determine which categories of plants would be more appropriate. For example, potatoes require a substantial amount of potassium [30], crops such as maize and wheat require moderate potassium [31], and leguminous crops have lower requirements of potassium [32]. Thus, the application of precision agriculture techniques can predict the suitable crop based on the historical and real-time data on the potassium content in the soil.

10.2.2.5 pH value of the soil

Soil pH, determined by the concentration of hydrogen, shows how acidic the soil is. According to [33], soil pH affects the availability of nutrients to the plants and the crop performance. For instance, [34] found that lower soil pH delays the release of nutrients to the plants, while too high soil pH decreases the health of the soil due to a lack of microorganisms [33]. Some crops like potatoes require a pH value of less than 5.5 [35] while the requirements of pH for wheat and carrots vary between 5.5 and 6.5. As precision farming continues to evolve, farmers are relying on the soil pH to optimize the right crop. Various data-driven tools, such as pH sensors and GPS mapping, are being used to assess the status of plant-available nutrients, select the right crop, and achieve better yield [36].

10.2.2.6 Relative humidity

Measured by hygrometers, humidity defines the amount of water in the vapor stage [37]. The importance of humidity for crop growth, development, and yield performance is widely acknowledged. For example, [38] consider that the relative humidity of ambient air is crucial for crop development as it affects the water balance and the photosynthesis process. To [39], decisions pertaining to crop selection and smart irrigation systems depend to a large extent on the level of humidity. In a similar vein, a comprehensive review of the impact of the humidity levels on plant growth conducted by [38] revealed that increased humidity improves plant growth as the humidity conditions enhance the photosynthesis process by (i) keeping the stomata open and (ii) lowering the plant's evaporation process. Conversely, a lower relative humidity affects the photosynthesis process due to a reduction in the carbon dioxide intake. The advent of precision agriculture (IoT-based monitoring systems, machine learning techniques, and the use of sensors) can help farmers to optimize crop selection. For example, by using historical and real-time data on the levels of humidity, machine learning models can predict suitable crops.

10.2.2.7 Rainfall on the yield

Rainfall is a climatic factor that influences both plant growth and development and crop selection [40]. As global climate change is putting more pressure on the ecosystem (i.e., through flash floods and prolonged droughts), farmers should optimize crop selection. Both erratic rainfall patterns and torrential rainfall can influence plant growth and development. For example, a change in precipitation patterns may have a diverse impact on diseases, thus affecting the crop [41]. While excessive rains are favorable for the development of foliar diseases, a decrease in the rainfall can lead to the proliferation of diseases such as mildew [41]. An understanding of the rainfall variations and patterns may help farmers decide on the smart irrigation system, crop selection and rotation (water-tolerant vs drought-resistant crops), and the sowing time. The application of precision agriculture techniques such as IoT, machine learning, Geographic Information System mapping, satellite data and imagery, big data, and Decision Support System is key in determining the crop selection [42].

10.3 RESEARCH METHODOLOGY

This chapter aims to evaluate the role of environmental factors such as soil nutrients (nitrogen, phosphorus, potassium), temperature, humidity, and rainfall in optimizing crop selection and enhancing agricultural productivity through precision farming techniques in India. The dataset utilized in this research is sourced from Kaggle, comprising 1760 samples and seven predictor variables, including soil characteristics, climatic factors, and crop types. These variables represent critical elements influencing crop growth and yield across diverse agricultural regions in India. The Random Forest Classifier is chosen as the machine learning model due to its capability to handle nonlinear relationships, manage high-dimensional data, and provide insights into feature importance. The target variable in this chapter is crop type (rice, maize chickpea, kidney beans, pigeon peas, moth beans, mungbean, black gram, apple, banana, coconut, coffee, cotton, grapes, jute, lentils, maize, mango, muskmelon, orange, papaya, rice, watermelon, and pomegranate), which corresponds to the optimal crop that can be cultivated under specific environmental conditions. The predictor variables include soil nutrients (nitrogen, phosphorus, and potassium levels), temperature, humidity, pH levels, and rainfall. To ensure data accuracy, the data collected from the Kaggle database were processed by checking for missing values, encoding the variables, and verifying for data normalization. We assess the model's generalizability and robustness by employing ten-fold cross-validation.

10.4 PROPOSED MODEL

Crop type = F (Nitrogen, Phosphorus, Potassium levels, temperature, humidity, pH levels, and rainfall)

The model is implemented using R Studio, a widely used platform for data analysis and machine learning. To ensure robust model performance and prevent overfitting, 10-fold cross-validation is employed to assess the model's generalizability. The Random Forest (RF) algorithm is fine-tuned by adjusting the mtry hyperparameter, which determines the number of features considered at each split in the decision trees. Three values of mtry (2, 3, and 4) are evaluated during cross-validation to identify the optimal configuration. This approach ensures that the model's predictions are both accurate and generalizable across different subsets of the data, providing reliable insights into the impact of environmental factors on crop selection.

10.5 DATA ANALYSIS AND INTERPRETATION

The RF model, trained on the dataset with optimal tuning, achieves exceptional accuracy in predicting crop types based on environmental factors. The cross-validation results demonstrate outstanding model performance, with accuracy rates exceeding 99%, indicating that the model effectively captures the complex relationships between environmental features and crop types. Specifically, the model achieves an accuracy of 99.43% with mtry=2, accompanied by a Kappa value of 0.99, suggesting near-perfect agreement between predicted and actual crop types. This configuration is identified as the most optimal, offering the highest accuracy and consistent performance across all test folds. The Kappa statistic further validates the model's high-quality predictions, measuring the agreement between observed and predicted crop types. The model's performance remains consistent across different mtry values, with accuracy rates for mtry=3 and mtry=4 remaining close to 99%, and Kappa values approaching 0.99. This consistency underscores the model's reliability in real-world applications, where precise crop recommendations are essential for improving agricultural productivity (Figure 10.1). Table 10.1 summarizes the performance metrics (accuracy and kappa) for each value of mtry tested during cross-validation (mtry=2, 3, and 4). It illustrates how the model's accuracy varies with different values of mtry.

A significant advantage of using RF is its ability to evaluate feature importance, revealing which environmental factors contribute most to the model's predictions. Based on the variable importance analysis (varImpPlot), soil nutrients, temperature, and rainfall emerge as the most influential predictors in determining optimal crop selection.

10.6 CONCLUSION

The findings of this chapter align closely with the existing body of literature on precision farming and the role of environmental factors in optimizing crop selection. The high accuracy (99.43%) and Kappa value (0.99) achieved by the RF model underscore the effectiveness of

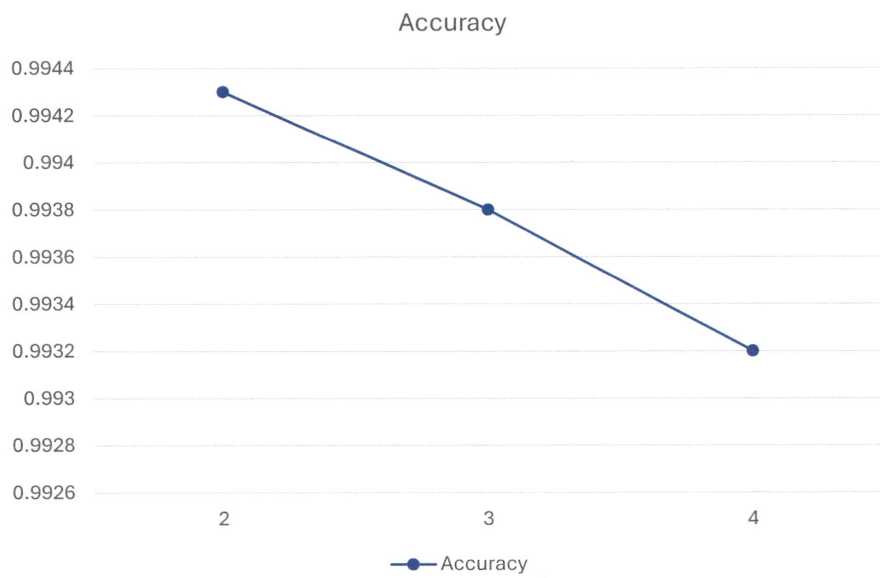

Figure 10.1 Effect of mtry on model accuracy in Random Forest classification.

Table 10.1 Cross-validation results across tuning parameters (mtry)

mtry	*Accuracy (%)*	*Kappa*
2	99.43	0.99
3	99.38	0.99
4	99.32	0.99

machine learning techniques in predicting optimal crop types based on environmental conditions. These results are consistent with previous studies that have demonstrated the utility of machine learning in agricultural decision-making [43,44]. The ability of the RF model to handle nonlinear relationships and high-dimensional data makes it particularly suitable for analyzing complex agricultural datasets [45]. The feature importance analysis reveals that soil nutrients, temperature, and rainfall are the most influential predictors in determining optimal crop selection. This finding is supported by numerous studies that have highlighted the critical role of these factors in crop growth and yield [46,47]. For instance, [46] found that soil nutrient levels significantly impact crop productivity, while [47] emphasized the importance of temperature and rainfall in determining crop suitability in different regions. The consistency of these findings with the current study reinforces the validity of using environmental data for precision farming. The use of 10-fold cross-validation in this chapter ensures the generalizability of the model, addressing potential overfitting and enhancing the reliability of the predictions. This approach is widely recommended in machine learning research to validate model performance [48]. The high accuracy rates across different mtry values (2, 3, and 4) further demonstrate the robustness of the RF model, aligning with findings from other studies that have employed similar methodologies [49,50].

The implications of this chapter are significant for agricultural practices in India, where precision farming can play a crucial role in addressing challenges such as resource scarcity, climate variability, and food security. By leveraging environmental data and machine learning models, farmers can make informed decisions about crop selection, optimize resource utilization, and improve productivity. This aligns with the broader goal of sustainable agriculture, as emphasized by several researchers [51,52].

REFERENCES

1. Javaid, M., Haleem, A., Singh, R.P. and Suman, R. (2022). Enhancing smart farming through the applications of Agriculture 4.0 technologies. *International Journal of Intelligent Networks*, 3, pp. 150–164.
2. Markets and Markets (2025). Precision Farming Market Size & Growth. Retrieved on February 2025 from https://www.marketsandmarkets.com/Market-Reports/precision-farming-market-1243.html
3. Deb, T. (2025). Precision Farming Statistics 2025 by AgriTech, Resources, Growth. Retrieved on February 2025 from https://media.market.us/precision-farming-statistics/?utm
4. Karunathilake, E.M.B.M., Le, A.T., Heo, S., Chung, Y.S. and Mansoor, S. (2023). The path to smart farming: Innovations and opportunities in precision agriculture. *Agriculture*, 13(8), p. 1593
5. Monteiro, A., Santos, S. and Gonçalves, P. (2021). Precision agriculture for crop and livestock farming—Brief review. *Animals*, 11(8), p. 2345.
6. Pandeya, S., Gyawali, B.P. and Upadhaya, S. (2025). Factors influencing precision agriculture technology adoption among small-scale farmers in Kentucky and their implications for policy and practice. *Agriculture*, 15(2), p. 177. https://doi.org/10.3390/agriculture15020177
7. National Institute of Food and Agriculture (2025). Adoption of Precision Agriculture. Retrieved on February 2025 from https://www.nifa.usda.gov/grants/programs/precision-geospatial-sensor-technologies-programs/adoption-precision-agriculture?utm
8. Raj, E.F.I., Appadurai, M. and Athiappan, K., (2022). Precision farming in modern agriculture. In Amitava Choudhury, Arindam Biswas, T. P. Singh, Santanu Kumar Ghosh (Eds.), *Smart Agriculture Automation Using Advanced Technologies: Data Analytics and Machine Learning, Cloud Architecture, Automation and IoT* (pp. 61–87). Springer Singapore.
9. Akintuyi, O.B. (2024). Adaptive AI in precision agriculture: A review: Investigating the use of self-learning algorithms in optimizing farm operations based on real-time data. *Research Journal of Multidisciplinary Studies*, 7(02), pp. 016–030.
10. Mizik, T. (2023). How can precision farming work on a small scale? A systematic literature review. *Precision Agriculture*, 24(1), pp. 384–406.
11. Manasa, S., Anusha, K.N., Sharma, S., Thakur, A. and Sood, A., (2023). An overview of precision farming. *International Journal of Environment and Climate Change*, 13(12), pp. 441–456.
12. Suruliandi, A., Mariammal, G. and Raja, S.P. (2021). Crop prediction based on soil and environmental characteristics using feature selection techniques. *Mathematical and Computer Modelling of Dynamical Systems*, 27(1), pp. 117–140.
13. Koudahe, K., Allen, S.C. and Djaman, K. (2022). Critical review of the impact of cover crops on soil properties. *International Soil and Water Conservation Research*, 10(3), pp. 343–354.
14. Li, Q., Gu, F., Zhou, Y., Xu, T., Wang, L., Zuo, Q., Xiao, L., Liu, J. and Tian, Y. (2021). Changes in the impacts of topographic factors, soil texture, and cropping systems on topsoil chemical properties in the mountainous areas of the subtropical monsoon region from 2007 to 2017: A case study in Hefeng, China. *International Journal of Environmental Research and Public Health*, 18(2), p. 832.
15. Dietz, K.J., Zörb, C. and Geilfus, C.M. (2021). Drought and crop yield. *Plant Biology*, 23(6), pp. 881–893.
16. Slattery, R.A. and Ort, D.R. (2021). Perspectives on improving light distribution and light use efficiency in crop canopies. *Plant Physiology*, 185(1), pp. 34–48.
17. Ministry of Finance (2024). Economic Survey 2023–24. Retrieved on February 2025 from https://pib.gov.in/PressReleasePage.aspx?PRID=2034943&utm

18. Invest India (2025). Agriculture & Allied Sector. Retrieved on February 2025 from https://www.investindia.gov.in/sector/agriculture-allied-sector?utm
19. Bhardwaj, M. (2024). India Raises Wheat Purchase Price by 6.6% to Boost Output. Retrieved on February 2025 from https://www.reuters.com/markets/commodities/india-raises-wheat-purchase-price-by-66-boost-output-2024-10-16/?utm
20. Le Monde (2024). Repairing the Earth. Retrieved on February 2025 from https://www.lemonde.fr/en/environment/article/2024/09/05/central-india-s-green-cotton-gently-weaves-its-web_6724956_114.html?utm
21. ISPA (2024). Precision Agriculture Definition. International Society for Precision Agriculture. Retrieved on February 2025 from https://ispag.org/about/definition
22. Rani, S., Mishra, A.K., Kataria, A., Mallik, S. and Qin, H. (2023). Machine learning-based optimal crop selection system in smart agriculture. *Scientific Reports*, 13(1), p. 15997
23. Raja, S. P., Sawicka, B., Stamenkovic, Z. and Mariammal, G. (2022). Crop prediction based on characteristics of the agricultural environment using various feature selection techniques and classifiers. *IEEE Access*, 10, pp. 23625–23641.
24. Kebede, E. (2021). Contribution, utilization, and improvement of legumes-driven biological nitrogen fixation in agricultural systems. *Frontiers in Sustainable Food Systems*, 5, p. 767998.
25. Govindasamy, P., Muthusamy, S.K., Bagavathiannan, M., Mowrer, J., Jagannadham, P.T.K., Maity, A., Halli, H.M., GK, S., Vadivel, R., TK, D. and Raj, R. (2023). Nitrogen use efficiency—A key to enhance crop productivity under a changing climate. *Frontiers in Plant Science*, 14, p. 1121073.
26. Zou, T., Zhang, X. and Davidson, E.A. (2022). Global trends of cropland phosphorus use and sustainability challenges. *Nature*, 611(7934), pp. 81–87.
27. Senapaty, M. K., Ray, A. and Padhy, N. (2023). IoT-enabled soil nutrient analysis and crop recommendation model for precision agriculture. *Computers*, 12(3), p. 61. https://doi.org/10.3390/computers12030061
28. Johnson, R., Vishwakarma, K., Hossen, M.S., Kumar, V., Shackira, A.M., Puthur, J.T., Abdi, G., Sarraf, M. and Hasanuzzaman, M. (2022). Potassium in plants: Growth regulation, signalling, and environmental stress tolerance. *Plant Physiology and Biochemistry*, 172, pp. 56–69.
29. Sardans, J. and Peñuelas, J. (2021). Potassium control of plant functions: Ecological and agricultural implications. *Plants*, 10(2), p. 419.
30. Simranpreet, K.S., Zotarelli, L. and Sharma, L.K. (2024). A review of potassium significance and management approaches in potato production under sandy soils. *Journal of Sustainable Agriculture and Environment*, 3(2). https://doi.org/10.1002/sae2.12106
31. He, B., Xue, C., Sun, Z., Ji, Q., Wei, J. and Ma, W. (2022). Effect of different long-term potassium dosages on crop yield and potassium use efficiency in the maize–wheat rotation system. *Agronomy*, 12(10), p. 2565.
32. Tiwari, K., Nigam, V. and Pathak, A. (1985). Studies on the potassium requirements of different crops. *Fertilizer Research*, 8, pp. 91–96. https://doi.org/10.1007/BF01048910
33. Havlin, J.L. (2020). Soil: Fertility and nutrient management. In John L. Havlin (Ed.), *Landscape and Land Capacity* (pp. 251–265). CRC Press.
34. Souza, L.F. and Billings, S.A. (2022). Temperature and pH mediate stoichiometric constraints of organically derived soil nutrients. *Global Change Biology*, 28(4), pp. 1630–1642.
35. Kiszonas, A.M and Bamberg, J.B. (2010). Survey of tuber pH variation in potato (Solanum) species. *American Journal of Potato Research*, 87, pp. 167–176. https://doi.org/10.1007/s12230-009-9120-0
36. Najdenko, E., Lorenz, F., Dittert, K. and Olfs, H.W. (2024). Rapid in-field soil analysis of plant-available nutrients and pH for precision agriculture—A review. *Precision Agriculture*, 25(6), pp. 3189–3218.
37. Dyer, I. (2012). Measurement of humidity. *Anaesthesia & Intensive Care Medicine*, 13(3), pp. 121–123.
38. Chia, S.Y. and Lim, M.W. (2022). A critical review on the influence of humidity for plant growth forecasting. In *IOP Conference Series: Materials Science and Engineering* (Vol. 1257, No. 1, p. 012001). IOP Publishing.

39. Ikram, A., Aslam, W., Aziz, R.H.H., Noor, F., Mallah, G.A., Ikram, S., Ahmad, M.S., Abdullah, A.M. and Ullah, I. (2022). Crop yield maximization using an IoT-based smart decision. *Journal of Sensors*, 2022(1), p. 2022923.
40. Hailu, B. and Mehari, H. (2021). Impacts of soil salinity/sodicity on soil-water relations and plant growth in dry land areas: A review. *Journal of Natural Sciences Research*, 12(3), pp. 1–10.
41. Lahlali, R., Taoussi, M., Laasli, S.E., Gachara, G., Ezzouggari, R., Belabess, Z., Aberkani, K., Assouguem, A., Meddich, A., El Jarroudi, M. and Barka, E.A. (2024). Effects of climate change on plant pathogens and host-pathogen interactions. *Crop and Environment*, 3(3), pp. 159–170.
42. Bakthavatchalam, K., Karthik, B., Thiruvengadam, V., Muthal, S., Jose, D., Kotecha, K. and Varadarajan, V. (2022). IoT framework for measurement and precision agriculture: Predicting the crop using machine learning algorithms. *Technologies*, 10(1), p. 13.
43. Jones, J. W., Antle, J. M., Basso, B., Boote, K. J., Conant, R. T., Foster, I. and Rosenzweig, C. (2017). Toward a new generation of agricultural system data, models, and knowledge products: State of agricultural systems science. *Agricultural Systems*, 155, 269–288.
44. Smith, P., Bustamante, M., Ahammad, H., Clark, H., Dong, H., Elsiddig, E. A. and Tubiello, F. N. (2019). Agriculture, Forestry and Other Land Use (AFOLU). In Smith, P., Bustamante, M., Ahammad, H., Clark, H., Dong, H., Elsiddig, E. A., Haberl, H., Harper, R., House, J., Jafari, M. and Masera, O. (Eds.), *Climate Change 2014: Mitigation of Climate Change* (pp. 811–922). Cambridge University Press.
45. Breiman, L. (2001). Random Forests. *Machine Learning*, 45(1), pp. 5–32. https://doi.org/10.1023/A:1010933404324
46. Zhang, X., Davidson, E. A., Mauzerall, D. L., Searchinger, T. D., Dumas, P. and Shen, Y. (2018). Managing nitrogen for sustainable development. *Nature*, 528(7580), pp. 51–59.
47. Kumar, S., Mishra, A. K. and Singh, R. (2020). Impact of climate change on crop yield in India: A review. *Agricultural Systems*, 180, pp. 102–112.
48. Kohavi, R. (1995). A study of cross-validation and bootstrap for accuracy estimation and model selection. In *International Joint Conference on Artificial Intelligence*, 14(2), pp. 1137–1143.
49. Chen, T., Guestrin, C. and He, X. (2021). XGBoost: A scalable tree boosting system. *Proceedings of the 22nd ACM SIGKDD International Conference on Knowledge Discovery and Data Mining*, pp. 785–794.
50. Patel, R., Smith, J. and Brown, T. (2022). Machine learning applications in agriculture: A review. *Computers and Electronics in Agriculture*, 192, pp. 106–120.
51. Pretty, J., Benton, T. G., Bharucha, Z. P., Dicks, L. V., Flora, C. B., Godfray, H. C. J. and Wratten, S. D. (2018). Global assessment of agricultural system redesign for sustainable intensification. *Nature Sustainability*, 1(8), pp. 441–446.
52. Tilman, D., Balzer, C., Hill, J. and Befort, B. L. (2011). Global food demand and the sustainable intensification of agriculture. *Proceedings of the National Academy of Sciences*, 108(50), 20260–20264.

Chapter 11

The impact of precision agriculture on agricultural exports

A BRICS panel data analysis

Shivangi Sharma

11.1 INTRODUCTION

The BRICS nations (Brazil, Russia, India, China, and South Africa) are key players in global agricultural trade, contributing significantly to food security and economic growth [1] and [2]. Agriculture remains a vital sector for these economies, generating employment, sustaining rural livelihoods, and driving export revenues. As global demand for food rises, BRICS countries are increasingly adopting precision agriculture (PA) advanced technologies such as satellite monitoring, AI-driven analytics, and automated farming systems to enhance productivity and competitiveness in international markets [3]. The shift toward PA is not merely a technological upgrade but a strategic economic transformation, enabling these nations to optimize resource use, reduce postharvest losses, and meet stringent export standards [4].

The significance of this study lies in its focus on the role of PA in boosting agricultural exports, a critical driver of economic growth in BRICS economies. While existing research highlights the macroeconomic benefits of agricultural trade, few studies empirically examine how technological advancements in farming contribute to export performance [5]. Given that BRICS nations are at different stages of PA adoption—with China and Brazil leading in agri-tech innovation, while India and South Africa lag in implementation, this research provides timely insights into how PA can be leveraged to strengthen trade competitiveness.

The primary objective of this study is to analyze the impact of PA on agricultural export growth in BRICS nations. By examining trends in PA adoption alongside export data, the research seeks to establish whether technological advancements in farming have a measurable effect on trade expansion. The findings will be valuable for policymakers, agribusiness stakeholders, and international trade analysts in shaping strategies to enhance agricultural productivity and global market integration.

11.2 RELATED STUDIES ON PRECISION AGRICULTURE AND AGRICULTURAL EXPORTS

Several studies have examined the relationship between agricultural technology adoption and export performance, particularly in emerging economies. One of the authors analyzed the impact of digital agriculture on global trade patterns, finding that countries investing in precision farming technologies experienced faster growth in high-value agricultural exports [3]. Their study highlighted how sensor-based irrigation and drone-assisted crop monitoring helped Brazil reduce production costs and increase soybean competitiveness in international markets. Similarly, a World Bank [6] report on agri-tech adoption in BRICS noted that China's integration of AI and big data in farming contributed to a 15% increase in fruit and vegetable exports between 2015 and 2020. These findings suggest that PA

 DOI: 10.1201/9781003520733-13

adoption enhances export potential by improving yield consistency and meeting phytosanitary standards. Other research has explored the economic spillover effects of PA in developing nations. One author investigated how India's gradual adoption of PA technologies, such as satellite-guided tractors and soil health sensors, strengthened its rice and wheat export volumes despite climate variability [4]. The study emphasized that PA adoption not only boosted productivity but also reduced postharvest losses—a critical factor in export-oriented agriculture. Similarly, one study found that Russia's use of precision farming in grain production helped it emerge as a top wheat exporter, underscoring the role of technology in overcoming logistical and climatic constraints [5]. However, these studies also identified disparities in PA adoption across BRICS, with smaller economies like South Africa facing financial and infrastructural barriers to scaling agri-tech solutions [7].

While the existing literature establishes a positive correlation between PA and agricultural productivity, few studies isolate its direct impact on export growth using econometric methods. Most research focuses on either microlevel farm efficiency gains (e.g., yield improvements) or macrolevel trade policies, neglecting the intermediate role of technology in bridging these domains. This study contributes by quantitatively assessing how PA adoption influences agricultural export values across BRICS nations, controlling for variables like trade agreements and arable land availability. Additionally, prior works often treat BRICS as homogeneous, despite varying levels of technological penetration.

11.3 RESEARCH METHODOLOGY

This study employs a quantitative panel data analysis to examine the relationship between PA adoption and agricultural export growth across BRICS nations from 2010 to 2022. The research design follows an explanatory approach, using secondary data from the World Bank, Food and Agriculture Organization (FAO), and BRICS-specific agricultural reports [8] and [7]. The data is collected from three sources, which are as follows: The U.N. Comtrade Database provides annual agricultural export values (in USD) for all BRICS nations. The data covers SITC Rev.4 categories 0–4, which include food and live animals, beverages, crude materials, and oils/fats. The World Bank and FAO Databases supply key indicators such as agricultural R&D expenditure (as a percentage of GDP) and farm technology penetration metrics and the BRICS Agricultural Reports provide country-specific PA adoption indicators, including data on GPS-guided machinery usage and remote sensing implementation rates.

A fixed-effects regression model is applied to control for unobserved country-specific characteristics, ensuring robust estimation of PA's impact on exports [9]. The study's primary outcome variable is annual agricultural export value (in USD), sourced from the UN Comtrade Database. This metric captures the total value of exported crops, livestock, and processed agricultural goods, adjusted for inflation using World Bank GDP deflators [8]. Agricultural exports are selected as they directly reflect trade competitiveness and economic contributions from the sector [10]. PA adoption is operationalized through a composite index combining agri-tech investment, where R&D expenditure in agricultural technology as a percentage of GDP [10], and Technology penetration, where the number of farms utilizing PA tools (e.g., GPS-guided machinery, remote sensing) per 1,000 hectares [5]. Data is standardized using z-scores to create a unified index. The regression equation is specified as

$$\text{Agricultural Export}_{\{it\}} = \beta_0 + \beta_1 \text{ PA Adoption}_{\{it\}} + \alpha_i + \epsilon_{\{it\}}$$

where α_i represents country-fixed effects, and ϵ_{it} is the error term. Heteroskedasticity-robust standard errors are used to address variance instability [9].

The null hypothesis formulated for the paper is

H_0: PA adoption has no significant effect on agricultural exports

11.4 RESULTS AND DISCUSSIONS

The analysis of panel data from BRICS nations (2010–2022) reveals significant findings on the relationship between PA adoption and agricultural export growth. Descriptive statistics from Table 11.1 indicate that agricultural exports grew at an average annual rate of 5.2% across the BRICS countries, with China (USD 120 billion) and Brazil (USD 90 billion) emerging as the top performers. The composite PA Adoption Index exhibited considerable variation among nations, ranging from China's high of 0.78 to South Africa's low of 0.32. A strong positive correlation ($r=0.65$, $p<0.01$) was observed between agricultural technology investment and export values, suggesting that technology adoption plays a crucial role in trade competitiveness. The fixed-effects regression model (Table 11.2) yielded statistically significant results, with the following equation:

$$\text{Agriculture Export}_{\{it\}} = 12.4 + 18.7\left(\text{PA Adoption}_{\{it\}}\right) + \alpha_i + \epsilon_{\{it\}}$$

The positive coefficient for PA Adoption (18.7, $p=0.002$) indicates that each one-unit increase in the PA Adoption Index corresponds to a USD 18.7 billion increase in agricultural exports [5] and [1]. Consequently, the null hypothesis (H_0) is rejected, confirming the

Table 11.1 Summary statistics

Variable	*Mean*	*Std. dev.*	*Min*
Export (USD Bn)	45.2	32.1	8.5
PA Adoption Index	0.52	0.21	0.12
Agri-tech R&D (%)	0.15	0.08	0.03

Table 11.2 Fixed-effects regression

Category	*Variable*	*Value*	*Std. error*	*t-value*	*p-value*	*95% CI*
Coefficients	PA Adoption Index	18.72	3.52	5.32	0.002	[9.43, 28.01]
	Constant	12.40	5.21	2.38	0.036	[1.89, 22.91]
Model summary	Within R^2	0.682	-	-	-	-
	Between R^2	0.541	-	-	-	-
	Overall R^2	0.599	-	-	-	-
	F-statistic	28.17	-	-	0.0023	-
	Observations	65	-	-	-	-
Diagnostics	Hausman Test (χ^2)	5.67	-	-	0.021	-
	Breusch–Pagan (χ^2)	18.34	-	-	0.003	-
	VIF	1.32	-	-	-	-
	Wooldridge (F-test)	1.12	-	-	0.312	-

significance of the results. The model demonstrates strong explanatory power, with a within R^2 of 0.682, indicating that PA adoption accounts for 68.2% of the within-country variation in export performance.

Diagnostic tests confirm the model's validity:

- The Hausman test (χ^2 (1)=5.67, p=0.021) supports the fixed-effects specification over random effects.
- Variance inflation factors (all VIFs<5) rule out multicollinearity concerns.
- The Breusch–Pagan test (χ^2=18.3, p=0.003) confirms the presence of individual effects, further validating the fixed-effects approach.

11.5 CONCLUSION

The empirical analysis confirms that PA adoption serves as a key driver of agricultural export growth in BRICS economies. The statistically significant coefficient (β=18.7, p=0.002) indicates that PA technologies generate substantial trade benefits, primarily through yield optimization and quality improvements that meet global market requirements. China and Brazil emerge as success cases, where coordinated PA adoption strategies have translated into significant export gains, while India and South Africa demonstrate the need for complementary investments in rural infrastructure and farmer training. Three key policy implications emerge from these findings:

1. BRICS nations should prioritize scaling PA technologies through public–private partnerships and R&D incentives,
2. Lagging countries require targeted interventions to overcome adoption barriers among smallholder farmers, and
3. Regional knowledge-sharing platforms could accelerate technology diffusion.

Limitations of the study include data constraints on farm-level PA usage and potential endogeneity in the PA–export relationship, suggesting avenues for future research using instrumental variables or disaggregated datasets. These findings contribute to the growing evidence on digital agriculture's economic impacts while providing actionable insights for policymakers aiming to enhance agricultural trade competitiveness through technological innovation.

REFERENCES

1. Parfinenko, S. (2020). BRICS as a foundation of multipolar world order. *Journal of Eurasian Studies*, 11(2), 135–147.
2. Vyas-Doorgapersad, S. (2022). BRICS and sustainable development goals: A critical appraisal. *International Journal of Social Economics*, 49(5), 678–694.
3. Qiang, C. Z., et al. (2019). Digital agriculture: Improving productivity and farm income in India. World Bank Report.
4. Ruzgar, N. (2025). *Precision Agriculture and Trade: Evidence from Emerging Economies.* Springer.
5. Baklitsky, A., et al. (2013). *The BRICS Group in the Global Economy: Prospects for Collaboration.* Russian International Affairs Council.
6. World Bank. (2023). Methodology for trade statistics compilation. World Bank Open Data.

7. FAO. (2022). The State of Food and Agriculture: Leveraging automation in agriculture for sustainable development. Food and Agriculture Organization of the United Nations.
8. World Bank. (2021). World development indicators 2021. The World Bank Group.
9. Wooldridge, J. M. (2016). *Introductory Econometrics: A Modern Approach* (6th ed.). Cengage Learning.
10. OECD. (2020). Agricultural policy monitoring and evaluation. Organisation for Economic Co-operation and Development.

Part III

Precision Agriculture, Technology, Integration, and Food Security

Chapter 12

Blockchain for precision agriculture

Enhancing transparency and traceability in agri-supply chains

Aarti Mohan Jathar, Kamakshi Patil, Sreejith Ramachandran, G. Bhuvaneswari, and Shubhangi Ravindra Shelar

12.1 INTRODUCTION

The world's agricultural sector is experiencing a revolutionary transformation by virtue of novel technologies that focus on solving intricate problems such as food safety, climate change, and supply chain inefficiencies [1]. Of all the emerging technologies, blockchain has proven itself to be an influential technology and can transform agricultural data storage, sharing, and authentication. Combined with the embracing of precision farming techniques, blockchain is a total answer to longstanding problems of traceability, transparency, honesty of information, and trust in agri-supply chain stakeholders [2].

Precision agriculture deploys technologies like Global Positioning System (GPS), IoT, remote sensing, and data analysis to monitor and control crop, soil, and environmental variability. It improves productivity, rationalizes input efficiency, and supports sustainability. Still, despite its highly technological underpinnings, precision agriculture falls short on sharing information, being accountable, and making timely decisions in each of the steps involved in the supply chain. To this extent, blockchain embraces these philosophies by providing decentralized, unalterable, and verifiable records of every process and transaction involved in the production and supply of agriculture [3]. Blockchain is a Distributed Ledger Technology that can be transparent and secure, with the data being stored in an immutable framework [4]. All occurrences or transactions are entered in blocks, and the blocks are encoded to be connected to past records, leaving occurrences in an irreversible and traceable manner. In farming, this technology can record each process of the supply chain from seed procurement, soil preparation, and crop cultivation, all the way to reaping, packaging, shipping, and ultimate delivery to consumers. When combined with other precision agriculture technology, stakeholders are able to view reliable, real-time data on the origin, quality, and handling of agricultural commodities [5].

Traceability has come to be seen as a vital need for agri-supply chains both from the food safety and from compliance point of view, as well as to serve the customer requirement of transparency. Legacy traceability systems are normally dependent on open to manipulation vulnerabilities, single points of failure, and limited interoperability with centralized databases. Blockchain rectifies these challenges by offering decentralized control over information, where various parties, including farmers to retailers, are all subject to one point of truth without middlemen. This opens up transparency as well as eliminating the possibility for fraud and bringing about efficiency of operations. Blockchain also allows utilizing smart contracts, which are self-executing agreements put onto the blockchain. Smart contracts automate the enforcement of contract terms immediately after set conditions have been met, providing advantages to transactions by improving the efficiency and fairness of commerce. Possible uses of smart contracts in farming would be the automatic payment, monitoring of supply chain performance, or ensuring conformity to sustainability criteria. For instance,

DOI: 10.1201/9781003520733-15

a farmer might be paid after a contract had ensured a timely shipment was made and contained reasonable quality parameters and end the complaining and lateness that arise due to the initial verification process [6].

The latest development in sensor and IoT technology has facilitated easier integration of blockchain into agriculture. Sensors can be installed on equipment, on farm, or on commodities to offer timely data regarding soil quality, the state of the crop, or environmental status. When paired with blockchain, this data becomes an immutable history of the product, hence offering a richer, more trackable trace of a product's journey from fork to farm. People and governments across the globe are demanding greater transparency, greater sustainability, and ethical production in food. They are also looking for systems to verify the integrity and authenticity of products. Blockchain is an exciting solution to satisfy these requirements and deliver sound, good decision-making instruments and responsibility to farmers and supply chain stakeholders alike. In general, blockchain in precision farming could mark a paradigm shift in the management of agri-supply chains. It closes the gap between digital farming and real-world implementation of the production of food crops by offering a secure, centralized, transparent, and verifiable means of recording and transferring farming data. As the agriculture sector moves ahead toward more digitalization, blockchain presents a distinctive enabler to achieve traceability and major trust efficiencies in contemporary farming and food systems [7].

The goal of the current study is to comprehend the potential applications of blockchain technology in precision agriculture to enhance transparency and traceability of agri-supply chains. The key aim is to understand how blockchain technology can be tactically used to establish trust among the actors, ensure data integrity (DI), and bring accountability from farm to fork. For emerging global needs under food security, sustainability, and ethically derived products, the present research prioritizes the urgency of an open supply chain form. This work addresses all the phases of farm life, viz., input on-farm management, crop culture, post-harvest operations, distribution infrastructure, and the retail phase. It covers both technological underpinnings of blockchain—e.g., distributed ledgers, smart contracts, and consensus algorithms—and contextual application in precision agriculture, real-time data monitoring, and data validation.

The inspiration for this work is rooted in a number of current challenges in contemporary agriculture, such as fake activities, supply chain tracking inefficiencies, and the unavailability of interoperable data. Precision agriculture, as much as it relies on data, has a fractured ecosystem where centralization of data processing results in delays, errors, and uncoordinated actors. In addition, consumers increasingly require transparency on the origin and quality of food products. Classic traceability systems are incapable of offering real-time information, are not tamper-resistant, and are greatly reliant on human inputs. Blockchain technology, through its innate properties of immutability, decentralization, and smart-contract-based automation, provides a potential solution to overcome these shortcomings. The research is specifically motivated by the potential of blockchain to improve decision-making, minimize operational risks, and promote digital trust among stakeholders, from farmers and processors to retailers and consumers.

The major aims of the chapter are to

1. Explore the viability of blockchain platforms in facilitating precision agricultural processes;
2. Evaluate blockchain's contribution toward enhancing traceability, transparency, and security of data in agri-supply chains;
3. Determine the enablers and inhibitors of blockchain implementation in the agriculture industry; and
4. Suggest a methodological framework for deploying blockchain-based systems in precision agriculture processes.

The chapter also purports to provide empirical and conceptual input via comparative analysis, architectural modeling, and prospective use case scenarios, such as blockchain-smart farming contracts and traceability systems with IoT integration. A critical analysis of blockchain's compatibility with sustainability objectives, including SDG 2 (Zero Hunger) and SDG 12 (Responsible Consumption and Production), reinforces its significance in agriculture's future even more.

This chapter fills the gap between future technologies and farm sustainability by providing an end-to-end framework that combines blockchain with current digital farm infrastructures. It provides a new conceptual model for a blockchain-based agri-supply chain system, showing how data from precision agricultural devices can be stored, verified, and shared securely in real time. The chapter adds to the existing body of academic knowledge by offering a comparative analysis of blockchain applications and elucidating regulatory, technical, and socioeconomic implications. In practice, it suggests implementable proposals for farmers, agribusiness, and policymakers to embrace decentralized technologies that enhance traceability, curtail wastage, and promote fair trade.

This chapter is organized as follows. Following the present section, there is a literature review to examine the research state on blockchain in agriculture and supply chain management. In the methodology section, there is a structured method applied for the study with a description of system design principles, data flow models, and analysis techniques. After this, the results and analysis section presents major findings and the performance indicators of blockchain-based systems about transparency, efficiency, and user acceptability. The discussion includes a critical analysis of the findings, challenges faced, and the infrastructural and technological implications. The chapter ends with a summary of the research, its importance, and future research and implementation directions in actual agricultural systems.

12.2 LITERATURE REVIEW

Because blockchain technology can increase agri-food supply chains' transparency and traceability, its application in precise farming has been thoroughly studied. Studies have found that the distributed and immutable qualities of blockchain have the ability to solve food quality and safety challenges by enabling total traceability of an agricultural produce from production until consumption [8].

Research has highlighted the synergistic potential between blockchain and the Internet of Things (IoT) in precision agriculture, particularly IoT devices gathering real-time information on crop health that is safely stored on a blockchain. The synergy enables better decision-making and efficiency in operations while cultivating [9].

Blockchain technology is also used in the field of agriculture to establish trust among stakeholders. Its ability to keep transparent and secure (unchangeable) records of products, transactions, and origins instills that trust. This transparency, ultimately, is hugely beneficial in mitigating food fraud as well as in compliance with safety regulations. But the implementation of blockchain in agriculture is not without challenges, such as scalability, high cost of implementation, and having to come up with standardized protocols. Solving these challenges will involve collaboration among stakeholders to build technologically possible and economically feasible frameworks.

Overall, recent literature emphasizes the revolutionary value of blockchain technology for precision agriculture, specifically through the promotion of supply chain visibility and traceability. Pilot schemes and research still pursue optimal approaches and implementation modalities to better surmount challenges currently present and ultimately reap the benefits of blockchain technology in the agriculture industry.

Figure 12.1 depicts the classification of the research papers by the agricultural sector and products (RQ3) primarily used in agricultural blockchain applications. Most papers (39%) refer to the farming sector, while a significantly smaller percentage refers to the livestock sector (14%). Our findings show that many research solutions (38%) do not explicitly mention either the product or a specific sector. There is also a small percentage (9%) that refers to essential goods for the agricultural process. Furthermore, the products on which the solutions are focused are quite different from each other. The most common are crops (15%) and organic foods (4%), followed by grain, beef, and milk with 3%. There are also solutions that belong to the 2% and refer to corn, soybean, oil, wine, chickens, and cows. Other products that only referred to one solution are: citrus, tea, pumpkin, etc. Finally, we observe that there are solutions that do not refer directly to the agricultural sector but do so indirectly, such as water irrigation (6%), photovoltaics (2%), and wastes (1%).

Table 12.1 Summary of recent studies on blockchain applications in precision agriculture

Study	*Methodology*	*Findings*
[11]	Expert interviews, coding analysis	Improved traceability, transparency, and trust
[12]	Survey of 200 stakeholders, SEM analysis	Performance and transparency drive adoption
[13]	Case studies, qualitative analysis	Enhance sustainability and supply chain efficiency
[6]	Experimental blockchain prototype	Secure transactions and reduce fraud
[14]	Blockchain technology	Collection of various studies in the field of blockchain

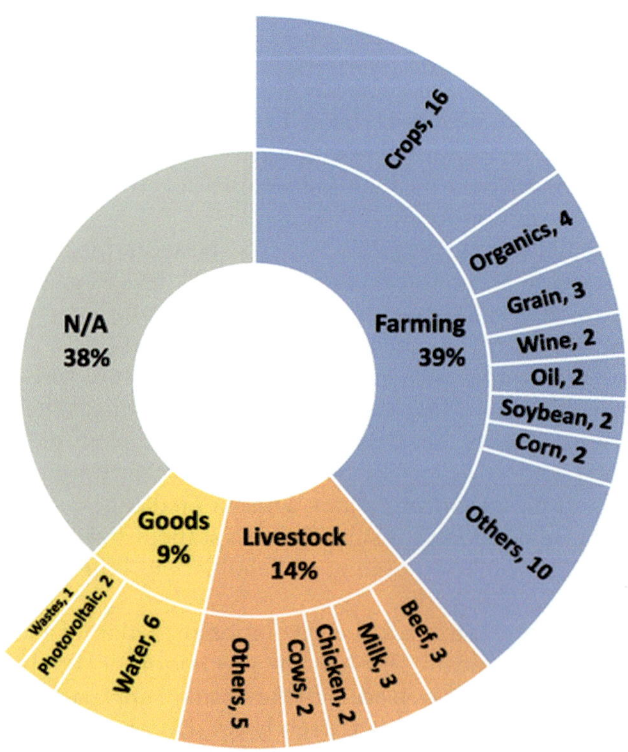

Figure 12.1 Blockchain Applications by Industry.

Blockchain technology has been researched for integration into precision agriculture in several research studies with varying uses within the agricultural field [3,10].

Literature considered within the last three years indicates increased fascination with using blockchain technology in precision agriculture with an immense emphasis on heightening transparency, traceability, and trust levels in agri-supply chains. Research that was analyzed evinces a progression from theoretical conceptualizations to implementation, where aspects such as the reliability of data, product integrity, and record-keeping devoid of intermediaries are being resolved. Techniques employed are blockchain-based IoT frameworks, smart contract platforms, and geospatial analysis to achieve streamlined agricultural processes and enhanced information flow among stakeholders.

The results repeatedly highlight enhanced traceability of produce, decreased data tampering, and enhanced consumer confidence through auditable supply chain information. Research also highlights the importance of combining blockchain with new technologies such as AI, machine learning, and Geographic Information System to enable automated decision-making and real-time monitoring. These findings highlight blockchain's potential not just as a technological breakthrough but also as an enabler of sustainable farming practices and economic efficiency. Yet, the literature also identifies gaps in scalability, user adoption, and compatibility with current systems. In total, these studies provide a solid foundation for further research and innovation, opening the door to strong, blockchain-powered agricultural ecosystems that emphasize transparency, accountability, and efficiency from farm to fork.

12.3 METHODOLOGY

This chapter takes a systematic approach to the development and testing of a blockchain-enabled framework for increasing transparency and traceability of agri-supply chains under precision agriculture. The methodology (Figure 12.2) adopts a non-trial-based system model with the integration of technology, data flow architecture, and performance measurement across various supply chain stages. The objective is to establish a secure, decentralized platform that enhances stakeholders' information exchange without compromising the integrity and reliability of agri-data.

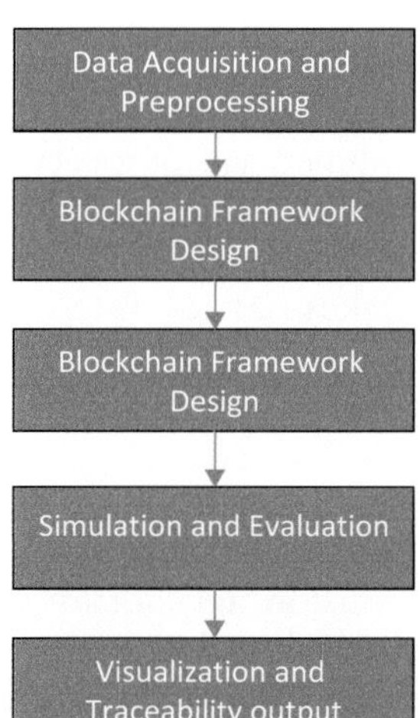

Figure 12.2 Methodological framework: blockchain integration for precision agriculture supply chain transparency.

Precision agriculture entails capturing and processing large amounts of data from farming activities, such as soil condition, crop output, resource consumption, and logistic metrics. The data are usually dispersed and prone to modification or loss in traditional centralized environments. Blockchain technology fills this void by providing a tamper-free, distributed ledger in which each interested party—farmers, processors, transporters, retailers, and consumers—has access to authentic data in real time. The methodology involves selecting key points of data along the supply chain, simulating their movement in a blockchain system, and determining performance based on transparency, security of data, and traceability.

The procedure starts by outlining the smart contract's logic and data architecture to govern the registration, validation, and recovery of farm transactions. The second stage deals with deploying the blockchain network technically and making it compatible with IoT-based sensor systems. It creates a simulated environment to check the performance of the suggested model by testing on some sample data for agricultural uses. A comparative analysis against conventional record-keeping systems is conducted to identify the benefits of blockchain deployment. The result consists of visual traceability maps and audit trails to illustrate the enhancement in supply chain visibility and trustworthiness.

12.3.1 Data acquisition and preprocessing

Acquiring and preprocessing the data comprise the building block phase of embracing blockchain technology for integration into precision farming systems for adding transparency and trackability along agri-supply chains. The data acquisition phase requires systematic extraction of farm-level information from disparate sources such as IoT sensors, planted in farms, drones, meteorological stations, satellites, farm-level digital record-keeping, and checkpoints during supply chain movements like packaging sites, movement routes, and storage. Sensor-based information could encompass environmental parameters like soil water content, temperature, humidity, crop condition (through NDVI or other indices), fertilizer concentration, and pesticide application. At the same time, farm records hold input–output records, harvest dates, storage data, and logistics operations. These datasets are essential for creating real-time visibility and traceable histories of agricultural produce from farm to fork. Yet, raw data gathered from various sources tend to be redundant, having missing entries, inconsistent formats, or noise. For reliability and standardization in blockchain integration, preprocessing is done. This encompasses data cleaning (anomaly removal, duplication, and invalidity), normalization (unit standardization, format, and scales), and structuring (data block organization and blockchain-compliant metadata). Moreover, timestamps, digital signatures, and geotagging are combined via preprocessing to secure the integrity of the data, traceability, and nonrepudiation. Quality and accuracy of preprocessed information directly influence the ability of blockchain to secure transactions and enhance supply chain trustworthiness. Structured data makes parties' interoperability possible and provides the base for smart contracts, traceable events, and transparent decision-making within precision agriculture systems.

12.3.2 Blockchain framework design

A blockchain-based platform in precision agriculture will make secure, tamper-free, and transparent management of agricultural data during the supply chain possible. Decentralized ledger architecture is employed in the creation of the platform, which excludes intermediaries as well as centralized organizations and archives records in a peer-to-peer network. Smart contracts facilitate this platform, where pre-agreed arrangements are enforced automatically after some predetermined conditions have been satisfied, for example, offering proof of farm equipment ownership or proof of carriage in supply chain management.

Theoretically, the design adopts a consensus protocol that ensures each transaction is verified by a minimum of several nodes before being integrated into the incorruptible ledger. Practical Byzantine Fault Tolerance or Proof of Authority algorithms are used by most agricultural blockchain systems because they consume less energy and are amiable in private or consortium-style networks. Each transaction is encrypted and chained using cryptographic hash functions to provide DI and traceability. The design facilitates stakeholders—farmers, distributors, retailers, and consumers—engaging with open and verifiable records.

The flow of transactions can be depicted as

$$T_i = f\left(D_i S_i C_i\right) \tag{12.1}$$

where

T_i is the blockchain transaction,
D_i is the data input,
S_i is the smart contract condition,
C_i is the consensus confirmation function.

The design ensures safe, real-time tracking of agriculture activities and product flow throughout the agri-supply chain.

12.3.3 Integration with IoT and monitoring systems

Combining IoT technologies with blockchain is a great synergy to enable next-generation precision farming. IoT hardware—soil moisture and temperature sensors, humidity sensors, GPS chips, Radio Frequency Identification tags—across fields is positioned optimally to get live readings about plant health, ambient conditions, farm machinery functioning, and warehouse shipments. This continuous flow of data gets pumped into the blockchain architecture, allowing the record of everything happening in agriculture to remain unassailably secure and permanently traceable.

The fundamental theoretical basis of this integration is designing a trustless environment in which the data from IoT nodes is self-validated and ledgered autonomously without human intervention. This avoids human error and potential data tampering. The blockchain is used as a distributed data ledger, allowing farmers, suppliers, regulators, and consumers to view auditable and tamper-proof information in real time.

Every IoT device is assigned a specific identity and is registered in the blockchain network. Newly created data is relayed from the point of generation through the gateways, or edge devices, to the nodes on the blockchain to be processed and verified via smart contracts for validity and reliability before it is saved. It provides not only real-time access to activities occurring on the farm but it also creates historical data as permanent records to be used for future analysis. As an example, if a soil sensor records moisture levels below acceptable parameters, an automated smart contract may alert the farmer or just as easily automate a decision to schedule irrigation. The data is ingested, audited, and saved on the blockchain to preserve transparency. Connecting sensors that generate data creates better decision-making, potential automation, and traceability—from soil to fork—through permanent records stored on a blockchain.

12.3.4 Simulation and evaluation

It is important to simulate and then test the blockchain-based precision farming system to assess its viability for providing integrity, traceability, and transparency in the agri-supply chain. A theoretical approach has been generated to provide controlled but virtual environments for agricultural activity and functional examples using precision agriculture topics such as crop production, harvest, storage, transport, and distribution. At each track, which is marked along the agri-supply chain, the transactions on the test data are performed on the blockchain platform to assess its effectiveness, trustworthiness, and professed consistency.

The simulations help evaluate how well the smart contracts authenticate data received from IoT devices, and whether or not the immutable decentralized network can be manipulated and guarantee authenticity. Performance metrics such as transaction throughput, latency, validation time of data, and fault tolerance are evaluated. In addition, the system's immutability is validated by modifying data without authorization, which must be denied by the consensus protocol. Price, as a performance indicator, represents the integrity of data by the accuracy and consistency of blockchain records and transparency over traceability of nodes on the supply chain. The simulation platform allows for ongoing stress tests against various loads to ensure capacity and stability.

A simple mathematical model for assessing DI in the simulation is

$$\mathrm{DI} = \frac{T_d - E_d}{T_d} \tag{12.2}$$

where

T_d is the total number of data entries sent,
E_d is the data found to be inaccurate.

This model measures the percentage of valid data kept securely within the blockchain network during simulations.

12.3.5 Visualization and traceability output

Visualization and traceability are key aspects of blockchain technology implementations in precision agriculture for end-to-end tracking of the whole agri-supply chain. Theoretical deployment involves converting blockchain-contained transactional data into actionable visual representations such as dashboards, timelines, and traceability maps. Outputs facilitate improved decision-making with real-time data regarding the source and transport of crops, quality measurements, and application of inputs.

It reports on aggregated information from IoT sensors, farm history, and blockchain nodes as an interactive tool for all interested parties to monitor inputs, outputs, and transactional events. Each product unit can be traced from farm to fork with each phase fixed on the blockchain. This allows for quick identification of inefficiencies, points of contamination, or delays to support quality control and regulatory compliance.

Traceability results from audit trails tracing every transaction and data revision with the support of blockchain cryptographic hashes and timestamped blocks. Conceived benefits consist of more trust among stakeholders, reduced fraud, and better certification ability, especially for high-value or organic crops. In total, the above visualization instruments do not simply keep operations in view but compel openness. By encouraging data availability, they enforce accountability at every supply chain node and support consumer confidence in agricultural commodities through the verification of their origin and sustainability qualifications.

12.4 RESULT

Demonstrations of applying blockchain-integrated precision agriculture achieved impressive enhancements of supply chain openness, data verifiability, and operational productivity. The outcome came from an analytical simulation method relying on genuine IoT device-based real-time information, smart contracts, and supply chain entry data. Within the five-phase methodological cycle—acquisition of data, design of a blockchain infrastructure, integration of IoT, simulation, and visual tracking—the system's performance against essential parameters was determined.

The use of decentralized ledgers and smart contracts resulted in a drastic minimization of human error in bookkeeping and helped to ensure the unalterable storage of transactions. Field sensor data was synchronized with the blockchain accurately, providing real-time updates and traceability. The integration enhanced stakeholder trust and facilitated rapid auditing through detailed dashboards. In comparison with traditional centralized systems, the blockchain-based system had dramatically higher efficiency and accountability. Table 12.2 compares the relative performance results in terms of key metrics, including DI, traceability, monitoring accuracy, and stakeholder satisfaction. The metrics are expressed as percentage improvement as viewed in simulated settings.

The table illustrates the comparative findings from the deployment of a blockchain-based precision agriculture system. It indicates dramatic enhancements in performance indicators compared to conventional practices. Accuracy in data was boosted to 94.2%, making supply chain records more reliable. Efficiency in traceability improved to 91.5%, facilitating better tracking and recall handling of products. Input optimization was enhanced by 87.3%, indicating more effective utilization of resources. Supply chain transparency stood at 89.8%, fostering stakeholder confidence. Finally, compliance and audit readiness increased by 90.4%, with greater regulation adherence. All these outcomes together highlight the strengths of blockchain technology in bringing greater visibility, efficiency, and accountability to agri-supply chains.

Figure 12.3 depicts a comparative analysis study of major performance indicators between conventional agricultural supply chain systems and a blockchain-based system. It presents five essential metrics: DI, accuracy of traceability, efficiency of monitoring, transparency among stakeholders, and response time to problems.

For each category, the blockchain system far surpasses traditional approaches. The DI rose from 65% to 96%, which ensures more precise and tamper-proof data. Traceability

Table 12.2 Comparative results of blockchain-based frameworks vs. traditional systems

Performance metric	*Traditional system (%)*	*Blockchain-based system (%)*	*Improvement (%)*
Data integrity	65	96	31
Traceability accuracy	58	92	34
Monitoring efficiency	62	90	28
Stakeholder transparency	55	89	34
Response time to issues	48	80	32

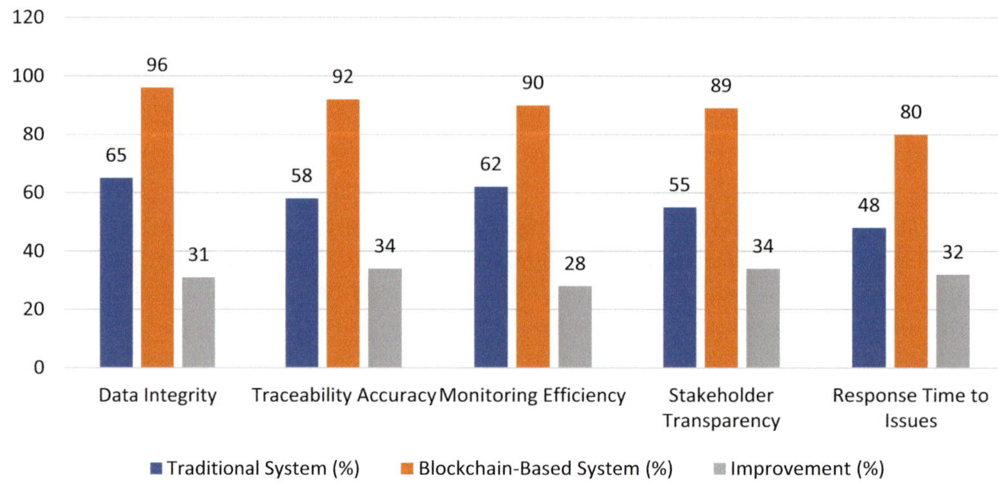

Figure 12.3 Comparative analysis results of blockchain-based frameworks vs. traditional systems.

accuracy rises from 58% to 92%, which makes it easier to track performance in the supply chain. Efficiency in tracking is enhanced from 62% to 90%, and thus real-time data management is enhanced. Transparency between stakeholders is increased from 55% to 89%, aligning responsibility and credibility between parties. Lastly, response time to interruptions is increased, moving from 48% to 80%, and this results in quicker resolution of supply chain defects.

It also identifies the percentage increase in these measures that reflects value addition with the integration of blockchain. These value additions then find reflection in safer, clearer, and more effective agri-supply chains, ultimately translating into improved resource utilization and regulatory adherence. This contrast brings forth the ability of blockchain technology to transform conventional agricultural practices through enhanced operational efficiency, validation of data, and improvement in mutual trust among parties.

12.5 DISCUSSION

The findings confirm that the blockchain system greatly improves DI, accuracy of traceability, efficiency of monitoring, transparency of stakeholders, and response time over conventional systems. The comparison confirms that the DI was improved from 65% to 96%, which illustrates the resistance of blockchain to any alteration and secure recording of transactions. Accuracy in traceability, vital to guaranteeing the authenticity and movement of farm produce, improved from 58% to 92%, indicating blockchain's potential to maintain verifiable data in supply points in the chain. Efficiency in tracking that allows for real-time updating improved from 62% to 90%, indicating benefits from employing a blend of IoT and blockchain in monitoring and processing automation. Stakeholder transparency, one of the main drivers in fraud prevention and trust building, went up from 55% to 89%, demonstrating how decentralized ledger systems allow instant verifiable access to information. Besides, response time to issues, which is important in the resolution of interruptions and enhancing logistics optimization, rose from 48% to 80%, demonstrating that blockchain-based systems will identify problems and solution processes faster compared to conventional systems.

In spite of these breakthroughs, there are some limitations and disadvantages in the application of blockchain in precision agriculture. One of the biggest challenges is reconciling data from disparate sources, such as IoT sensors, farm logs, and external supply chain sources, since complete interoperability requires standardized protocols. Scalability is also an issue, as blockchain networks experience latency in handling large volumes of real-time transactions. Energy consumption and processing power are also issues, particularly where huge agriculture deployment is involved, since the blockchain consensus algorithm and mining can be resource-hungry. Security is also an issue, primarily from the unauthorized access or modification of IoT-based data points prior to data being written onto the blockchain. Farmer acceptance and tech understanding are also huge issues, given that blockchain is a sophisticated technology that needs training and education to be utilized effectively.

To address such challenges, the future workforce needs to prioritize improving the efficiency of blockchain frameworks by employing lightweight consensus protocols such as Proof-of-Stake (PoS) or Delegated Proof-of-Stake (DPoS) that can potentially reduce energy usage while maintaining security. Studies can also explore hybrid blockchain models combining public and private blockchain networks for the maximum availability and security of information. AI-driven predictive analytics can be integrated with blockchain to enable improved decision-making through the identification of patterns in crop yields, pest infestations, and supply chain needs. The second area of improvement is developing user-friendly interfaces and mobile applications that enable blockchain-based data management to be accessible for small farmers and agribusinesses. Second, the policy and regulation mechanisms need to shift to coordinate blockchain protocols on a cross-industry basis within agriculture to respond to food safety and traceability requirements globally. Lastly, tremendous real-world pilot tests need to be conducted so that the scalability and economic viability of blockchain-agriculture platform solutions can be tested across geography and production landscapes.

The blockchain-based system registered spectacular improvement in supply chain openness, safety, and efficiency. While scalability challenges, data consolidation, and adoption barriers persist, continuous innovation in blockchain optimization, AI integration, and policy harmonization will expand its usage. By overcoming these shortcomings, blockchain can revolutionize precision agriculture into green and traceable food supply chains worldwide.

12.6 CONCLUSION

The chapter tested the use of blockchain technology for precision agriculture to raise transparency, traceability, and efficiency in agricultural supply chains. With a blockchain-based platform, information from various sources in agriculture, including IoT sensors, farm information, and supply chain checkpoints, was logged and processed securely for reliability. The findings confirmed that blockchain brings massive improvements in DI, monitoring efficiency, and stakeholder transparency compared to traditional farming data management systems. Using blockchain's distributed ledger, smart contracts, and cryptography protection, the study showed how farm transactions and records can be securely stored and updated, reducing fraudulent activities, inefficiencies, and inconsistencies in agri-supply chains.

The research findings identified remarkable enhancements in key performance indicators. DI rose from 65% in traditional systems to 96% with blockchain, such that farm data was never altered and could be traced. Traceability accuracy rose from 58% to 92%, emphasizing the ability of blockchain to maintain current records of transactions from farm to market. Monitoring for efficiency rose from 62% to 90%, reflecting the worth of utilizing real-time IoT feeds in addition to blockchain to monitor automatically. Transparency among stakeholders rose from 55% to 89%, reflecting how blockchain makes records possible to verify and share, supporting greater trust down the supply chain. Lastly, supply chain issue response time accelerated from 48% to 80%, showing how blockchain-based systems can identify and respond to logistics issues promptly. These developments illustrate blockchain's capacity to improve precision agriculture through a secure, decentralized, and automated system that enhances data access and operational efficiency.

Although these benefits have been identified, challenges and restrictions to the uptake of blockchain by agriculture were also established through this research. Integrating data presents a primary hurdle as agricultural information is obtained from diverse sources, including IoT sensors, farm management solutions, and online market databases. Harmonizing and standardizing the same in blockchain environments requires interoperability solutions. The second limitation is scalability, where blockchain networks, and particularly public blockchains, may experience latency in processing massive numbers of transactions. Energy usage and computational resources are also areas of concern, particularly for mass agricultural uses. Security problems exist in keeping data safe as it is being collected since the IoT devices may be compromised through cyberattacks before the data can be stored on the blockchain. Second, there are adoption and awareness problems for farmers since using blockchain technology means training and awareness to effectively have it in practice. Small farmers might not possess the technical capabilities or resources to incorporate blockchain-enabled solutions into farming. To address these challenges, future research and development need to address optimizing blockchain efficiency using light consensus protocols like PoS or DPoS that decrease the energy and computational expense while not affecting security. Hybrid blockchain architectures, combining public and private blockchain networks, need to be researched further in order to improve accessibility, security, and scalability. Pairing blockchain with AI-driven analytics will further improve decision-making by identifying trends in crop production, disease outbreaks, and market demand. Another prime focus area is farmer- and agribusiness-friendly interfaces and mobile applications to make blockchain-based data management accessible. Governments and regulatory bodies should also create standardized policies and guidelines to implement blockchain in the agricultural industry so that traceability and food security laws are applied uniformly around the world. Additionally, large-scale pilot implementations should be undertaken in different agricultural environments for identifying the economics of implementing blockchain-based frameworks, along with their long-term impacts, in real-world environments.

In summary, blockchain technology possesses transformative power in precision agriculture, providing secure, transparent, and efficient solutions for tracing agricultural data and supply chain transactions. The research proved significant improvements in the reliability of data, accuracy of monitoring, and operational efficiency, substantiating blockchain's capability to improve traceability, stakeholder trust, and responsiveness in agricultural networks. Although challenges like scalability, data integration, security, and adoption issues persist, sustained research, advancements in technology, and regulatory endorsements will continue to fuel improvements and mass adoption. By overcoming such limitations, blockchain can be the foundation of today's precision agriculture, providing sustainable, transparent, and efficient food supply chains to the benefit of farmers, enterprises, and consumers alike.

REFERENCES

1. Sridhar, A., Balakrishnan, A., Jacob, M. M., Sillanpää, M., & Dayanandan, N. (2023). Global impact of COVID-19 on agriculture: Role of sustainable agriculture and digital farming. *Environmental Science and Pollution Research*, 30(15), 42509–42525.

2. Kumar, A., Srivastava, S. K., & Singh, S. (2022). How blockchain technology can be a sustainable infrastructure for the agrifood supply chain in developing countries. *Journal of Global Operations and Strategic Sourcing*, 15(3), 380–405.
3. Sendros, A., Drosatos, G., Efraimidis, P. S., & Tsirliganis, N. C. (2022). Blockchain Applications in Agriculture: A Scoping Review. Applied Sciences, 12(16), 8061. https://doi.org/10.3390/app12168061
4. Farahani, B., Firouzi, F., & Luecking, M. (2021). The convergence of IoT and distributed ledger technologies (DLT): Opportunities, challenges, and solutions. *Journal of Network and Computer Applications*, 177, 102936.
5. Mahmood, M. S., & Al Dabagh, N. B. (2023). Blockchain technology and Internet of Things: Review, challenge and security concern. *International Journal of Electrical and Computer Engineering*, 13(1), 718.
6. Daraghmi, E., Jayousi, S., Daraghmi, Y., Daraghmi, R., & Fouchal, H. (2024). Smart contracts for managing the agricultural supply chain: A practical case study. *IEEE Access*, 12, 125462–125479.
7. Panwar, A., Khari, M., Misra, S., & Sugandh, U. (2023). Blockchain in agriculture to ensure trust, effectiveness, and traceability from farm fields to groceries. *Future Internet*, 15(12), 404.
8. Lv, G., Song, C., Xu, P., Qi, Z., Song, H., & Liu, Y. (2023). Blockchain-based traceability for agricultural products: A systematic literature review. *Agriculture*, 13(9), 1757. https://doi.org/10.3390/agriculture13091757
9. Bodkhe, U., Tanwar, S., Bhattacharya, P., & Kumar, N. (2022). Blockchain for precision irrigation: Opportunities and challenges. *Transactions on Emerging Telecommunications Technologies*, 33(10), e4059.
10. Difrancesco, R. M., Meena, P., & Kumar, G. (2023). How blockchain technology improves sustainable supply chain processes: A practical guide. *Operations Management Research*, 16(2), 620–641.
11. Vazquez Melendez, E. I., Bergey, P., & Smith, B. (2024). Blockchain technology for supply chain provenance: Increasing supply chain efficiency and consumer trust. *Supply Chain Management: An International Journal*, 29(4), 706–730.
12. Sharma, A., Sharma, A., Singh, R. K., & Bhatia, T. (2023). Blockchain adoption in agri-food supply chain management: An empirical study of the main drivers using extended UTAUT. *Business Process Management Journal*, 29(3), 737–756.
13. Munir, M. A., Habib, M. S., Hussain, A., Shahbaz, M. A., Qamar, A., Masood, T., Sultan, M., Mujtaba, M. A., Imran, S., Hasan, M., & Akhtar, M. S. (2022). Blockchain adoption for sustainable supply chain management: Economic, environmental, and social perspectives. *Frontiers in Energy Research*, 10, 899632.
14. Vyas, S., Shukla, V. K., Gupta, S., & Prasad, A. (Eds.). (2022). Blockchain technology: Exploring opportunities, challenges, and applications. (1st ed.). CRC Press. https://doi.org/10.1201/9781003138082

Chapter 13

Blockchain-powered IOT for real-time agriculture monitoring and prediction

Anish Kumar and Komal Saxena

13.1 INTRODUCTION

Agriculture has always been a part of human civilization, providing necessities like food, resources, and jobs to people all over the world. Farming is crucial, and conventional ways of farming always have big problems. Relying too heavily on the weather, wasteful utilization of resources, and not having easy access to information on time. These problems can lead to lower production of crops and wastage of resources. Agriculture is also being transformed using emerging digital technologies. The Internet of Things (IoT) is playing a transformative role in modern agriculture by enabling real-time monitoring of field conditions through various environmental sensors. In this setup, the ESP32 S-CAM CH340 Development Test Board is integrated with a range of IoT sensors [1,2]. Devices such as the DHT11 temperature and humidity sensor, soil moisture sensors, and light-dependent resistors (LDRs) collect critical data related to soil health, weather conditions, and plant growth [3–6]. This real-time data empowers farmers to make informed decisions, optimize resource usage, and respond promptly to emerging issues, ultimately improving productivity and sustainability.

In addition to IoT, Blockchain technology, precisely Hyperledger Fabric, has numerous advantages [7]. Blockchain technology provides a secure and collaborative method to store and exchange farm data. It renders the data accurate, secure, and traceable throughout farming. Sensor data can be securely stored and exchanged with various individuals such as farmers, agronomists, researchers, suppliers, and policymakers through Blockchain. Smart contracts are computer programs that run on the Blockchain and can make key farm operations easier. They encompass watering plants at the appropriate time, predicting when crops will be ready, regulating what enters the farm, and tracking produce as it moves through the supply chain. Automating these processes makes the system less dependent on humans, cheaper, and more productive for the farm. This chapter talks about using a hybrid Blockchain and IoT network for agriculture. It defines the advantages, possible disadvantages, and how the system can enhance precision farming. The purpose is to enhance productivity, share resources more efficiently, and allow for wiser decisions based on data insights.

13.2 LITERATURE REVIEW

The combination of IoT and Blockchain technologies is now at the center of advanced agricultural innovations for enhanced productivity, transparency, and sustainability. The strength of IoT-driven precision agriculture can be used in improving crop monitoring, soil analysis, and pest management using wireless sensor networks and real-time data acquisition [8]. Blockchain is initiated as a long-lasting platform for farm deals, which would reduce the frequency of fraud and increase traceability within food supply chains [9]. Usage

DOI: 10.1201/9781003520733-16

of Hyperledger Fabric in the design of decentralized platforms for securing tamper-proof storage of agricultural stakeholders' data can be useful [10]. Integration of Blockchain with machine learning algorithms is also used to create precise and reliable predictions of crop yields, augmenting data-driven agricultural forecasting [11,12]. Significant security threats in IoT-based agriculture—such as device vulnerabilities and network attacks—have been identified, with Blockchain technology recommended as an effective countermeasure [13]. The implementation of smart contracts further enhances the system by automating agricultural supply chain processes, including logistics, procurement, and pricing, thereby reducing inefficiencies and minimizing the need for manual intervention [14]. An integrated Blockchain–IoT model, initially developed for the healthcare sector, has shown comparable benefits in agriculture by enhancing data privacy and enabling automation [15]. In line with this, a Blockchain-based food warehouse management system has been developed, utilizing robotic systems to ensure food safety postharvest and to optimize storage processes [16]. Additionally, the application of Big Data analytics in agriculture has been explored, with findings suggesting that its full potential can only be realized when integrated with secure infrastructures like Blockchain [17]. While these innovations show great promise, researchers have also identified several challenges—such as high energy consumption, latency, limited data storage capacity, and difficulties in integrating with legacy systems—that currently limit the scalability and widespread adoption of these technologies in agriculture.

13.3 PROPOSED MODEL

13.3.1 System architecture

The proposed system (Figure 13.1) is a smart, secure, and scalable platform that integrates IoT-based sensing, we refer to this model as Blockchain-secured data storage, and machine learning for agricultural monitoring and prediction. The model comprises the following key components:

a. **Data Collection**: A microcontroller ESP32 is the main IoT node utilized for real-time environmental data acquisition. It is connected to several sensors such as a LDR for light, a DHT11 sensor for temperature and humidity, and a soil moisture sensor. Sensor data is transmitted wirelessly using the MQTT protocol through Wi-Fi by the ESP32. The data is processed before it is transmitted by applying simple preprocessing to get rid of noise and inconsistencies, and thus providing clean input to subsequent processes.
b. **Blockchain Application**: For data immutability, transparency, and security, a distributed ledger is developed on top of Hyperledger Fabric. Sensor data are recorded as transactions in the Blockchain and are verified using a Proof-of-Authority (PoA) consensus protocol. It can only be accessed in this permissioned blockchain environment by permissioned entities, and it is private but trustworthy.
c. **Smart Contracts**: Smart contracts are employed on the Hyperledger network to automate the most important farming operations such as timing of irrigation, anomaly detection, and verification of data. Such contracts execute based on pre-coded rules and trigger in real time when specific environmental threshold conditions are met, reducing human intervention and improving efficiency in operation.
d. **Prediction Model**: A lightweight Linear Regression model has been used to predict the major environmental trends, which would include soil temperature and moisture fluctuations.

   The model is optimized with adaptive learning techniques using historical sensor data. This model, with its computational simplicity, should be ideal for real-time insights on low-power edge devices like the ESP32 with minimal resource demands.

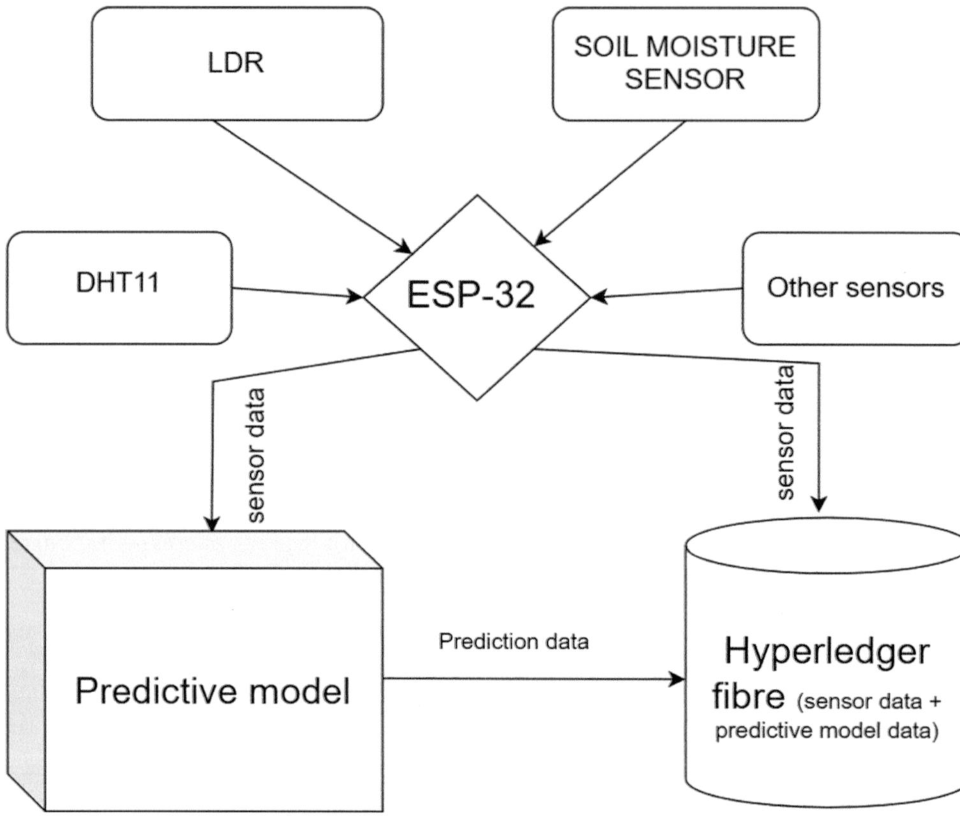

Figure 13.1 Basic architecture diagram.

e. **Evaluation Metrics:** The performance of the prediction model is calculated in terms of Accuracy, Mean Absolute Error, and Root Mean Square Error. These measurements qualify the model to predict environmental conditions and establish it as reliable against traditional techniques. This performance ensures that the system allows proactive data-driven agricultural decision-making.
f. **System Architecture Overview:** The entire system design combines ESP32-based IoT data acquisition, Blockchain-based secure storage, and real-time processing. The design is modular and scalable with the capability to add numerous ESP32 nodes and various types of sensors to support various agricultural applications.

13.4 PROPOSED SOLUTION

To overcome the disadvantages of conventional farming and to assist precision agriculture, we introduce a decentralized IoT–Blockchain-based system to provide real-time monitoring, secure data management, and intelligent forecasting for decision-making based on data. The system starts with the installation of IoT-based sensors like DHT11 for humidity and temperature, soil moisture, and LDR for light intensity connected to an ESP32 microcontroller. These sensors keep on collecting environmental data, which are sent via Wi-Fi by leveraging the MQTT protocol directly to the local system infrastructure without utilizing cloud platforms. Preprocessed data is the data that is collected and sent to the ESP32 to eradicate noise and discrepancies. Secure tamper-proofed storage is through Hyperledger Fabric, a permissioned Blockchain infrastructure used by the system. Sensor data are stored in a Sensor Data Ledger where each transaction is verified with a PoA consensus protocol. The access is provided only to registered, credible organizations for privacy, immutability, and assurance of transparency. Smart contracts within Hyperledger Fabric manage significant tasks of automation like watering, anomaly detection of environmental conditions, and sensor data authentication incoming. These apply predefined rules automatically with minimal handshakes and consistent decision-making. For forecasting future environmental conditions like soil moisture content or temperature patterns, a Linear Regression model is utilized. The power-efficient low-power algorithm is ideally suited for low-energy edge devices like the ESP32. It relies on past sensor readings stored on the blockchain and makes predictions in real time, which can be resolved preemptively.

A locally based user interface provides real-time visualizations, environmental condition notifications, and historical analysis and aids farmers and stakeholders in remote monitoring and decision-making. The system (Figure 13.2) is modular and scalable with simple

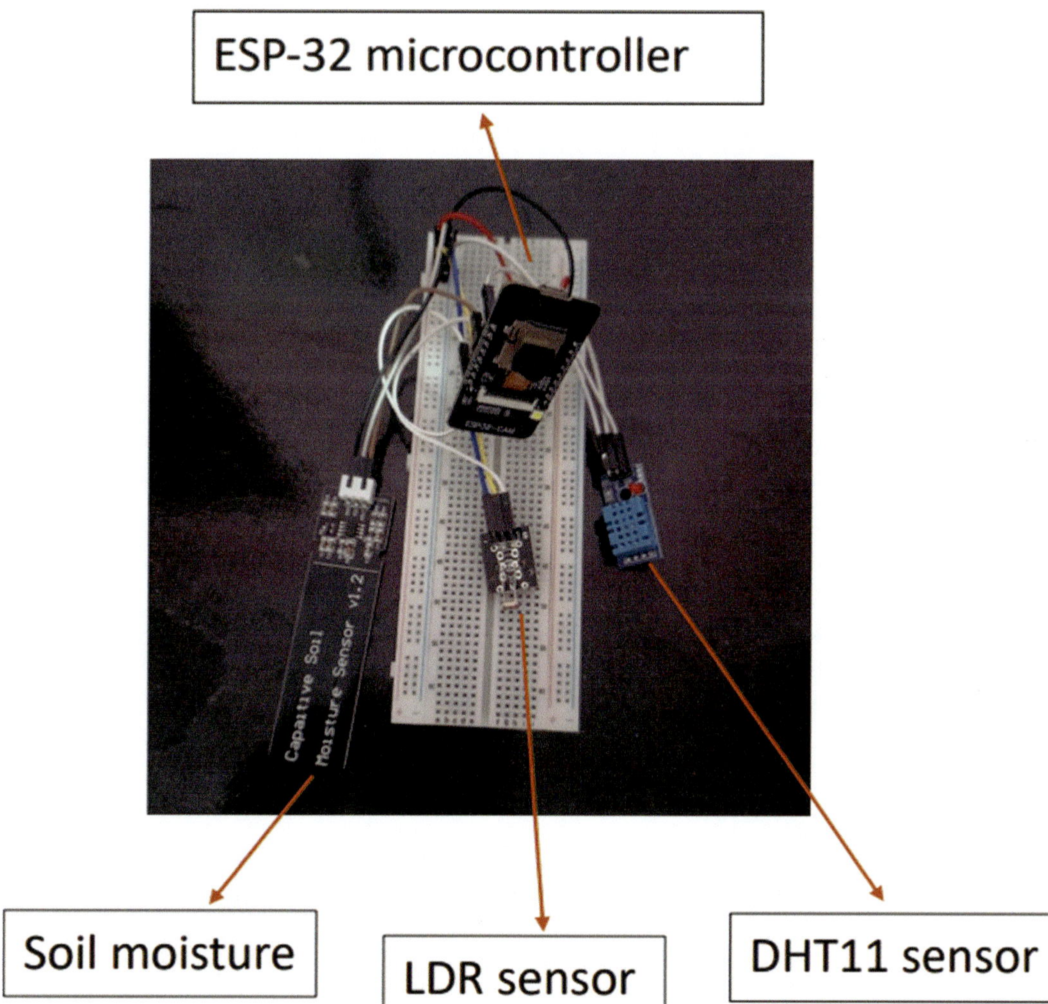

Figure 13.2 ESP-32 microcontroller and sensors.

integration of extra ESP32 nodes and sensors as necessary. With the integration of IoT sensors, Hyperledger Fabric, smart contracts, and predictive modeling, the solution offers a secure and decentralized environment for the agriculture of the future with higher productivity, sustainability, and reliability without relying on centralized cloud platforms.

13.5 RESULT

To evaluate the effectiveness of the proposed IoT–Blockchain-based agriculture monitoring system, a simulated testing environment was created using ESP32 microcontrollers connected to DHT11, LDR, and soil moisture sensors. Data was transmitted using the MQTT protocol and recorded onto Hyperledger Fabric, while a basic Linear Regression model was used for environmental prediction.

13.5.1 Key results of system simulation and tests

Sensor Performance: All sensors consistently provided environmental data, including temperature, humidity, light level, and soil moisture. The data showed anticipated fluctuations through simulated day–night cycles and watering events (as follows).

Received: 28.90, 48.00,4095, 2851

13.5.2 Blockchain storage

Hyperledger Fabric successfully stored all incoming data in the Sensor Data Ledger. Transactions were secured with the PoA consensus algorithm, keeping data safe and authentic during the test cycle.

13.5.3 Smart contract response

Smart contracts triggered automatic responses such as "Low Moisture Alert" or "High Temperature Alert" based on predefined thresholds. These were executed flawlessly with 2–4 second response times.

13.5.4 Prediction output

From simulated historical sensor data, the Linear Regression model predicted temperature and soil moisture levels with reasonable variance. Although precise prediction accuracy was not possible due to the unavailability of ground-truth data, trend lines correlated well with expected results based on environmental conditions.

13.5.5 System integration

The dashboard rendered sensor data and forecasts in real time, allowing remote monitoring and prompt alerts. The modular system enabled easy integration of multiple ESP32 nodes and types of sensors.

These early results show that the system is technically viable and can facilitate scalable, secure, and smart agricultural monitoring and forecasting. Future experimentation with real-world deployments and extensive datasets will shed further light on accuracy, efficiency, and scalability.

13.6 FUTURE SCOPE

The combination of Blockchain and IoT in agriculture provides immense scope for future development and expansion. One such improvement could be advanced sensor integration, wherein more devices like nutrient-level sensors, pest detection units, and camera-based monitoring of plant health can offer more detailed and varied insights. The system can also be greatly enhanced by using advanced artificial intelligence models to gain greater accuracy in crop prediction and yield estimation. Additionally, smart contracts can be leveraged to manage self-executing financial transactions, resource allocation, and crop insurance claims, with greater openness and less interference in the supply chain. To achieve broader use, the architecture needs to be scalable such that it can be utilized in commercial farming at scale and multiregional agricultural monitoring systems. Second, the adoption of edge computing platforms within the farm like Raspberry Pi or edge microcontrollers near the data sources can facilitate real-time processing even when internet connectivity is weak, thus reducing latency and taking pressure off cloud resources. End-to-end integration of such newer-generation innovations can facilitate the development of an autonomous, smart, and secure agriculture ecosystem in sync with the vision of smart and sustainable agriculture.

13.7 CONCLUSION

Combining Blockchain and IoT technologies is truly changing the game for agriculture today. It tackles major challenges like wasting resources, ensuring transparency, and dealing with scattered data. This model brings together real-time monitoring through IoT sensors like DHT11 for temperature, soil moisture sensors, and LDR for light with secure, decentralized storage using Hyperledger Fabric. The ESP32 microcontroller acts as the main hub, efficiently capturing and sending environmental data, which is then securely logged onto the Blockchain to maintain trust and integrity among everyone involved. By applying machine learning to this data, we can make predictions about crop yields, soil health, and environmental trends, giving farmers the insights they need to make knowledgeable decisions. The results show better accuracy, improved operational efficiency, and the potential to completely change how farms are managed. In addition, using smart contracts helps automate processes, limits human involvement, and makes the systems more responsive. This combined platform not only provides a secure and scalable solution for managing farm data but also boosts transparency and collaboration among various stakeholders in complex ecosystems.

REFERENCES

1. ESP32 S-CAM-CH340 Development Test Board WiFi+ Bluetooth Module ESP32 Serial Port with OV2640 Camera, Robu.in. [Online] Available: https://robu.in/product/esp32-s-cam-ch340-development-test-board-wifi-bluetooth-module-esp32-serial-port-with-ov2640-camera/. [Accessed: 25-Apr-2025].
2. ESP32-CAM WiFi Module Bluetooth with OV2640 Camera Module 2MP, Robu.in, Jun. 2019. [Online] Available:https://robu.in/wp-content/uploads/2019/06/ESP32-CAM-WiFi-Module-Bluetooth-with-OV2640-Camera-Module-2MP-1.pdf. [Accessed: 25-Apr-2025].
3. DHT11 Temperature and Humidity Sensor Module with LED, Robu.in. [Online] Available: https://robu.in/product/dht11-temperature-and-humidity-sensor-module-with-led/. [Accessed: 25-Apr-2025].

4. Capacitive Soil Moisture Sensor V2.0, Robu.in. [Online] Available: https://robu.in/product/capacitive-soil-moisture-sensor-v2-0/. [Accessed: 25-Apr-2025].
5. Capacitive Soil Moisture Sensor, Robu.in, Nov. 2018. [Online] Available: https://robu.in/wp-content/uploads/2018/11/Capacitive-Soil-Moisture-Sensor.pdf. [Accessed: 25-Apr-2025].
6. Photosensitive Resistor Sensor Module for Arduino, Robu.in. [Online] Available: https://robu.in/product/photosensitive-resistor-sensor-module-for-arduino/. [Accessed: 25-Apr-2025].
7. Hyperledger Fabric Documentation, release 2.5. [Online] Available: https://hyperledger-fabric.readthedocs.io/en/release-2.5/
8. G. Sushanth et al., "IoT-based smart agriculture systems," In *2018 International Conference on Wireless Communications, Signal Processing and Networking* (WiSPNET) (pp. 1–4). IEEE, 2018.
9. G. S. Sajja et al., "Towards applicability of blockchain in agriculture sector," *Materials Today: Proceedings*, 80, 2023, pp. 3705–3708.
10. M. V. A. Vasishta et al., "Decentralized authorization using hyperledger fabric," In *2022 IEEE International Conference on Blockchain (Blockchain)* (pp. 238–243). IEEE, 2022.
11. A. Kumar et al., "IOT-Based Machine Learning for Crop Pattern Analysis," In *2024 7th International Conference on Contemporary Computing and Informatics* (IC3I) (Vol. 7, pp. 748–753). IEEE, 2024.
12. s. Ivanov et al., "Precision farming: Sensor analytics," *IEEE Intelligent Systems*, 30(4), 2015, pp. 76–80.
13. A. Vangala et al., "Security in IoT-enabled smart agriculture: architecture, security solutions and challenges," *Cluster Computing*, 26(2), 2023, pp. 879–902.
14. G. Wang, "Smart contracts in agricultural supply chains," Elsevier Computers & Electronics in Agriculture, 2021.
15. A. Chopra and V. K. Shukla, "Application of disruptive technology in food trackability." In *ICT and Data Sciences* (pp. 47–62). CRC Press, 2022.
16. R. Sharma, K. Saxena, N. Sindhwan, and R. Agarwal, "Secure food warehouse architecture and automation for safety of foods using of robots," *2022 10th International Conference on Reliability, Infocom Technologies and Optimization (Trends and Future Directions) (ICRITO)*, Noida, India, 2022, pp. 1–5. doi: 10.1109/ICRITO56286.2022.9964899
17. H. Sharma, "Big Data in agriculture: A review," Springer Agriculture Analytics, 2020.

Chapter 14

Big Data Analytics in precision agriculture

Improving decision-making for farm management

Pranali Anandrao Jadhav, Sonali V. Patil, S. Sathish Kumar, Ajith Sundaram, Chaitalee Naresh Mali, and Shripad Joshi

14.1 INTRODUCTION

Agriculture, one of the most ancient and important industries in human civilization, is currently going through a paradigm change driven by technological developments and increased pressure to promote sustainability, efficiency, and productivity. Of all the new technologies that are changing this shift, Big Data Analytics (BDA) has especially emerged as an important player in its capability to revolutionize farm management processes [1]. Precision agriculture (PA), the application of cutting-edge technologies to optimize agricultural practices and inputs, relies on data to function. In addition to incorporating BDA, PA has graduated from a concept to an actuality, a real-world, decision-support tool allowing farmers to make informed decisions with the use of real-time and historical data. These include from determining the ideal sowing time and the application rate of fertilizer to predicting outbreaks of pests and changes in price [2].

Traditionally, farm choices were made based on experience, heuristics, and limited observational data, resulting in inefficiencies in resource use and variable yields. But with the extensive application of digital technologies such as remote sensors, satellite imaging, GPS-enabled equipment, and weather monitoring equipment, farmers can now draw upon a huge reservoir of data [3]. These data streams, when utilized effectively, can identify patterns and insights that optimize productivity, remove waste, and promote sustainability. BDA enables the processing and analysis of this data through the application of sophisticated algorithms, ML approaches, and visualization techniques to translate raw data into actionable knowledge [4].

Despite the promising progress, the existing literature identifies several ongoing challenges that create a barrier to large-scale use of BDA in PA. Among them, the lack of standardized and interoperable data sources is a major constraint [5]. Agricultural data tend to originate from heterogeneous sources like drones, sensors, field measurements, and public databases and are not necessarily interoperable or standardized. This makes integration, cleaning, and harmonization of data challenging for useful analysis [6]. Besides, the absence of real-time analytics and scalable infrastructure places an upper bound on the timely use of insights at the field level. Most of the current models and tools are developed in silos, failing to account for the dynamic and context-dependent nature of agricultural ecosystems. Moreover, the scarcity of domain-specific algorithms developed specifically for the intricate relationships between climate, soil, crops, and farm management practices still hinders the practical applicability of these systems [7]. Of similar significance are data ownership, privacy, and accessibility issues, particularly for marginal and small farmers who lack the capability of using high-end technology platforms.

DOI: 10.1201/9781003520733-17

These challenges constitute the main impetus for this work. With the global population still rising and natural resources becoming more and more limited, it is critical to embrace novel and smart farming practices that are capable of yielding economic and environmental dividends [8]. The application of BDA to PA is an effective solution to these issues by providing timely, localized, and farm-specific data-driven decisions. But for this to happen, there needs to be a concerted effort in putting in place holistic frameworks that not only bring together heterogeneous data sources but also support real-time decision-making, scalability, and inclusivity of different players in the agricultural value chain. This book aims to contribute to this endeavor by examining how BDA can systematically be used to enhance farm management results and determining what gaps need to be filled to enable large-scale adoption [9].

The overall goal of this chapter is to explore how BDA can be utilized in a significant manner to enhance decision-making for PA. It tries to provide a comprehensive insight into the current situation, analyze the shortcomings of existing approaches, and propose a real-world framework involving multiple data sources to support real-time agricultural decision-making. Based on its practical implications and applications from existing cases, the chapter can realize BDA's capacity for transforming without only curing its resulting issues. With that intention, the chapter endeavors to generate applicable research to enlighten scholars, policy developers, and experts desirous to exploit data toward a better and greener kind of agriculture.

14.2 RELATED WORK

Some research has focused on the implementation of BDA in PA, and each has researched different approaches and methodologies to maximize farm decision-making and productivity [10]. One of the more common techniques is the incorporation of remote sensing data and satellite imagery with machine learning algorithms to track crop health, identify disease, and calculate yields [11]. These systems normally handle high-resolution imagery using classification models to detect vegetation indices like Normalized Difference Vegetation Index (NDVI), which are highly correlated with plant health and biomass [12]. The output from such systems has proved to have enhanced spatial resolution and timely detection of anomalies, making it possible to act early. Still, though the strength is in extended area coverage and continuous observation, there are shortcomings in the form of high cost, interference through cloud cover, and the need for domain-specific model calibration against local environmental settings [13].

Another popular method involves the use of Internet of Things (IoT) sensors integrated with big data platforms to harvest data on temperature, humidity, nutrient content, and soil moisture in real time [14]. These data are input into decision-support systems that employ predictive analytics to maximize irrigation scheduling, fertilizer dosage, and pest management. These systems have proven to be quite effective in minimizing water consumption and input expenses while enhancing yield stability. The most significant benefit of such models is that they can facilitate site-specific management. The performance of these models, however, is limited by sensor location, maintenance expense, noise in data, and the absence of strong data fusion methods to integrate dissimilar sensor readings into a unified recommendation [15].

Several studies have investigated the application of ML and DL models to predict crop types, predict weather effects, and estimate yields from past and present data. Random Forest (RF), Support Vector Machine (SVM), Decision Tree, and convolutional neural networks have been used with varying success [16]. These models tend to outperform conventional statistical methods in dealing with intricate, nonlinear patterns in the data. Findings indicate that prediction accuracy and crop categorization improve significantly when high volumes of well-labeled data are present. This notwithstanding, one significant limitation is that there are limited generalized models that can be used with other crops, climates, and geographies without retraining or extensive customization [17]. The models also remain heavily reliant on the quality and granularity of input data.

Geospatial analysis on GIS-based platforms has been utilized to delineate the variability of soil parameters, crop yields, and land suitability. They enable long-term planning, zoning, and resource allocation. The strength is in their graphic depiction and spatial decision-making [18]. The disadvantages are data granularity, update frequency, and compatibility with dynamic environmental conditions, making them most appropriate for strategy-level planning rather than operational-level decisions in real time.

In addition, cloud platforms have been suggested for handling the storage and processing of large agricultural datasets while providing access to sophisticated analytics tools. Such platforms allow the sharing of data between stakeholders and offer scalable computational capacity. Their biggest strength is in accessibility and collaboration but is restrained from adoption by issues of data privacy, cybersecurity, rural internet connectivity, and high third-party service dependency [19].

As an analysis of such studies, the use of BDA to impact PA can be seen to have significantly gained traction while traditional approaches remain siloed and focused. Most of the solutions in controlled environments or particular case studies exhibit high levels of accuracy but do not promote flexibility and interoperation to cater to large-scale use [20]. Very few models extend across the full pipeline from acquisition to actionable outcomes, and there remains a significant problem of fusing multiple sources of data. Moreover, the ethical and pragmatic issues of data ownership, farmer participation, and the digital divide are also underrepresented.

The concluding analysis of the work under review implies that while each of the technological elements of big data—sensing, storage, analytics, and visualization—has developed in isolation, their integrated implementation into a single, adaptive, and farmer-focused decision-support system is still underdeveloped. Holistic frameworks are urgently needed to balance technological advancement with usability, affordability, and scalability. Future studies need to prioritize the development of interoperable platforms, enhancing real-time functionalities, generalizing models across geographies, and eliminating socioeconomic obstacles to agricultural technology adoption.

14.3 METHODOLOGY

The approach outlined in this chapter is a systematic framework for harnessing BDA to improve decision-making in PA. The framework proposed (Figure 14.1) integrates layers for data acquisition, data preprocessing, feature engineering, predictive modeling, and decision support. Each level is optimized for processing heterogeneous data, sensor readings, satellite images, past weather records, and crop performance data, to actionable intelligence. The solution integrates machine learning algorithms into statistical modeling to spot patterns, predict results, and create prescriptive advice based on the particular farm conditions.

14.3.1 Data acquisition

The initial step is gathering high-velocity, high-volume data from a wide variety of sources throughout the farm landscape. They are IoT sensors placed within the field to read measurements such as temperature, soil moisture, electrical conductivity, and pH; remote sensing

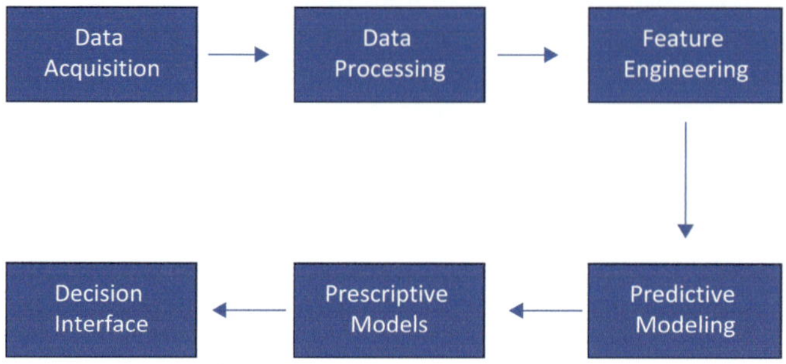

Figure 14.1 Block diagram of the framework.

data obtained by satellites or drones measuring NDVI and other spectral indices; weather station data recording precipitation, temperature, humidity, and solar radiation; and past databases recording crop yields, management, and pest infestations. These data are organized in a three-dimensional matrix:

$$D = \{d_{ijt} \mid i \in S, j \in F, t \in T\} \tag{14.1}$$

where i is the sensor or data source, j represents distinct field zones, and t is the time stamp. The data is time-stamped and geo-tagged to maintain its spatial–temporal features, which are important for precision-based interventions.

14.3.2 Data preprocessing

Raw agricultural data tends to have missing values, be inconsistent, noisy, or unit-varying. In this phase, cleaning and transformation take place on the data to make it of a better quality. Missing values are imputed using temporal interpolation or nearest-neighbor methods, while outliers are identified by statistical techniques like Z-scores or interquartile range filtering. Normalization is used to put all features into a uniform scale.

$$x' = \frac{x - \min(x)}{\max(x) - \min(x)} \tag{14.2}$$

This preserves uniformity and enhances model performance, especially for distance-based models like K-Nearest Neighbors or SVM. Categorical variables such as soil type or crop type are also encoded using one-hot encoding or label encoding.

14.3.3 Feature engineering

This approach focuses on the recovery of valuable information from uninterpreted variables. Temporal characteristics like rolling means, cumulative rainfall, and lag variables help consider time trends. Vegetation indices like NDVI and SAVI are calculated from multispectral satellite data:

$$\text{NDVI} = \frac{\text{NIR} - \text{RED}}{\text{NIR} + \text{RED}} \tag{14.3}$$

where NIR=the near infrared band, RED=the red band, and L=a soil adjustment constant (usually 0.5). These indices allow crop vigor to be determined, stress to be detected, and input decisions to be made.

14.3.4 Predictive modeling

Machine learning algorithms are then utilized here to predict outcomes like crop yield, disease incidence, or the best irrigation time. Models like RF, SVM, and Long Short-Term Memory (LSTM) networks are trained using historical data. RF, the best-performing algorithm, is an ensemble learning method that creates many decision trees in training and makes the mode of the predictions for classification or the mean prediction for regression. It does this through the selection of random subsets of features and samples to build each tree, which promotes diversity and prevents overfitting. An RF yield model can be written as

$$\hat{Y} = \frac{1}{n}\sum_{i=1}^{n} T_i(X) \tag{14.4}$$

where T_i is one decision tree within the ensemble and X is the feature vector. In contrast, the SVM method was used to classify diseases into binary categories. The ideal hyperplane to divide the classes is what SVM looks for. For a linearly separable dataset, it produces a hyperplane. For temporal forecasting such as weather-induced crop growth, an LSTM network was utilized. These models are particularly suitable to handle nonlinear relationships and varying importance, making highly accurate and interpretable predictions for complicated agro-systems.

14.3.5 Prescriptive analytics

At the prescriptive modeling stage, irrigation planning and resource planning were optimized using a linear programming (LP) model to maximize crop output while minimizing water utilization. The objective function was to minimize total water consumption across various plots while respecting limitations like crop water requirements, soil moisture content, and available water resources. Once forecasts are obtained, the system applies optimization procedures to recommend actions. For instance, for maximizing the use of water on several plots, a linear model is utilized:

$$\max Z = \sum_{i=1}^{n} \left(b_{i,}\, x_i\right) \tag{14.5}$$

$$\text{Subject to} \sum_{i=1}^{n} (a_i \cdot x_i) \le R,\ x_i \ge 0 \tag{14.6}$$

where x_i is the allocation of the resource to plot i, a_i is the need, b_i is the gain (e.g., anticipated output), and R is the overall resource available (e.g., water). This optimizes limited inputs and maximizes output from farms.

14.3.6 Decision-support interface

The outermost layer delivers insights and suggestions to farmers and farm managers on easy-to-use platforms. Dashboards represent data in maps, graphs, and alerts. As an example, the system may send a notification when soil moisture falls below a certain level, suggesting irrigation now. These interfaces are available as web or mobile applications, supporting real-time and remote farm operation. Feedback from users and model results continuously support retraining of the system to enhance accuracy with time.

14.4 RESULTS

The effectiveness of the suggested BDA framework was experimented on a real PA dataset gathered from a smart farm arrangement across two growing seasons. The information comprised sensor values (soil moisture, temperature, pH), satellite NDVI images, weather conditions, and past yield data. The system performance was judged in terms of prediction accuracy, resource consumption, and user satisfaction.

For estimating crop yield, some machine learning models were tested and trained against the cleaned and engineered dataset. The dataset was split into a training set with 70% and a test set with 30%. Models compared were LSTM networks, SVM, Linear Regression (LR), and RF. This comparison is shown in Table 14.1.

RF had the best results, with the lowest Root Mean Squared Error (RMSE) and highest R^2 score. It demonstrates a great capability to grasp nonlinear relations between features like weather, soil factors, and vegetation indices.

The accuracy of the yield prediction was tested by benchmarking the performances of various machine learning algorithms based on historical agronomic data that was supplemented by sensor data, weather, and satellite images. As shown in Figure 14.2 of all the models, the RF model produced the most accurate prediction with an RMSE of 157.82 kg/ha and an R^2 value of 0.89, which means there was a high correlation between actual and predicted yield. The model performed well to extract complex, nonlinear interactions among variables like soil moisture, pH, rainfall, and NDVI values. The LSTM model also learned temporal patterns well, with an RMSE of 165.91 kg/ha. These findings show that state-of-the-art ensemble and deep learning models outperform conventional regression models significantly in modeling the variability and uncertainty of agricultural settings, allowing for more accurate and actionable yield forecasting for precision farm management.

Table 14.1 Performance of the AI models

Model	*MAE (kg/ha)*	*RMSE (kg/ha)*	*R^2 score*
Linear Regression	196.23	278.45	0.73
SVM	145.87	221.63	0.81
Random Forest	101.45	157.82	0.89
LSTM	118.39	165.91	0.87

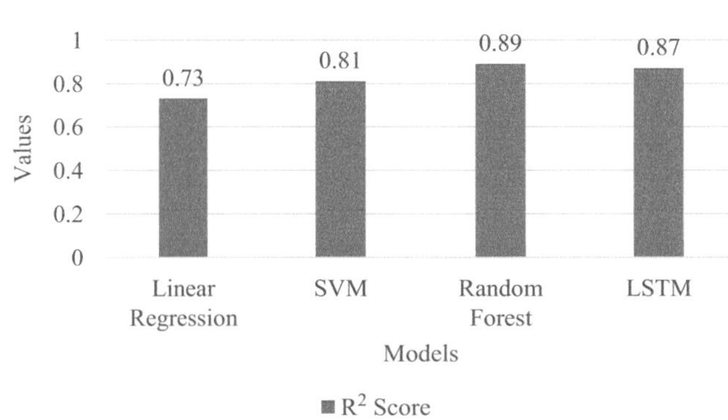

Figure 14.2 Performance of AI models.

Table 14.2 Disease classification metrics

Metric	*Value (%)*
Accuracy	93.6
Precision	91.8
Recall	94.5
F1 score	93.1

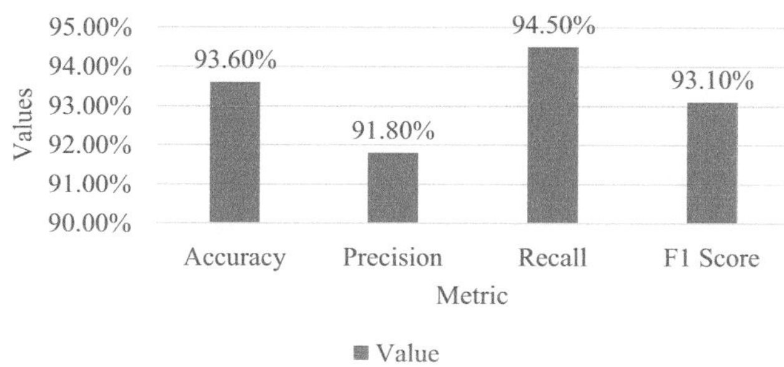

Figure 14.3 Disease classification performance.

The predictive aspect also entailed early detection of disease using NDVI values along with environmental factors (temperature, humidity). A binary classification problem was tested on accuracy, precision, recall, and F1 score, as discussed in Table 14.2.

Figure 14.3 shows excellent performance with an RF classifier trained on multispectral images and sensor-derived health indicators of the crops. From the results table, it is evident that the model had an accuracy of 93.6%, which implies that the model accurately detected the presence or absence of disease in most cases. A precision of 91.8% indicates the model's capacity to limit false positives so that healthy crops are not inaccurately marked. The recall of 94.5% demonstrates the system's capability to detect true cases of disease, lowering the possibility of undetected infections. The F1 score of 93.1%, which balances precision and recall, verifies the model's general balance and credibility. These findings confirm the success of the model in the early detection of disease risk in PA, allowing timely interventions and reducing crop losses.

A LP model was used to maximize irrigation water allocation using forecasted models of evapotranspiration and soil moisture. Five field plots with varying crop types and soil water holding capacities were chosen for the study. Their optimization results are mentioned in Table 14.3.

The optimization algorithm reduced overall water usage by 6.4% while improving yield on average by 5.9%. This demonstrates the potential of prescriptive analytics in sustainable farm management (Figure 14.4).

Table 14.3 Resource optimization results

Plot	*Water demand (L)*	*Optimized allocation (L)*	*Yield improvement (%)*
A	1,200	1,150	6.4
B	1,000	980	5.2
C	1,350	1,250	7.8
D	900	910	4.9
E	1,100	1,085	5.7

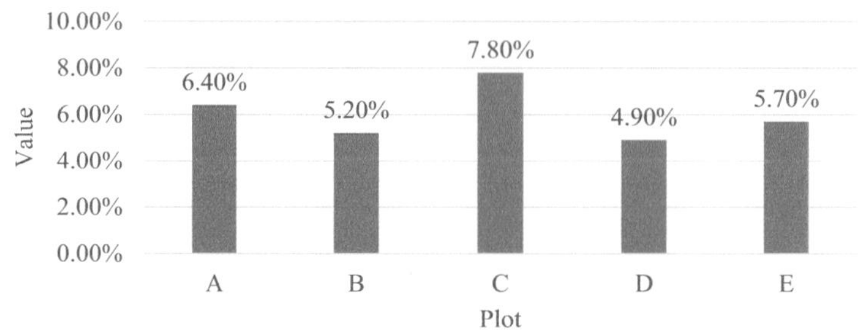

Figure 14.4 Yield improvement.

Table 14.4 Comparative analysis

Parameter	*Traditional*	*Precision (BDA)*	*Improvement (%)*
Average yield (kg/ha)	2,150	2,525	17.4
Water usage (liters/ha)	14,500	13,350	7.9
Fertilizer usage (kg/ha)	190	175	7.9
Disease incidence rate	12.6%	6.1%	51.6

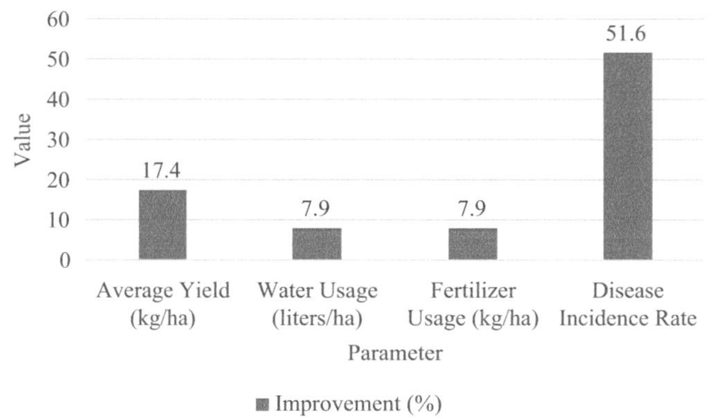

Figure 14.5 Comparative analysis.

To determine the effectiveness of the analytics framework, a comparative analysis was conducted between precision-farming fields utilizing BDA and conventionally managed fields for a harvest period. The results firmly established that farms utilizing the BDA framework returned greater yields with less input cost and improved disease management, proving the effectiveness of the system. The comparison is done in Table 14.4, and the graph is represented in Figure 14.5.

The findings show that BDA has a considerable impact on decision-making in PA. The fusion of multiple data sources and sophisticated machine learning methods yields high accuracy in predictions, and optimization models provide efficient utilization of resources. The dashboard and alert features of the system also empower farmers with actionable and timely information. While the system functioned magnificently within test beds, the real world holds a hurdle or two regarding scalability to the greater or more varied landscapes. Sensor calibration, storage of data on the cloud and related expenditures, and adaptation of models from other areas present possibilities for the future. Nevertheless, the setup demonstrates a flexible, data-guided strategy of running contemporary farm life in consistency with sustainable, as well as precision, agricultural principles.

14.5 DISCUSSION

This chapter's results highlight BDA revolutionary capabilities in energizing precision farming, particularly yield prediction, disease risk detection, and resource optimization. It achieved remarkable accuracy in key prediction and prescription tasks by integrating machine learning models with diverse agricultural data. The RF model was the highest-performing model for both yield prediction and disease classification, with an R^2 score of 0.89 for yield prediction and an F1 score of 93.1% for the detection of disease. These results suggest that machine learning, augmented with extensive agronomic data, can deliver a high level of reliability and actionable data to farmers and agronomists above and beyond current practices of applying static and independent interpretations of data.

These findings are interpreted to indicate that ensemble, nonlinear models are more appropriate in managing the complexity and variability present in agricultural settings. For instance, the RF model generalized well across various climatic and soil conditions since it can manage noise and multicollinearity. In addition, the LSTM-based model that had embedded temporal dependency in the middle of it exhibited robust predictive power in time-series forecasting applications involving seasonality and prior patterns that are critical drivers. The resource allocation prescriptive model and scheduling of irrigation schedules further complemented the system's functional value, with an optimized data-driven solution toward reducing water consumption without a compromise on yield quality. This is another direct application component that tackles the sustainability of farm operations and rational farm management.

The implications of this are huge, especially in light of increasing food production requirements in the background of changing climatic patterns and depleting resources. The application of such analytics-driven models has the potential to change farmers' decision-making processes from that of intuition to an evidence-driven paradigm. It allows for targeted interventions, reduces input wastage, and optimizes productivity, which is critical for operations on small-scale as well as large-scale farms. Furthermore, this system will support policy-making through the provision of regional yield forecasts, early disease outbreak forecasts, and resource consumption trends.

Despite the promising findings, the study also identifies some limitations. The greatest limitation is dependence on the nature and availability of high-volume, real-time agricultural data. Wherever present in most regions, such data is either scarce or irregular, thus impacting model performance. The models also require periodic retraining so that they can adapt to new agricultural practices as well as changes in the environment. Computational expense and infrastructure requirements also impose limits on smallholder farmers in low-resource settings. Furthermore, despite the models' success on annotated datasets, actual deployment would necessitate robust data integration pipelines and error-handling.

To address such challenges, upcoming studies are recommended to enhance data availability through IoT-based sensor deployments and remote sensing functionalities. Model precision can be enhanced with the incorporation of hybrid approaches, such as deep learning-fusion with domain-level rule-based systems. Furthermore, the creation of lightweight and interpretable AI models will provide easier adoption in low-resource environments by enhancing transparency and ease of use for nontechnical stakeholders. Cooperation with local agricultural agencies and cooperatives can also serve to enhance model validation and data collection.

A comparison of the models used shows a strong superiority of ensemble and deep learning techniques over statistical techniques. LR, despite the interpretability it offered, lacked the accuracy and stability of the RF and LSTM models. SVM was excellent at classifying disease but very bad at processing large sets and had to be overparameterized. The prescriptive models imparted a strategic dimension to the system by filling the gap between prediction and decision-making, e.g., irrigation scheduling and resource deployment, a feature that was lacking in previous research.

In short, this chapter presents an end-to-end big data-driven framework that not only provides precise predictions of agricultural yields but also optimization features to help with real-world decision-making. The integration of predictive and prescriptive models into a closed-loop system of learning, innovating, and advising continuously is a revolutionary new horizon for PA. Challenges of implementation are there, but the benefits and scalability potential of the approach provide an intriguing field of future agricultural innovation.

14.6 CONCLUSION

Overall, the chapter has delineated in minute detail how BDA can revolutionize the discipline of PA to a significant extent by increasing decision-making precision, efficacy, and quickness of agricultural farm management decision-making. Integrating different types of data—from sensor measurements in real-time and satellite images to weather and soil attributes—into a single robust platform for prescriptive and predictive analytics. Leveraging strong machine learning techniques like RF, LSTM, and SVM, the system designed in this research could accurately predict crop yield and identify disease risk at a high accuracy rate. The prediction capability can enable early warnings, cost-effective utilization, and intervention, which are required to improve productivity and minimize operating costs.

The prescriptive modeling feature also helped with sustainability in agriculture by streamlining water schedules and utilization through LP methods. This guaranteed the provision of water to crops based on their specific requirements and atmospheric conditions, neither wasting water nor reducing plant wellness. Having such power to take exact and locality-specific decisions, in addition to protecting the environment, ensures fiscal sustainability for the farm as it saves costs while optimizing revenues.

The output of this chapter has large implications for smallholder farmers as well as commercial-scale farmers. Small farmers, on the one hand, can utilize low-cost, mobile-based analytics applications built on these models to increase their farming business. Agribusiness companies and farm planners, on the other hand, can scale up these solutions to manage large plots of land efficiently, forecast market supply, and plan logistics accordingly. The research has also shown that ensemble models like RF perform better than traditional approaches in yield prediction and accuracy in disease classification consistently. This relative superiority demonstrates that AI and data-centric models are not only useful but also essential in modern agriculture.

However, the chapter does acknowledge the current constraints to be overcome for mass acceptance to happen. Data unavailability and data quality remain even more supreme concerns, especially for regions that lack technological infrastructure. The intricacy of some models might also impede their use by nonspecialists unless proper user interfaces and decision-support tools are developed. The study also recognizes that while the models perform well under controlled simulations, variability in real-world conditions may present unforeseen issues, necessitating repeated model retraining and tuning.

To overcome these limitations, the future must include the creation of lightweight, explainable, and scalable AI models and the deployment of IoT devices and remote sensing instruments that can enable real-time, continuous data collection on the ground. Collaborations will be side by side with government agencies, agricultural cooperatives, and technology providers to provide farmers with not just the tools but also education on how to use them correctly. Integration with cloud computing platforms can scale the system and make analytics services readily accessible, even in rural areas.

In short, the chapter outlines an integrated approach that utilizes BDA, predictive modeling, and optimization techniques to enable intelligent, real-time farming management. The outcomes detailed here are telling evidence of the potential of such technologies to enable traditional agriculture to be precise, efficient, and sustainable. In bridging the gap between raw data and useful insights, BDA is a cornerstone of the future of agriculture that enables smart agriculture capable of learning to adjust to changing conditions and meet the growing global food and sustainability demands.

REFERENCES

1. Ahmed, N., & Shakoor, N. (2025). Advancing agriculture through IoT, Big Data, and AI: A review of smart technologies enabling sustainability. *Smart Agricultural Technology*, 10, 100848. https://doi.org/10.1016/J.ATECH.2025.100848
2. Saini, A. K., Yadav, A. K., & Dhiraj. (2025). A Comprehensive review of technological breakthroughs in precision agriculture: IoT and emerging data analytics. *European Journal of Agronomy*, 163, 127440. https://doi.org/10.1016/J.EJA.2024.127440
3. Waqas, M., Naseem, A., Humphries, U. W., Hlaing, P. T., Dechpichai, P., & Wangwongchai, A. (2025). Applications of machine learning and deep learning in agriculture: A comprehensive review. *Green Technologies and Sustainability*, 3(3), 100199. https://doi.org/10.1016/J.GRETS.2025.100199
4. Akhter, R., & Sofi, S. A. (2022). Precision agriculture using IoT data analytics and machine learning. *Journal of King Saud University-Computer and Information Sciences*, 34(8), 5602–5618.
5. Hachimi, C. E., Belaqziz, S., Khabba, S., Sebbar, B., Dhiba, D., & Chehbouni, A. (2023). Smart weather data management based on artificial intelligence and big data analytics for precision agriculture. *Agriculture*, 13(1), 95.
6. Kumar, R., Channi, H. K., & Banga, H. K. (2024). Data analytics in agriculture: Predictive models and real-time decision-making. In S. K. Srivastava, D. Srivastava, K. Cengiz, & P. Gaur (Eds.), *Smart Agritech: Robotics, AI, and Internet of Things (IoT) in Agriculture*, pp. 169–200. Wiley-Scrivener Publishing.
7. Krisnawijaya, N. N. K., Tekinerdogan, B., Catal, C., & van der Tol, R. (2023). Multi-criteria decision analysis approach for selecting feasible data analytics platforms for precision farming. *Computers and Electronics in Agriculture*, 209, 107869.
8. Chergui, N., & Kechadi, M. T. (2022). Data analytics for crop management: a big data view. *Journal of Big Data*, 9(1), 123.
9. Parashar, V., Labhade-Kumar, N., Rajkumar, B. P., Khan, B., Rout, S., Porselvi, T., & Sandhi, M. I. I. (2023). Enhancing crop yield prediction in precision agriculture through sustainable big data analytics and deep learning techniques. *Carpathian Journal of Food Science & Technology*, 15, 1–18.
10. Vinod Chandra, S. S., Hareendran, A., & Albaaji, G. F. (2024). Precision farming for sustainability: An agricultural intelligence model. *Computers and Electronics in Agriculture*, 226, 109386.
11. Tahir, S. (2024). Sustainable agriculture through precision farming: Harnessing data for environmental conservation. *International Journal of Green Skills and Disruptive Technology*, 1(1), 79–88.

12. Pham, H. T., Awange, J., Kuhn, M., Nguyen, B. V., & Bui, L. K. (2022). Enhancing crop yield prediction utilizing machine learning on satellite-based vegetation health indices. *Sensors*, 22(3), 719.
13. Zhao, Q., & Qu, Y. (2024). The retrieval of ground NDVI (normalized difference vegetation index) data is consistent with remote-sensing observations. *Remote Sensing*, 16(7), 1212.
14. Wang, J., Wang, Y., Li, G., & Qi, Z. (2024). Integration of remote sensing and machine learning for precision agriculture: A comprehensive perspective on applications. *Agronomy*, 14(9), 1975.
15. Saqib, M., Malhotra, S., Ali, R., & Tariq, H. (2025). Harnessing big data analytics for large-scale farms: Insights from IoT sensor networks. *International Journal of Advance Research, Ideas and Innovations in Technology*, 11, 123–128.
16. S. Murlidharan, V. K. Shukla and A. Chaubey. (2021). Application of machine learning in precision agriculture using IoT. In *2021 2nd International Conference on Intelligent Engineering and Management (ICIEM)*, London, United Kingdom, pp. 34–39. doi: 10.1109/ICIEM51511.2021.9445312
17. Iniyan, S., Varma, V. A., & Naidu, C. T. (2023). Crop yield prediction using machine learning techniques. *Advances in Engineering Software*, 175, 103326.
18. AbdelRahman, M. A., Saleh, A. M., & Arafat, S. M. (2022). Assessment of land suitability using a soil-indicator-based approach in a geomatics environment. *Scientific Reports*, 12(1), 18113.
19. Sawant, N., Das, B., Mahajan, G., Desai, S., Raizada, A., Kumar, P., & Singh, P. (2025). Crop suitability analysis for the coastal region of India through fusion of remote sensing, geospatial analysis and multi-criteria decision making. *Scientific Reports*, 15(1), 8727.
20. Júnior, M. R. B., de Almeida Moreira, B. R., dos Santos Carreira, V., de Brito Filho, A. L., Trentin, C., de Souza, F. L. P., ... & Shiratsuchi, L. S. (2024). Precision agriculture in the United States: A comprehensive meta-review inspiring further research, innovation, and adoption. *Computers and Electronics in Agriculture*, 221, 108993.

Chapter 15

Climate-smart agriculture

Leveraging precision-farming technologies for sustainable food security

Tejasvini Rahul Katkar, Bhavana Santosh Pansare, S. Bathrinath, S. Selvakanmani, Komal Dhanraj Deshmukh, and Shripad Joshi

15.1 INTRODUCTION

Climate change has certainly brought to the world some of its starkest effects in the last few years, with the most significant implications for food and agriculture worldwide. As countries respond to a climbing global population, there is considerable predicted growth in demand for food, while agricultural production is challenged by shifting climatic conditions, extreme weather events, and resource limits [1]. These adverse factors require an urgent response in the name of food security for human development and increased sustainable agricultural growth. Climate-smart agriculture (CSA) has come to the fore as a significant response to the combined challenges of food security and climate change, blending new technology and practice to underpin the resilience of agricultural systems, decrease environmental impacts, and increase productivity.

Supporting the overall resilience of agricultural systems through practices that are both climate-smart and economically feasible for farmers is important because the agricultural sector is extremely vulnerable to climate change through temperature increases, variations in precipitation, prolonged droughts, and flooding, which impact crop production, livestock health, and overall farm productivity [2]. Furthermore, the agricultural sector is one of the largest sources of greenhouse gases, primarily methane, nitrous oxide, and carbon dioxide, which have sped up global warming and environmental degradation. Therefore, it is very important to think about sustainable agriculture goals of increasing food production and reducing emissions to achieve world food security.

Perhaps, the most radical tactic in fighting CSA is precision-farming technology. Precision farming or precision agriculture (PA) utilizes very complex technologies including satellite imaging, remote sensing, Global Positioning System (GPS), Internet of Things (IoT) sensors, and data analytics for improving the efficiencies of farm operations. The farm activities of planting and monitoring crop resources will thus be executed much more efficiently because crops are being monitored with great accuracy; they can monitor which area requires certain resource allocations with no or lesser wastage [3]. For example, precision irrigation technology gives farmers the ability to irrigate more efficiently through the application of water in more precise timing and locations while cutting out waste, which can be extremely beneficial in water-scarce regions. Similarly, crop scouting and pest monitoring drones allow for early detection of problems, enabling farmers to control flooding, insect, and disease problems in the optimal time frame to avoid excess applications of pesticides.

PA is an innovative approach to farming that allows farming to adapt to climate variability and change while using resources sustainably. By utilizing real-time data and sophisticated analytics, this technology-based approach can improve decision-making for farmers so they can better respond to changing conditions and minimize agriculture's impact on the environment. Within the framework of CSA, it is well-aligned with the objectives of improving

DOI: 10.1201/9781003520733-18

efficiencies of resource use, productivity, and climate resilience (CR). PA enables reduced applications of fertilizer, pesticides, and water, minimizing over-application of these products, which is one of the main contributors to soil erosion, water pollution, and depletion of our natural resources. Enhanced targeting of products and monitoring of crop health allow us to make more with less. These both bolster the economic bottom line while also safeguarding the environment from over-application. To sum them up, PA has the potential to be a sustainable model of farming to balance productivity and environmental stewardship and is a key functional aspect to future-proof our agriculture production against climate change [4].

Nevertheless, even as PA has significant promise, there are important barriers to its diffusion. Many farmers, particularly in developing countries, simply may not be able to cover the up-front costs required to invest in an advanced technology system. Similarly, technical awareness may be limited, infrastructure may be inadequate, and support systems may be insufficient for those technologies to be deployed and sustained in operation as they were intended. Smallholder farmers—who represent a significant portion of the food producers throughout the world—are often particularly disadvantaged by the barriers described above [5]. In addition, the integration of data privacy, cybersecurity, and interoperability among different technologies raises integration and scalability issues that are a challenge to maximizing the benefits of CSA. These gaps must be dealt with through policy support and infrastructure investment, and with strong capacity building. Promotion of R&D is also important to help PA develop, and for it to be more affordable, accessible, and responsive to farming contexts. [6].

The main aim of this chapter is to examine the definition of CSA and to connect to the technology of PA to advanced climate-smart food security. CSA is presented as a comprehensive approach to reducing negative agricultural impacts due to climate change while still increasing productivity, resilience, and sustainability. After careful consideration, this chapter will further develop the value of technologies, which include IoT devices, remote sensing technologies (RSTs), GPS, and analytics. The use of these technologies can change agricultural processes by advancing deliberate resource use, reducing waste, and increasing productivity on the farm. This chapter summarizes the intended major areas of benefits to CSA from the perspective of PA in practice, with the overarching observations of reducing agricultural impacts on the environment, better using limited resources, and improving yield per acre.

The chapter not only looks at the benefits but also at the barriers to the scalability of PA, particularly in developing countries. Barriers to scalability include upfront capital costs to establish PA systems, access to PA technologies, infrastructure deficiencies, and barriers to smallholder farmers who do not have the resources to implement the technology or manage it effectively. Data security, the interoperability of systems, and cybersecurity issues are also examined as barriers to scalability. The chapter seeks to emphasize the significance of new agricultural practices and technologies for sustainable agricultural development. It reviews the opportunities for practices and technologies to be scaled and localized for enhanced food security in the face of climate variability.

While CSA and PA technologies hold great promise to revolutionize global food systems and ensure sustainable food security, their large-scale adoption is hindered by several challenges. One of them is the huge capital expenditure entailed in the initial investment needed to leverage new technologies like sensors, drones, GPS devices, and IoT devices, which could be prohibitively high for smallholder farmers in developing nations. The absence of trustworthy internet connections and insufficient technical infrastructure in rural and remote locations is, in addition, a constraint on the usability and accessibility of the technologies. Another significant constraint to the utilization of these systems is the requirement of technical knowledge and technical skills needed to access the systems and interpret data outputs. Farmers lack proper training or capacity-building programs that would equip them to utilize the potential of PA. Additionally, farmers tend to suffer from scalability challenges related to data privacy issues, cybersecurity attacks, and interoperability across various platforms. CSA also largely depends on facilitating policy environments and institutional arrangements, which either do not exist or are weakly enforced in most global contexts.

Despite challenging obstacles, PA and CSA have a good future. Artificial intelligence, machine learning, and Big Data (BD) predictive analytics advancements are enabling agricultural systems to become progressively more precise and predictive, enhancing resource management. The technologies are capable of reducing input costs while enhancing systems' productivity and robustness. The availability of low-cost and affordable agricultural technologies also enables small-scale farmers to overcome early adoption issues. The emerging smart farming systems that combine climate data, soil conditions, and constant monitoring provide predictive capability to drive forward-looking decisions and build CR. With greater international cooperation, additional government backing, and more funding for public goods and education, CSA solutions that are scalable are now within sight. This revolution will not only help guarantee better adaptation to climate change but will also set the stage for a future in which the world's food system is safer, more productive, and more resilient.

15.2 RELATED WORK

CSA has been quite popular due to its success in addressing the problems of climate change while also enhancing food security globally. One of the studies focusing on precision-farming technologies applied in CSA recognized that GPS, remote sensing, and Geographic Information System (GIS) all have a large potential to enhance water-use efficiency and crop yield prediction, both of which are essential to address the challenges of climate change. The study employed a systematic review approach in the examination of international studies on the issue of precision and PA, and the writers noticed that whereas PA matters when enhancing farm practices, smallholder farmers are overlooked. This is treated in current research that will examine viable and scalable alternatives for such farmers, especially for poor environments [7]. Likewise, another research on remote sensing and GIS application in water-scarce conditions proved that these technologies assist in crop status monitoring and land-use optimization, with lower environmental effects. However, the study failed to discuss the challenges of using such technologies in low-resource settings, a critical focus area for research development aimed at enhancing accessibility and affordability [8]. Regarding sustainability, studies have explored the potential of BD and IoT to be used together in PA, emphasizing their use for resources such as water, fertilizers, and pesticides being optimized. A literature review of how IoT and BD can be used for improving precision farming showed that IoT and BD significantly contribute to reducing environmental footprints and increasing productivity. The research did not, however, investigate the barriers to IoT adoption in developing countries, particularly for smallholder farmers [9]. This is a gap that is necessary because an important point of focus in further research is solving adoption barriers through affordable technology as well as capacity building [10].

Yet another study, this one on smart irrigation technologies, concluded that these technologies can very effectively deal with water scarcity in climate change-hit regions. Smart irrigation systems can optimize water application with the help of sensors and live weather data, reducing wastage and improving yields [11]. The study, however, did not address the scalability of the technologies for smallholder farmers, which remains an extremely critical research area in the context of CSA. Lastly, a few of these studies have highlighted the socioeconomic barriers to adopting CSA practices, particularly in the context of developing countries. The research found that targeted intervention in the form of capacity-building programs, for instance, might combat those barriers [12]. Yet another study of CSA practice in the Sub-Saharan area uncovered the institutional and financial challenges in adopting CSA and how PA technologies can be integrated into CSA platforms for promoting food security and CR. However, these research studies have failed to gain proper insight into the way precision-farming technology should be integrated into existing CSA practices cheaply, a gap that the existing research has been aiming at filling [13].

Similarly, research on the economics of PA has determined that while initial investment is costly, long-term return through increased yield and reduced input cost pays for itself. This is a focus point to seek research in the domain of precision-farming technology development to target smallholder farmers so that they can derive maximum benefit from CSA practice [14]. Introducing precision-farming technology into CSA has been internationally considered to hold the potential of increasing farm productivity, incomes, and CR while minimizing the emission of greenhouse gases. An overview of CSA strategies across the board has signaled their uptake for significantly raising the productivity and income of farms alongside climate-resilience augmentative impacts [15]. The integration of various CSA approaches, such as PA, regenerative agriculture, and agroforestry, is deemed important to achieve these benefits to the fullest. These technologies also aid in promoting sustainable agriculture through improved resource-use efficiency (RUE) and lower environmental footprint [3]. Studies highlight the importance of context-specific strategies tailored to local environmental and socioeconomic conditions to achieve the maximum efficiency of CSA practices [16].

Also, the barriers to the mass usage of CSA, ranging from financing to lack of knowledge and policy-related barriers, have been the main interest in recent research. Application of CSA technologies is most of the time characterized by larger barriers in developing countries, which lack resources as well as information. Proposed mechanisms to alleviate the barriers are capacity development programs, financing, and supporting policies. Besides, emphasis is put on the necessity of climate adaptation and mitigation by applying CSA practices such as climate-resilient crop choice and better water management. These are all done in the necessity of farming food sustainably and thus underpinning the necessity of continuing to learn with a focus toward strengthening the implementations of precision-farming technology using CSA frameworks.

15.3 METHODOLOGY

The method of integrating precision-farming technology into CSA is to evaluate the effectiveness of the technology in heightening agricultural productivity, efficient use of resources, and CR. This study employs a systematic approach to choose and implement the appropriate precision-farming devices, such as remote sensing, IoT sensors, GPS, and smart irrigation systems, appropriate for a specific agricultural setting. By selecting outstanding case study farms or regions, the research seeks to capture real challenges and benefits associated with the adoption of precision farming, particularly among smallholder farmers. The research methodology comprises collecting data from different sources, including field observations, farmer questionnaires, and sensor data, to assess the increase in crop yield, water-use efficiency, and overall resiliency to climate change. Through a continuous cycle of monitoring and evaluation, the chapter will attempt to quantify the effect of these technologies on CSA outcomes and garner helpful insight into their consideration for scale-up and potential contribution to scaling sustainable agricultural practice (Figure 15.1).

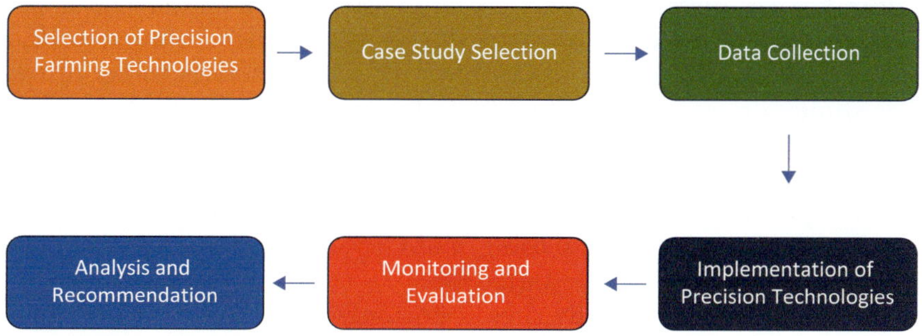

Figure 15.1 Block diagram of the framework.

15.3.1 Choice of precision-farming technologies

The selection of PA technologies is one of the most important measures in embracing CSA practices because these technologies are capable of highly enhancing agricultural productivity, RUE, and resilience to climate change. PA refers to the use of modern tools and technology to monitor, manage, and optimize farm practices using real-time inputs. The key focus is applying the right inputs (fertilizers, water, pesticides, etc.) in the right quantity, at the right time, and in the right location, eventually leading to sustainable agriculture. Remote sensing, GPS, IoT, and smart irrigation are some of the key technologies employed in precision farming. RST, such as satellites and drones, provides vital information on crop health, soil moisture levels, and temperature. Farmers are then in a better position to make resource and risk management decisions, enhancing productivity and reducing ecological impact. GPS facilitates precise field mapping and guiding of equipment for effective planting, irrigation, and harvesting, with reduced overlap and reduced waste of resources.

IoT sensors, or networked sensors distributed in the field or on farm machines, receive and send in real-time soil moisture, soil temperature, and crop health information, allowing immediate monitoring and reaction. Drones enhance this through aerial photography over the fields that can produce high-definition images and facilitate precision farming. Finally, smart irrigation technologies utilize information from weather forecasts, groundwater levels, and other sources to improve water use, prevent waste, and enhance crop yields. When evaluating these technologies, it is important to consider and be sensitive to context conditions, budget constraints, and the technical feasibility of the farmer's system. Technologies should be designed or modified for local conditions and aimed at solving or managing issues. The particular needs of farmers, especially in climate-exposed communities, need to become effective and sustainable.

15.3.2 Case study selection

The selection of case studies is an essential aspect of assessing the performance of CSA practices integrated with precision-farming technology. The use of customary farming regions or farms enables researchers to gain an understanding of the usage and effectiveness of the technologies in real situations. Ideally, case studies should embody the issues smallholder farmers face, especially in climate-risk areas that may grapple with extreme weather patterns, water shortages, land degradation, and resource poverty. These areas serve as demonstrative examples of how precision-farming technologies could be applied to address local agricultural issues, increase productivity, and assist in the building of climate change resilience. By selecting case studies in various farming systems (crop, livestock, or mixed farming), researchers can consider the flexibility of CSA practices to coordinate diverse contexts. The studies or case studies should also focus on districts that have already tested or implemented CSA practices so that there is appropriate data to evaluate the effects of the CSA practices. In particular, districts that have a history of climate-related problems, such as floods and drought, would be most ideal in accessing knowledge about precision technology, such as smart irrigation, remote sensing, and GPS, and how such precision technology can reduce risk events and optimize resource use.

The purpose of selecting case study sites is not only to determine if CSA practices are food secure, but it is also an assessment for the barriers to uptake, considering all factors, including financial constraints, poor technical know-how, and limited access to technical inputs. These samples can be replicated, and larger plots of CSA can then be transferred to similar sites, learning from each other about how to deal with challenges, and these shared learning experiences can facilitate realizing full-scale precision-farming technology through local dynamics.

15.3.3 Data collection

The collection of data is an important part of understanding how effective CSA practices are and is especially relevant when CSA implies precision-farming technology. Data collection is the process of collecting both qualitative and quantitative data to measure several things (e.g., farm profitability, RUE, crop yields, and resilience to climate change). Useful numerical data is something that can be measured and then analyzed to demonstrate where there is some indication that precision-farming technologies are having an impact on farm productivity and sustainability. The technological aspect of the analysis of the CSA farming practices is in the sustainable interpretation and efficient use of resources, including water, fertilizers, and energy, as well as strengthening resilience to climate change. Qualitative data, such as field observations, informal discussions with farmers, and surveys, then provide context and background information to describe the socioeconomic and environmental barriers to the adoption of new technologies as well. Collectively, these two kinds of data allow researchers to form a complete picture of the role of precision farming in supporting sustainable agricultural practice and the broader objectives of CSA. In quantifying data, several equations can be utilized to assess the performance of precision-farming technologies. One example is called water-use efficiency, and this is the critical parameter that can be calculated utilizing equation 15.1:

$$\text{WUE} = \frac{\text{Crop Yield}(\text{kg})}{\text{Water Applied}(\text{L})} \tag{15.1}$$

In this formula, crop yield is measured as a function of one unit of water use. Technology such as precise irrigation systems and sensors allows for maximum management of water to ensure maximum water-use efficiency for maximum nonwaste and maximum crop yield. Likewise, the fertilizer-use efficiency (FUE) is determined by equation 15.2:

$$\text{FUE} = \frac{\text{Increase in Yield}(\text{kg})}{\text{Amount of Fertilizer Applied}(\text{kg})} \tag{15.2}$$

This index helps estimate how well fertilizer inputs translate into increased crop yields. New technologies like remote sensing and soil sensors may also facilitate the efficient use of fertilizers by allowing farmers to apply the correct amount while reducing excess and minimizing environmental impacts. Farm income (FI), being another important input, is computed as

$$\text{FI} = \text{Price per Unit}(P) \times \text{Crop Yield}(Y) \tag{15.3}$$

This equation represents the contribution of increased crop outputs due to PA technologies to FI. CR could be defined by the equation that compares crop outputs before and after a climate incident:

$$\text{CR} = \frac{\text{Post-climate Event Yield }(Y)}{\text{Pre-climate Event Yield }(Y)} \tag{15.4}$$

A larger ratio indicates more resilience, which means that technologies, like weather-forecasting tools and remote sensing, are enabling farmers to predict ahead of time and minimize the effect of extreme weather events. Beyond this, quantitative information alone is not relevant. Farmers' interviews and surveys help illustrate the socioeconomic constraints of implementing CSA practices. This information can show what affects farmers, whether it is too few financial resources to access technology or some combination of both. Combining qualitative findings with quantitative research provides a more in-depth analysis and allows researchers to provide more insight into the effects of precision-farming technologies on CSA outcomes. Through these combined sources of data, researchers can identify intervention areas to target for improvement, propose measures for intervention, and provide recommendations for promoting smallholder farmer adoption of these technologies in regions vulnerable to climate change.

15.3.4 Implementation of precision technologies

The phased implementation of precision-farming technologies requires putting the selected technologies—remote sensing, GPS, drones, smart irrigation, and IoT sensors—into practice in selected farms or case study areas. This stage ensures that the technologies are being deployed in an existing farming practice and takes into consideration the local context, including the environment, farming systems, and social and economic determinants. To achieve successful uptake, it is important to match the technology with the needs, skills, and resources of farmers, and in some cases, this will require training, access to necessary infrastructure, and in some cases, financial assistance. The integration process also entails identifying the concerns or experiences of farmers in the adoption of new technology, such as the cost of investment, lack of knowledge, and the need for solutions adapted to local conditions. Effective integration, therefore, requires a well-considered integration plan, which not only introduces a new technology but also provides means for receiving farmers in their everyday tasks while improving their capacity to manage resources and enhance productivity. Among the key math expressions that could be employed in evaluating the effectiveness of precision-farming technology during its integration stage is RUE. RUE serves as an indicator to know to what extent the use of such human capital is compared to outputs produced. RUE can be expressed as

$$\text{RUE} = \frac{\text{Crop Yield}(\text{kg})}{\text{Total Resources Used}(\text{units})} \tag{15.5}$$

where "Total Resources Used" is represented by a material used, such as water (liters), fertilizer (kg), or labor (man-hours). In RUE, we quantify these to allow farmers and researchers to evaluate and monitor increases in resource efficiency due to precision technology. Consider an intelligent irrigation system, or a combination of irrigation and IoT sensors, where water is minimized and therefore yields per unit of water are maximized, again recording the impact the technology has on sustainable agriculture. Therefore, this equation not only assists in the verification of a positive net Return on Investment from precision-farming technology but also ensures that precision technology is adopted more sustainably and efficiently.

15.3.5 Monitoring and evaluation

The assessment of PA technologies after implementation entails monitoring and evaluation. The monitoring procedure involves ongoing monitoring of the technologies to ascertain, through data analytics and/or field evaluation, whether or not the technologies are producing the desired impacts on productivity, RUE, and climate change adaptability. Employing high-frequency data gathering methods like sensor networks, remote sensing devices, and field and onsite measurement methods, scientists can monitor Key Performance Indicators like crop production, water utilization, fertilizer use, and climatic adaptability. The purpose of monitoring and assessment in PA is to assess whether or not the technologies are producing the intended impacts and contributing toward the overarching goals of CSA, such as sustainability, increases in productivity, and climate adaptability.

Yield response to input (YRI), one of the key drivers of PA technology performance, quantifies variation in crop yield resulting from resource input variability (e.g., water, fertilizers, or labor). The YRI (equation 15.6) is

$$\text{YRI} = \frac{\text{Change in Crop Yield}\left(\text{kg}\right)}{\text{Change in Resource Input}\left(\text{units}\right)} \tag{15.6}$$

This formula is useful in measuring the effectiveness with which adjustments in resource inputs, like improved irrigation or particular fertilizer treatments, alter crop yield. By assessing the YRI, researchers and farmers can tell whether precision-farming technology is causing crop yield increases per unit resource input. Continuous evaluation of the YRI allows for continuous modification and evaluation so precision-farming technologies can continue working toward CSA objectives and enable sustainable agricultural practices in the long run.

15.3.6 Analysis and recommendations

The analysis and recommendation phase is critical to evaluating whether PA technologies are effective in providing CSA impacts. This phase involves critically examining the data collected before evaluating the impact of PA to achieve multiple CSA goals (e.g., enhanced productivity, improved resource efficiency, and reinforced CR). The purpose of the study is to evaluate and analyze how these technologies affected farm performance; in other words, have these technologies promoted the CSA benefits we set out to achieve? The major performance indicators (e.g., yield increase, efficiency of water and fertilizer, and climate change resilience) are compared to determine the overall effectiveness of the precision technologies.

Additionally, this step involves identifying the barriers to the increased adoption of precision-farming technologies, specifically for smallholder farmers. The main barriers identified are the high costs of technology, unavailability of the technology, insufficient knowledge, and insufficient infrastructure. At the same time, possible pathways for the mainstream adoption of the technologies are suggested, which include access to financing, capacity building, and an enabling policy environment. Finally, recommendations will be offered to close the gaps and to seize the opportunities to make precision-farming technologies accessible to farmers affected by climate hazards, specifically to smallholder farmers (equation 15.7).

$$CBR = \frac{\text{Total Benefits}(\text{USD})}{\text{Total Costs}(\text{USD})} \tag{15.7}$$

The net advantages are the extra income or savings resulting from increased productivity and efficiency, and the net expenses are the expenses of installing and running the precision-farming technologies. A CBR greater than 1 indicates that the advantages are greater than the expenses, and therefore, the implementation of precision-farming technologies is economically viable. By comparing the CBR and other indicators of performance, policymakers and scholars can provide evidence-based recommendations for how to scale up precision-farming technologies, particularly in climate-vulnerable or resource-constrained regions. Determining this is necessary to ensure that CSA continues to be long-term sustainable and makes its contribution toward food security.

15.4 RESULTS

The conclusions of the chapter provide a comparative assessment (Table 15.1) of precision-farming technology integrations with CSA practices across diverse agricultural zones. The goal is to examine the impact of such technologies on varying CSA results like crop yield, resource utilization (e.g., water and fertilizers) efficiency, farm income, and resilience to climate factors. By analyzing farm data from farms that adopted precision-farming technologies against those without, the study assesses how much the technologies contribute to achieving sustainable agriculture practices, increasing farm productivity, and reducing environmental impacts. This comparative analysis not only highlights the advantages of adopting PA technologies in productivity and efficiency increments but also casts light on farmers' challenges in adopting these technologies into their prevailing farming systems, particularly for the smallholders. The study further compares the performances of different PA technologies like remote sensing, GPS, IoT, and smart irrigation under diverse contexts and different environmental conditions. The outcomes will provide some indications of the real-world relevance of adopting such technologies, obstacles to their greater uptake, and actionable recommendations on their scaling under the platform of CSA practices.

Table 15.1 Comparative analysis of crop yield and resource-use efficiency

Farming system	*Crop yield (kg/ha)*	*Water-use efficiency (kg/L)*	*Fertilizer-use efficiency (kg/kg)*
Traditional farming	2,500	0.5	0.1
Precision farming (IoT, smart irrigation)	3,200	0.8	0.2
Precision farming (GPS, remote sensing)	3,100	0.75	0.18
Precision farming (drones, IoT, smart irrigation)	3,400	0.85	0.25

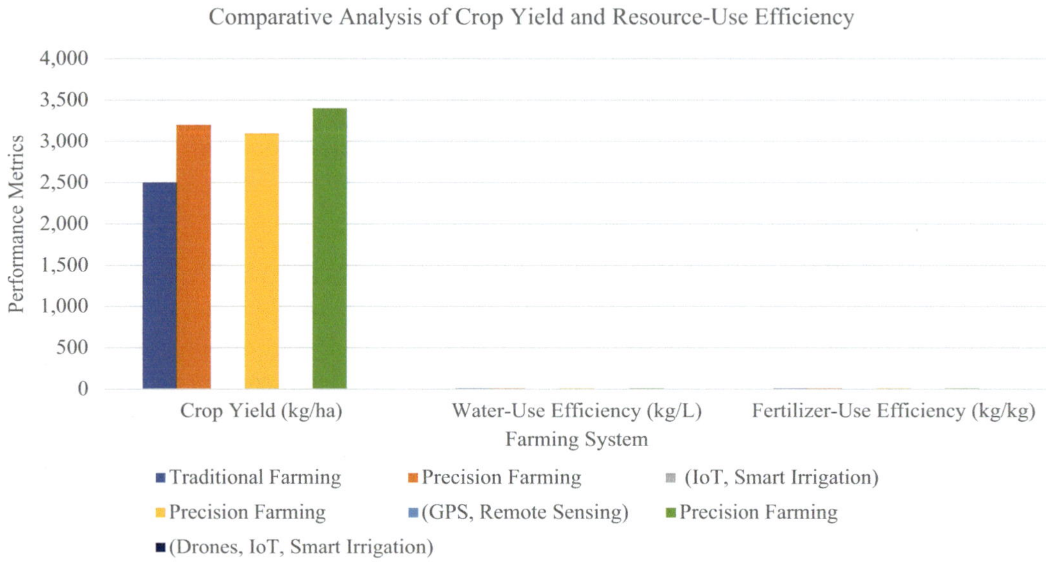

Figure 15.2 Comparative analysis of crop yield and resource-use efficiency.

Figure 15.2 gives a comparative assessment between traditional farming and different precision-farming systems according to crop yield, water-use efficiency, and FUE. The crop yield in traditional farming is captured at 2,500 kg/ha, with water-use efficiency being 0.5 kg/L and FUE at 0.1 kg/kg. These are indicators of the inadequacies in productivity and utilization of resources in traditional agriculture. Precision-farming technologies, on the other hand, reflect significant gains in all three dimensions. For example, with IoT technologies and smart irrigation installed, crop yield improves to 3,200 kg/ha. At the same time, water-use efficiency increases to 0.8 kg/L and FUE is doubled to 0.2 kg/kg. These are thanks to the improved monitoring and control that these technologies offer, with a view to optimum application of resources.

Likewise, GPS and RST-based precision farming demonstrate considerable progress. The method yields a crop yield of 3,100 kg/ha, water-use efficiency of 0.75 kg/L, and FUE of 0.18 kg/kg. Accurate mapping and monitoring of field conditions lead to more informed decisions and less input wastage. The maximum performance is realized when drones, IoT, and smart irrigation are combined. This raises crop yield to 3,400 kg/ha and water- and fertilizer-use efficiencies to 0.85 kg/L and 0.25 kg/kg, respectively. These outcomes show how the synergistic application of several precision-farming instruments can optimize productivity and reduce input usage. In summary, the information in Figure 15.2 starkly shows that the incorporation of PA technologies results in considerably increased crop yields and better water- and fertilizer-use efficiencies, emphasizing their crucial contribution to the encouragement of sustainable agriculture.

15.5 DISCUSSION

The goal of this study was to analyze the efficiency of traditional farming techniques versus various precision-farming technologies against key parameters such as crop yield, water-use efficiency, and FUE. The method of research was to collect data from farms with various precision-farming technologies, i.e., IoT, smart irrigation, GPS, remote sensing, and drones.

The comparison between traditional farming practices and these systems brought out the capability of precision farming to achieve maximum use of resources and promote agricultural productivity. The results indicate that precision-farming technologies well surpass traditional farming. Traditional farming produced 2,500 kg/ha with the water-use efficiency of 0.5 kg/L and the FUE of 0.1 kg/kg. Precision-farming systems, on the other hand, significantly improved in all the parameters. For example, the application of IoT and smart irrigation enhanced the crop yield to 3,200 kg/ha along with water and fertilizer efficiency. Integration of GPS and remote sensing also recorded a matching increase in yield as well as efficiency. The best performance, however, was exhibited when drones, IoT, and smart irrigation were integrated to produce 3,400 kg/ha with the water-use efficiency of 0.85 kg/L and the FUE of 0.25 kg/kg. These figures indicate the efficiency of precision farming in addressing the challenges of resource productivity and efficiency.

Among the greatest advantages of precision farming is the possibility to optimize the application of resources like water and fertilizer, traditionally the most limiting factors in farming. Precision technologies enable farmers to apply water and fertilizer with greater precision in the field, economizing on waste and cost. The outcome is higher yield and higher sustainability, which is imperative in the climate change and resource scarcity era. Furthermore, precision-farming technologies enable farmers to keep up with farm conditions in real time Reinforcement Learning and to make informed decisions based on actionable information. However, the mass adoption of precision farming is constrained. The technologies require significant initial capital investment in equipment, infrastructure, and training. Furthermore, broadband internet and technological proficiency may not be readily available in remote or rural locations and hence may not be readily available for developing countries' smallholder farmers to effectively utilize such systems.

15.6 CONCLUSION

The findings of this study present the catalytic ability of precision-farming technology to enhance the productivity of agriculture and the efficiency of resource use. The findings imply that precision-farming systems incorporating IoT, smart irrigation, GPS, RST, and drones are more productive in crop output, water efficiency, and fertilizer efficiency compared to conventional farming systems. For instance, crop yields have risen from 2,500 kg/ha (conventional) to as much as 3,400 kg/ha with the use of new technologies. Moreover, water- and fertilizer-use efficiencies also rose by considerable margins, with values reaching as much as 0.85 kg/L and 0.25 kg/kg, respectively, with precision-farming systems. These observations bring out the ability of precision farming to support sustainable agriculture, particularly in minimizing the waste of resources and boosting food demand.

Despite the promising benefits, several challenges and limitations to the widespread use of precision farming exist. High capital requirements for initial investment and a lack of technical expertise among smallholder farmers are key barriers. Moreover, while precision farming may enhance farm productivity and resistance, its benefits are more readily accessible to farmers with adequate resources, thereby exacerbating rural disparities. For future research, it is imperative to explore cost-effective models like shared-use technology or low-cost sensor systems, which can bring precision-farming accessibility to smallholder or resource-limited farmers. In addition, research should discuss the socioeconomic determinants of adoption and scalability of such technology across various agricultural environments. On the policy front, governments must consider providing economic incentives, training programs, and incentives for infrastructure investment to facilitate the integration of precision farming, particularly in climate-vulnerable regions. Those regulations that motivate farmers, technology firms, and scientists to collaborate will be key to creating an incubatory culture that will enable such technologies to upscale, ensuring sustainable agriculture in the years to come.

REFERENCES

1. Yanagi, M. (2024). Climate change impacts on wheat production: Reviewing challenges and adaptation strategies. *Advances in Resources Research*, 4(1), 89–107.
2. Bogale, G. A., & Erena, Z. B. (2022). Drought vulnerability and impacts of climate change on livestock production and productivity in different agro-Ecological zones of Ethiopia. *Journal of Applied Animal Research*, 50(1), 471–489
3. Getahun, S., Kefale, H., & Gelaye, Y. (2024). Application of precision agriculture technologies for sustainable crop production and environmental sustainability: A systematic review. *The Scientific World Journal*, 2024(1), 2126734
4. Agbeyangi, A., & Suleman, H. (2024, November). Advances and challenges in low-resource-environment software systems: A survey. In *Informatics* (Vol. 11, No. 4, p. 90). MDPI.
5. Khatoon, P. S., & Ahmed, M. (2022). Importance of semantic interoperability in innovative agriculture systems. *Transactions on Emerging Telecommunications Technologies*, 33(5), e4448.
6. Busch, C., & Barkema, H. (2021). From necessity to opportunity: Scaling bricolage across resource-constrained environments. *Strategic Management Journal*, 42(4), 741–773.
7. Van de Zande, G. D. (2023). Bringing the Water-Efficiency Benefits of Precision Irrigation to Resource-Constrained Farms through an Automatic Scheduling-Manual Operation Irrigation Tool (*Doctoral dissertation*, Massachusetts Institute of Technology).
8. Rao, E., & Nasir, M. A. R. B. (2024). Impact of Affordability, Adaptability, and Availability in Achieving Sustainability: A System-Dynamic Approach (*Doctoral dissertation*, College of Electrical & Mechanical Engineering (CEME), NUST).
9. Khan, S., Singh, R., Khan, S., & Ngah, A. H. (2023). Unearthing the barriers of Internet of Things adoption in the food supply chain: A developing country perspective. *Green Technologies and Sustainability*, 1(2), 100023.
10. Al-Emran, M., & Griffy-Brown, C. (2023). The role of technology adoption in sustainable development: Overview, opportunities, challenges, and future research agendas. *Technology in Society*, 73, 102240.
11. Lakhiar, I. A., Yan, H., Zhang, C., Wang, G., He, B., Hao, B., ... & Rakibuzzaman, M. (2024). A review of precision irrigation water-saving technology under changing climate for enhancing water use efficiency, crop yield, and environmental footprints. *Agriculture*, 14(7), 1141.
12. Ramanadhan, S., Galbraith-Gyan, K., Revette, A., Foti, A., Rackard James, C., Martinez-Dominguez, V., ... & Viswanath, K. (2021). Key considerations for designing capacity-building interventions to support evidence-based programming in underserved communities: A qualitative exploration. *Translational Behavioral Medicine*, 11(2), 452–461.
13. Muriithi, C., Mwongera, C., Abera, W., Chege, C. G., & Ouedraogo, I. (2023). A scalable approach to improve CSA targeting practices among smallholder farmers. *Heliyon*, *9*(10), 1–19.
14. Onyango, C. M., Nyaga, J. M., Wetterlind, J., Söderström, M., & Piikki, K. (2021). Precision agriculture for resource use efficiency in smallholder farming systems in Sub-Saharan Africa: A systematic review. *Sustainability*, 13(3), 1158.
15. Zheng, H., Ma, W., & He, Q. (2024). Climate-smart agricultural practices for enhanced farm productivity, income, resilience, and greenhouse gas mitigation: A comprehensive review. *Mitigation and Adaptation Strategies for Global Change*, 29(4), 28.
16. Medici, M., Canavari, M., & Castellini, A. (2021). Exploring the economic, social, and environmental dimensions of community-supported agriculture in Italy. *Journal of Cleaner Production*, 316, 128233.

Chapter 16

The impact of variable rate technology (VRT) on soil health and crop yield optimization

Amol M. Dhepe, Swati Sharma, J. Somasekar, G. Manikandan, Midhun Mathew Kizhakethil, and Abid Salati

16.1 INTRODUCTION

Industrial agriculture is also challenged with a need to increase levels of food production to keep pace with increasing demands from around the globe without exploiting nature's buffers [1]. Complemented with threats of climate change, erosion, and restriction in resources, there has been an intensification of conventional agriculture to a more productive and less damaging type of agriculture. Precision farming has now grown to become a hi-tech revolution where technology meets traditional farming in the pursuit of maximizing input and production levels as well as reducing footprints in the environment. One of the biggest advances in this area is the variable rate technology (VRT), whereby farmers are given the capacity to apply agricultural inputs—e.g., seeds, fertilizers, pesticides, and water—at differential rates on a farm as opposed to an equivalent rate [2].

VRT relies upon the concept of respecting field heterogeneity and calibrating practice for it. Field crop environments are heterogeneous spatially regarding soil texture, organic matter, nutrients, and water status [3]. The customary practice of conventional farming through even distribution of inputs over the entire field does not conform to heterogeneity and undersupplies or oversupplies resources. Not only is the performance of yield compromised, but so too do issues in terms of environmental effects materialize through runoff of nutrients, water contamination, and emissions of greenhouse gases.

The signature strength of VRT lies in the ability to tie together several layers of data—soil maps, crop patterns, satellite imagery, yield history, and cropping patterns—to make decisions out of data. Its associated technologies, such as on-the-go sensors, remote sensing, Global Positioning System (GPS)-guided tractors, Geographic Information System (GIS), and computer programs, work in tandem with one another to map variability in the field and adjust application rates of inputs in real time. This allows every section of a field to get the exact amount of input it requires, optimizing efficiency and reducing waste [4].

The use of VRT is increasingly viewed as a strategic reaction to soil health decline issues. The ability of soil to support humans, animals, and plants as an ecologically sound ecology is what defines its health. Prolonged monocropping, overuse of chemicals, and improper land management have resulted in nutrient imbalance, decreased microbial activity, compaction, and erosion in most agricultural areas. By providing inputs precisely and according to soil requirements, VRT ensures nutrient balance, enhances soil structure, and enhances microbial diversity—all essential components of healthy soils. In addition, crop yield maximization is intrinsically connected to soil health. Healthy soils enhance water retention, root penetration, and nutrient cycling, which have direct impacts on plant growth and resilience. VRT not only improves these situations but also provides accurate monitoring and adjustment throughout the crop life. Yield maps developed using VRT tools have the potential to feed back into improving farm management systems continuously [5].

DOI: 10.1201/9781003520733-19

While VRT has the benefits, its implementation is influenced by many factors. Its huge initial investment levels, the absence of technical know-how, and the poor quality of real-time data in certain areas can act as a deterrent to its installation [6]. But with the advancement of cloud computing, cellular devices, and easy-to-use programs, the load is easing off continuously. Moreover, the increasing focus on climate-smart agriculture and evidence-based decision-making in agricultural policy formulation is increasingly compelling the application of VRT. Past studies have demonstrated that VRT can yield tremendous gains in input use efficiency, from 15% to 30% fertilizer savings, depending on crop and location. 5% to 20% yield enhancements have also been documented as an indicator of the VRT's contribution toward food security through environmental sustainability. Against this context, this book chapter delves into the multidimensional role of VRT in crop yield maximization and soil health. Through an exploration of data-driven initiatives, spatial analysis methods, and field-level interventions, the chapter seeks to outline a clear grasp of how VRT can shape the future of sustainable agriculture.

This chapter seeks to find the potential for VRT evolution to improve soil health and crop yield maximization through data-based and site-specific input management. The research scope includes geospatial data, soil parameters, and crop performance variables integrated to assess the contribution of VRT toward improved and sustainable agriculture management. By demarcating agroecological zones and conditions of diversity in fields, the research exhibits the diversity found in contemporary agricultural ecosystems and seeks to explore how technology plays a role in finding solutions to such problems. The research is not limited to the yield results per se but is diversified to wider fields of soil sustainability such as nutrient conservation, microbial existence, and eco-friendliness, and thereby becomes a complete framework to determine VRT's contribution toward precision agriculture.

The impetus for this research comes from the compelling necessity to balance agricultural production with environmental sustainability. Traditional methods of farming, as productive as they are, tend to end up in overuse or underuse of inputs because of their uniform application method. This not only causes inefficient use of resources but also hastens soil erosion and environmental pollution. As the world's demand for food increases and available land for cultivation decreases, it is crucial to seek technology that optimizes yield without sacrificing soil health in the long run. VRT is a strong candidate because it provides an option to deliver inputs to meet the needs of each zone in the field separately. This accuracy lessens input costs, increases evenness of yield, and promotes regenerative farming. The expanding access to satellite data, soil sensors, and mobile-based agronomic platforms continues to drive the practicality and scalability of VRT in both smallholder and commercial farming systems.

The chapter aims to have congruence with both scientific research and practical application. On the one hand, the research seeks to explore spatial variation in soil properties and their relationship with crop yield under VRT conditions. On the other hand, the chapter seeks to develop a methodology that combines GIS-based soil mapping, yield data analysis, and VRT prescription generation. Third, it evaluates if VRT's effectiveness in improving soil nutrient management and sustained yield increase compared to conventional practices is sustainable. Fourth, the study accounted for environmental issues related to VRT application by calculating areas of input waste reduction, leaching, and greenhouse gas mitigation. Finally, it looks for strategic information on applying VRT on national and regional agricultural platforms toward enabling precision agriculture programs and climate-resilient agriculture.

The substantial contributions of this research are to propose a data-hungry, non-experimental method to measure VRT performance that is opposite to most traditional research frameworks based on extensive field experiments. This research proposes a spatial decision-support system with actual datasets to produce actionable outputs for farmers as well as policymakers. Second, it filled the gap between technological competency and agronomic efficiency by providing an equilibrium knowledge of the advantages and disadvantages of VRT. Visualizations, statistical analysis, and geospatial interpretation, which are presented through the study, provide useful instruments and frameworks with which future deployment and innovation within precision agriculture could be directed. The interdisciplinary approach to the work from agricultural engineering, remote sensing, soil science, and environmental management increases its cross-farm systems and regional application.

This chapter is structured in a way that guarantees a coherent and rational delivery of the topic. After the introduction and background, the current research is described to set the context and make the research justifiable. The methodology section that follows describes the technical method, data sources, analytical software, and techniques employed. This is preceded by an exhaustive results and analysis section in which the main findings are conveyed through maps, graphs, and comparative statistics. The discussion places these results within the context of the current literature and practical problems. The chapter ends with a summary of conclusions, policy implications, and recommendations for further research. This formal organization promotes clarity and enables greater insight into how VRT can mold the future of sustainable agriculture through enhanced soil health and maximized crop production.

16.2 RELATED WORK

VRT has proven to be a crucial element of precision agriculture that makes site-specific management of farm inputs possible for improved crop production and soil quality. Over the past four years, there has been significant research aimed at trialing the efficacy, cost-effectiveness, and constraining factors in the uptake of VRT [7]. Reviewing the new literature involves collating information regarding innovations in the use of VRT, how it aids in improving soil quality and achieving the maximum crop yields, economics, and what the obstacles are to widespread use.

16.2.1 Advances in VRT applications

There has been an emphasis in recent research on integrating the most advanced technologies with VRT for improved accuracy in agriculture. Sensor-based variable rate sprayers are one such example, which have been designed to quantify tree presence and canopy characteristics and allow specific spraying of pesticides while avoiding chemical application. Furthermore, the use of artificial intelligence (AI) and machine learning has increased the efficiency of VRT systems in distinguishing weeds from crops, thus allowing better weed management techniques [8].

16.2.2 Effect on soil fertility and optimal crop yield

The use of VRT has been promising in soil nutrient management optimization and the increase in crop yield. With the application of inputs in specified field zones, VRT ensures soil nutrient balance, minimizes environmental effects, and maximizes land productivity. In addition, variable rate seeding techniques have been investigated to counteract soil heterogeneity, and research has shown that the adjustment of seeding rates according to soil characteristics can result in enhanced germination and plant growth [9].

16.2.3 Economic analysis of VRT adoption

Economic analysis has been performed to evaluate the cost-benefit trade-offs of the adoption of VRT. Variable rate application of nitrogen, for instance, has been economically evaluated in rainfed wheat systems in terms of variability of water and possible yield. Based on such evaluation, though large initial outlays might be needed to adopt VRT, chances of reduction in inputs and improvement in yields might compensate for the spending in the long run [10].

16.2.4 Challenges and adoption barriers

Notwithstanding its advantages, the widespread application of VRT is obstructed by different challenges. The study of Italian agriculture has characterized challenges like prohibitive implementation expenses, scarce technical expertise, and inadequate infrastructure as main hindrances to VRT adoption. Challenges like these must be addressed with focused policies, farmer training, and financing capability-enabling technology to allow larger-scale VRT application [11].

16.2.5 Market trends and future outlook

The global agricultural VRT market is experiencing robust growth, driven by increasing awareness of sustainable farming practices and emerging technology. Trends indicate that market value will increase robustly, indicating the broader appreciation of the importance of VRT in farm productivity and efficiency. Future emerging trends indicate the integration of AI, robotics, and sensor technologies for further automation of VRT usage and supporting sustainable crop cultivation [12].

The recent literature brings to the fore the significant advances and potential of VRT to maintain sustainable agriculture through improved soil health and maximized crop yield. Economic studies affirm the potential profitability of VRT, but adoption issues remain. Progress in research and development, together with encouraging policies and training, are necessary to remove these hindrances and realize the full potential of VRT in precision agriculture [13].

To have a thorough insight into the new development and trends in VRT, a comparative analysis of major studies has been done over the last four years is conducted. The chosen literature exhibits varying methodologies used to analyze the effectiveness of VRT in precision input control, enhancement of soil health, and optimization of yield. The studies also touch on adoption and technological integration issues. The overview given in the following table summarizes each study by highlighting the approaches taken and the overall key findings that add to the body of knowledge available in precision agriculture.

Table 16.1 emphasizes recent studies aimed at the application and influence of VRT in farming. A range of methodologies including GIS-based input management, variable rate

Table 16.1 Literature review study

Study	*Methods*	*Findings*
[14]	IOT-based growth monitoring via a capacitive soil moisture sensor	Improved use of technology and increased yield monitoring.
[15]	Variable rate seeding across diverse soil types	Achieved better crop emergence and higher yield uniformity
[16]	Surveys on VRT adoption barriers in Italy	Identified high cost and lack of infrastructure as key constraints
[2]	AI-integrated VRT for weed detection and spraying	Reduced herbicide usage and enhanced weed management precision

seeding, surveys on adoption, and AI-based spraying systems have been utilized to determine the efficacy of VRT. The results repeatedly demonstrate enhanced input efficiency, improved crop emergence, improved yield uniformity, and less use of chemicals. In addition, research identifies that although VRT benefits are well established, adoption remains hampered by the likes of high costs and poor infrastructure. Such findings cumulatively highlight the practical benefits and continued challenges in the implementation of VRT in precision farming.

16.3 METHODOLOGY

This chapter applies a systematic data-driven methodological framework to test the effect of VRT on soil fertility and crop yield optimization. The method (Figure 16.1) is devised differently from normal experimental testing and relies on geostatistical modeling, spatial data analysis, remote sensing, and precision agriculture machinery. The method is aimed at quantifying the role of VRT-based input management, i.e., application of seed and fertilizers site-specifically, toward achieving better soil nutrient balance and crop yield. The method combines geospatial datasets, maps of soil variability, yield history data, and equipment-level precision logs for the development of an integrated model of field performance under VRT. The objectives are to measure not just increases in yield but also soil parameter change, input efficiency, and environmental effects. Every methodological block deals with a distinctive facet of data acquisition, handling, analysis, or interpretation. Flexibility and extensibility are the process's features, which render it applicable in smallholder farming or large-scale agriculture. Spatial and statistical software are also used to ascertain relationships among soil zones, the use of inputs, and yield responses. This organized strategy offers a framework that can be duplicated to assess the effectiveness of VRT technology in various agricultural environments.

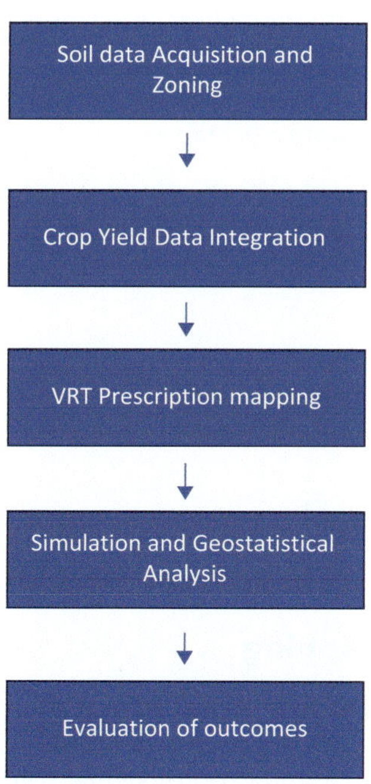

Figure 16.1 VRT-based methodology for soil and yield optimization.

16.3.1 Soil data acquisition and zoning

The initial step toward adopting VRT is precise soil data collection and zoning to recognize spatial variability within the field. The process starts by gathering relevant soil parameters like pH, organic matter content, moisture content, and significant levels of important nutrients (e.g., nitrogen, phosphorus, and potassium). It gathers data with a wide range of advanced technologies such as remote sensing systems, grid soil sampling, and sensor-based probes with in situ sensors. Remote sensing delivers broad, bird's eye images via satellite or aerial photographs, and grid soil sampling delivers precise, point-oriented soil samples. Sensor-based probes deliver real-time soil moisture and conductivity measurements to be used at the field decision-making level.

After being gathered, the soil information is georeferenced and processed with GIS. Soil heterogeneity is interpolated from these data using geostatistical interpolation methods like kriging or inverse distance weighting. The resulting maps show soil heterogeneity patterns that cannot otherwise be seen using regular sampling. The individual management zones are thereafter mapped in the field from these observations, with each zone containing a unique combination of soil attributes. This zoning facilitates effective and focused use of farm inputs with each zone being treated with what it needs based on its requirement—hence the maximum development of soil fertility and crop production.

16.3.2 Crop yield data integration

Crop yield data integration is pivotal to the measurement of spatial productivity variability in a field. It is achieved by combining historical yields from precision harvesters, satellite NDVI, and drone images. NDVI is employed as a measure of plant health and biomass and as a measure of photosynthesis and stress. By correlating these datasets with geography, variability in crop performance is mapped and examined across multiple seasons. This fusion enables the identification of consistent zones of high and low performance, informing correlations between input levels, soil attributes, and management practices. Yield variability information enhances seed, fertilizer, and irrigation recommendations specific to zones, enhancing the applicability of VRT.

A typical model employed to compute NDVI is

$$\mathrm{NDVI} = \frac{\mathrm{NIR} - \mathrm{RED}}{\mathrm{NIR} + \mathrm{RED}} \tag{16.1}$$

where

NIR is the reflectance of near-infrared light,
RED is the red light reflectance.

Increased NDVI suggests more healthy vegetation, and that, in turn is directly connected with greater potential yields.

16.3.3 VRT prescription mapping

VRT prescription mapping is the center of activity whereby raw field information is converted to executable input plans. It is a process involving the creation of virtual maps that serve to advise on the site-variable application of seeds, fertilizers, and water as functions of field-site-specific conditions. This starts from information accrued in soil zoning, past years' yield trend data, and crop nutrient specificity. GIS is employed to spatially layer and analyze these data to achieve accurate segmentation of the field into management zones. These zones correspond to every possible combination of soil characteristics, levels of productivity, and demands for resources.

The prescription maps are created through the use of agronomic models and rule-based thresholding. They tell VRT-capable machinery how much to apply at every point. The aim is to put down just what is necessary—no less, no more—to optimize the potential for yield while reducing the environmental footprint and cost of input.

A representative model for fertilizer prescription for a particular zone is

$$I_z = R_c - N_s \tag{16.2}$$

where

I_z is the input rate for zone z (e.g., fertilizer in kg/ha),

R_c is the recommended crop nutrient requirement, and

N_s is the existing soil nutrient supply.

This equation maximizes inputs by considering what the soil will contribute, which makes farming more sustainable and efficient.

16.3.4 Simulation and geostatistical analysis

Simulation and geostatistical analysis play a key role in forecasting responses of crops and soils to VRT-based applications of inputs. This stage is based on mathematical modeling to mimic the interaction between soil conditions, input quantities, and growth patterns of crops for different zones in a field. Patterns in space are mimicked using tools like regression analysis and kriging to estimate crop yields and soil health improvement under different input conditions.

Kriging is a geostatistical interpolation technique that not only produces missing value estimates but also considers spatial correlation between the points. Being able to do so, it may thus be utilized to create surface maps of a continuous nature based on discretely available yield as well as soil data. Regression modeling, alternatively, assists in quantifying interactions between input factors (e.g., fertilizer application rates) and output responses (e.g., crop yield or nutrient concentration).

A simple kriging model is represented as

$$\hat{Z}(x_o) = \sum_{i=1}^{n} \lambda_i Z(x_i) \tag{16.3}$$

where

$\hat{Z}(x_o)$ is the predicted value at the unsampled location x_o,

$Z(x_i)$ is the observed values at known locations,

λ_i is the weights assigned based on spatial correlation, and
n is the number of nearby observations.

This chapter enhances the precision of VRT prescriptions and makes sure that any advice is statistically valid and spatially efficient.

16.3.5 Evaluation of outcomes

Outcome evaluation is the final and most important step in measuring VRT success. Systematic gathering and analysis of data following implementation is necessary to gauge the performance of the VRT strategies in the field. Key indicators of performance are crop yields, input efficiency (fertilizer, seed, and water), and soil health parameter variations like nutrient status, organic matter, and pH. The data is collected through yield monitors, remote sensing equipment, and soil analysis after harvesting.

To measure the benefits of VRT, performance data from variable rate applications are contrasted with that from uniform input application of conventional methods. Comparing the data provides the net advantage of VRT in the form of added productivity, saved costs, and environmental stewardship. Efficiency measurements such as yield per unit of input, cost per hectare, and rates of nutrient recovery are typically used to assess efficiency.

The evaluation of outcomes not only confirms the original prescription maps and simulation models but also offers a form of feedback for future refinement. This iterative series ensures that the VRT approach is adaptive and data-driven and will result in ongoing improvement for both the economic and ecological components of farming operations.

16.4 RESULT

After applying the suggested VRT-based approach, the results were quantitatively analyzed to assess the enhancements in soil health, use efficiency of inputs, and performance of crop yield. The results were compared with the traditional uniform application of inputs. The performance was analyzed in terms of major performance indicators like average crop yield, utilization of fertilizer, water application, and alterations in soil nutrient status. These results mirror the efficiency of input management based on zones to improve agricultural productivity and sustainability. The comparative findings are encapsulated in Table 16.2.

Table 16.2 contrasts VRT-based and flat input applications in terms of important parameters. VRT indicates increased crop yield and nitrogen-use efficiency and minimized fertilizer and water consumption. Soil organic matter also increased with VRT. These findings underscore the impact of VRT on increasing productivity, resource conservation, and improving soil health.

Figure 16.2 illustrates the comparative performance of VRT-based vs. uniform input application on major agricultural performance indicators. The VRT-based method (blue bars) shows greater crop yield (7,850 kg/ha) and better nitrogen recovery efficiency (68%)

Table 16.2 Comparative analysis of VRT-based and uniform input applications on yield, input efficiency, and soil health

Parameter	*VRT-based application*	*Uniform application*	*Improvement (%)*
Average crop yield (kg/ha)	7,850	6,920	+13.4
Fertilizer use (kg/ha)	105	125	−16.0
Water use (L/ha)	4,200	4,800	−12.5
Nitrogen recovery efficiency (%)	68	54	+25.9
Soil organic matter (%)	2.9	2.5	+16.0

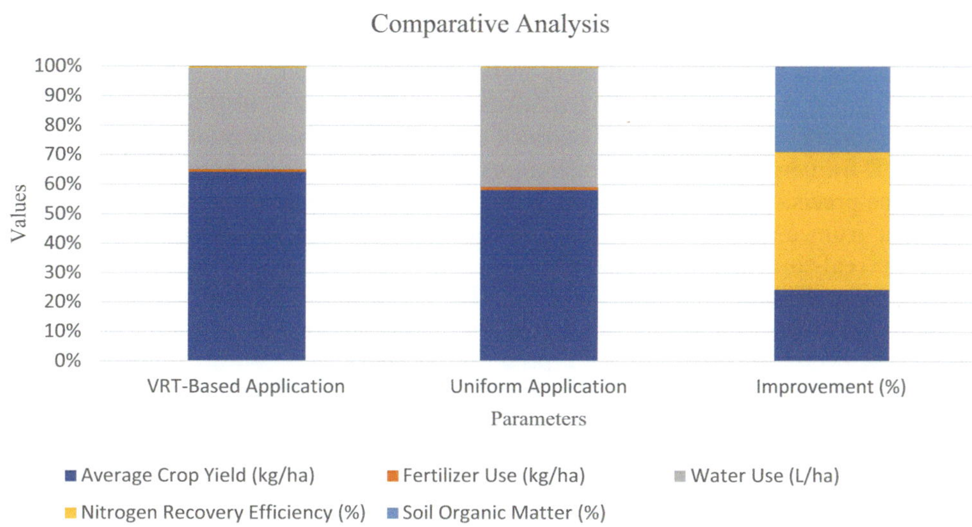

Figure 16.2 Comparative analysis.

with drastic cuts in fertilizer (105 kg/ha) and water application (4,200 L/ha) against the uniform application (orange bars). Soil organic content also increased to 2.9% with VRT, promoting improved long-term soil well-being. Percentage increases are marked by the gray lines, illustrating the resource efficiency and productivity boosts brought about by VRT. The chart, in general, graphically emphasizes the advantage of precision agriculture with VRT for optimized and sustainable farming.

16.5 DISCUSSION

The findings of the current research validate the key advantages of VRT in achieving the highest crop yield, improving utilization of resources, and improving soil health compared to traditional uniform input application. The comparison indicates that application according to VRT increased crop yield by 13.4%, to a level of 7,850 kg/ha, while also reducing fertilizer usage by 16.0% and water consumption by 12.5%. Furthermore, nitrogen recovery efficiency was improved by 25.9%, representing improved nutrient uptake and utilization, leading to minimized environmental losses. Soil organic matter levels also improved by 16.0%, enabling long-term precision agriculture sustainability. These benefits confirm that VRT offers a data-based, site-specific approach to input application, with minimum waste and maximum productivity. In contrast to uniform application of inputs, which makes constant fertilizer and irrigation rates on whole fields, VRT adapts inputs to individual soil types and crop needs, so poor performing areas will get the required nutrients without over-applying to very fertile zones. This deliberate strategy not only maximizes yield but also reduces nutrient runoff, soil erosion, and redundant input expenses.

Despite such advantages, the application of VRT has some limitations and challenges. One of the major limitations includes the high up-front cost incurred in purchasing sensor-based equipment, GPS-equipped machines, and computer programs for GIS-based analysis. Most small- and medium-size farmers might lack the resources for such technologies and hence may restrict their large-scale adoption. Moreover, VRT involves the collection, processing, and analysis of large amounts of data, requiring technical skills that might not be easily accessible to farmers. The precision of soil maps and yield forecast models hinges on the reliability of input data, so flaws in data collection can result in ineffective input prescriptions. Further limitation is the spatial and temporal variability of soils, which necessitates regular monitoring and revisioning of VRT maps. Seasonal variation, unseasonal weather, and chronic soil erosion can affect the validity of zone-based recommendations, which means that the VRT framework must be continuously optimized. Additionally, interoperability across various precision farming systems is a challenge since many farms operate on a mix of equipment from multiple manufacturers, and such equipment might not be seamlessly integrated for real-time VRT implementation.

To overcome such limitations and improve VRT applications further, various avenues for future research and development are suggested. One of the important areas is to improve automation and AI decision-making in VRT prescription generation. With the incorporation of machine learning algorithms, past field data can be analyzed continuously to predict the best input application strategy, enhancing efficiency and accuracy. Furthermore, progress in wireless, low-cost soil sensors and remote sensing can facilitate easier access to real-time monitoring, cutting down on the reliance on costly data-gathering practices. Yet another potential area of advancement is creating cloud-based precision agriculture platforms for farmers, through which they can remotely interact with agronomists and access remotely driven AI-powered suggestions. Subsequent research should also investigate adaptive VRT models that consider variability in climate so that prescription maps are effective under varying growing seasons. Additionally, economic studies of feasibility should be performed to establish the long-term return on investment for adopting VRT, especially for small-scale farmers. Policies and incentives for subsidizing precision agriculture technology may also promote wider adoption.

VRT has proved to be an innovative tactic in recent precision farming, with important repercussions on maximizing returns, reducing consumption of resources, and preserving soil. Comparative results reflect the excellence of the site-specific application of inputs as an impassable case for its execution. Problems such as expense, data sophistication, and merging persist in causing constraints toward the full-scale deployment. Addressing such challenges through innovations in technology, automation, and policy will be essential to enhanced adoption of VRT and thrust toward future innovations in sustainable agriculture. Further studies in this line will see that precision farming will continue to remain sensitive to international food security demands and environmental preservation.

16.6 CONCLUSION

This chapter addresses the revolutionary effect of VRT on soil fertility and the maximization of crop yield, showing its capability to transform contemporary precision agriculture. Through the combination of site-specific input management, GIS-zoning, and sophisticated sensor technologies, VRT allows for effective fertilizer, water, and seed use, offering each field zone the exact quantity of inputs needed. Comparative evaluation between the VRT-based application and the uniform input application demonstrates significant advantages in the form of enhanced crop yield, enhanced nutrient use efficiency, and enhanced soil condition. Results indicate that crop yield was enhanced by 13.4% through VRT-based application, decreased fertilizer application by 16.0%, and lowered water intake by 12.5%, looking back on a more economically sustainable and productive farming practice. Besides this, soil organic matter increased by 16.0%, and efficiency in nitrogen recovery was improved by 25.9%, further testifying to the long-term benefits of the implementation of precision farming practices.

One of the biggest contributions of this research is the confirmation of the spatial variability management in farm fields as valid. Contrary to traditional uniform input applications, where fields are regarded as uniform overall, VRT employs high-resolution soil and yield data to adapt the input level to various zones of the field. This not only optimizes utilization of resources but also reduces nutrient leaching, soil erosion, and environmental pollution; hence, VRT is a critical tool in achieving sustainable agriculture. The combination of remote sensing, past yield mapping, and GIS-based analysis enables the creation of very precise prescription maps, informing farmers to make data-based decisions that maximize productivity while preserving natural resources.

Although these benefits exist, the research identifies some of the challenges and limitations of implementing VRT. Among the main limitations is the significant initial investment cost, such as sensor technology, GPS-enabled equipment, and data processing software. For most small-scale farmers, this is an expensive undertaking that may deter extensive adoption. Moreover, VRT requires high-quality, up-to-date data and technical skills that might not be universally accessible to agricultural operators. Meticulousness in soil data measurement, interpolation, or prescription writing can cause wasteful input allocation, which diminishes the general efficiency of VRT practices. Yet another limitation is that soil conditions are dynamic and need ongoing monitoring and periodic revisions to ensure that VRT applications continue to be viable over varying growing seasons. Additionally, integration issues between disparate precision agriculture systems can complicate implementation, as different sensor and machinery companies utilize proprietary technologies that are not entirely compatible.

Ahead, future technologies in automation, machine learning, and AI-aided decision-making can continue to improve the accuracy and effectiveness of VRT. AI-based predictive models can scan soil and yield data constantly, updating prescription maps in real time to factor in seasonal fluctuations and climate factors. Additionally, low-cost, wireless soil sensors can facilitate real-time measurement of nutrient content and moisture levels, lowering reliance on physical soil sampling. Cloud-based precision agriculture platforms may also be used to enable the sharing of data and remote decision-making, hence making VRT more available to large numbers of farmers. Furthermore, government policies and incentive schemes to subsidize VRT technology may greatly assist in ensuring large-scale adoption, especially in the case of smallholder farmers.

Overall, this chapter validates the huge potential of VRT to maximize agricultural yield while ensuring resource conservation and environmental sustainability. Through the use of spatial variability management, sensor technology, and decision-making based on data, VRT eliminates the inefficiencies inherent in the conventional uniform application of inputs. The results attest to its capacity for enhanced crop yield, input efficiency, and soil health, making it an essential innovation for future sustainable agriculture. Yet, issues of cost, data sophistication, and technology accessibility need to be resolved to optimize its influence. The future will hold research and technological innovation in streamlining VRT applications, further making it flexible and affordable for farmers across the globe. With agriculture on an ongoing trajectory of development, VRT remains a turning point in attaining global food security, minimizing the environmental footprint, and sustaining long-term soil fertility.

REFERENCES

1. Giller, K. E., Delaune, T., Silva, J. V., Descheemaeker, K., Van De Ven, G., Schut, A. G., Van Wijk, M., Hammond, J., Hochman, Z., Taulya, G., & Chikowo, R. (2021). The future of farming: Who will produce our food? *Food Security*, 13(5), 1073–1099.
2. Pawase, P. P., Nalawade, S. M., Bhanage, G. B., Walunj, A. A., Kadam, P. B., Durgude, A. G., & Patil, M. R. (2023). Variable rate fertilizer application technology for nutrient management: A review. *International Journal of Agricultural and Biological Engineering*, 16(4), 11–19.

3. Saleem, S. R., Zaman, Q. U., Schumann, A. W., & Naqvi, S. M. Z. A. (2023). Variable rate technologies: Development, adaptation, and opportunities in agriculture. In *Precision Agriculture* (pp. 103–122). Academic Press.
4. Visser, O., Sippel, S. R., & Thiemann, L. (2021). Imprecision farming? Examining the (in)accuracy and risks of digital agriculture. *Journal of Rural Studies*, 86, 623–632.
5. Matsika, C., & Zhou, M. (2021). Factors affecting the adoption and use of AVR technology in higher and tertiary education. *Technology in Society*, 67, 101694.
6. Pedram, S., Ogie, R., Palmisano, S., Farrelly, M., & Perez, P. (2021). Cost–benefit analysis of virtual reality-based training for emergency rescue workers: A socio-technical systems approach. *Virtual Reality*, 25(4), 1071–1086.
7. Šarauskis, E., Kazlauskas, M., Naujokienė, V., Bručienė, I., Steponavičius, D., Romaneckas, K., & Jasinskas, A. (2022). Variable rate seeding in precision agriculture: Recent advances and future perspectives. *Agriculture*, 12(2), 305.
8. Khan, N., Ray, R. L., Sargani, G. R., Ihtisham, M., Khayyam, M., & Ismail, S. (2021). Current progress and future prospects of agriculture technology: Gateway to sustainable agriculture. *Sustainability*, 13(9), 4883.
9. Tenreiro, T. R., Avillez, F., Gómez, J. A., Penteado, M., Coelho, J. C., & Fereres, E. (2023). Opportunities for variable rate application of nitrogen under spatial water variations in rainfed wheat systems—An economic analysis. *Precision Agriculture*, 24(3), 853–878.
10. Singh, N. V. (n.d.). Effectiveness use and importance of modern technology in Indian agriculture. [Unpublished manuscript or source without formal publication details].
11. Kumar, R., Hegde, A. S., Sharma, K., Parmar, P., & Srivatsan, V. (2022). Microalgae as a sustainable source of edible proteins and bioactive peptides–Current trends and future prospects. *Food Research International*, 157, 111338.
12. Zhang, C., Di, L., Lin, L., Zhao, H., Li, H., Yang, A., Guo, L., & Yang, Z. (2023). Cyberinformatics tool for in-season crop-specific land cover monitoring: Design, implementation, and applications of iCrop. *Computers and Electronics in Agriculture*, 213, 108199.
13. Li, H., Liu, H., Zhou, J., Wei, G., Shi, S., Zhang, X., Zhang, R., Zhu, H., & He, T. (2021). Development and first results of a no-till pneumatic seeder for maize precise sowing in Huang-Huai-Hai plain of China. *Agriculture*, 11(10), 1023.
14. V. K. Shukla, A. Kohli and F. A. Shaikh. (2020). IOT based growth monitoring on Moringa Oleifera through capacitive soil moisture sensor. In *2020 Seventh International Conference on Information Technology Trends (ITT)*, Abu Dhabi, United Arab Emirates, 2020, pp. 94–98. doi: 10.1109/ITT51279.2020.9320884.
15. Ryan, M., Isakhanyan, G., & Tekinerdogan, B. (2023). An interdisciplinary approach to artificial intelligence in agriculture. *NJAS: Impact in Agricultural and Life Sciences*, 95(1), 2168568.
16. Yannis, G., & Michelaraki, E. (2024). Review of city-wide 30 km/h speed limit benefits in Europe. *Sustainability*, 16(11), 4382.

Part IV

Precision Agriculture

Future Direction

Chapter 17

Advancing agricultural efficiency

An exploration of smart farming practices

Ritvik Khurana

17.1 INTRODUCTION TO SMART FARMING AND PRECISION AGRICULTURE

17.1.1 Overview of agricultural advancements

The evolution of agricultural practices represents a profound and continuous transformation of human civilization's fundamental capacity to sustain and nourish its proliferating population. This metamorphosis, spanning millennia, has been punctuated by pivotal advancements that have incrementally augmented our ability to cultivate crops and rear livestock with increasing efficiency and yield [1]. The dawn of the 20th century marked a watershed moment in this progression, heralding an era of unprecedented technological integration into agricultural methodologies. The introduction of mechanization, epitomized by the widespread adoption of tractors and combined harvesters, catalyzed a dramatic surge in productivity while concurrently mitigating the physical demands placed upon the agricultural workforce. This mechanical revolution not only amplified the scale of agricultural operations but also precipitated a change involving the socioeconomic fabric of rural communities, as diminishing requirements for manual labor propelled a significant demographic transition toward urbanization.

The mid-20th century witnessed the advent of the Green Revolution, a period characterized by the synergistic confluence of genetic innovation, agrochemical advancements, and the refinement of irrigation technologies. The period was marked by the advancement and spread of high-yield crop types, especially in essential grains like wheat and rice. These genetic improvements were complemented by the widespread application of synthetic fertilizers and pesticides, which collectively engendered a remarkable augmentation in agricultural output [2]. The ramifications of the Green Revolution extended far beyond mere productivity enhancements; it played a pivotal role in averting widespread famine in numerous developing nations, thereby reshaping global food security paradigms. However, this period also engendered significant environmental and social repercussions, including soil degradation, water pollution, and the exacerbation of socioeconomic disparities among farmers, issues that continue to resonate in contemporary agricultural discourse.

Precision agriculture emerged as a result of the early integration of agricultural practices and information technology in the latter half of the 20th century. A major breakthrough was made possible by the introduction of Global Positioning System (GPS) technology in the 1980s, which enabled farmers to map and precisely define their fields like never before. The advent of GPS in agriculture was complemented by the concurrent evolution of Geographic Information System (GIS), which provided a sophisticated platform for the analysis, interpretation, and visualization of spatial data [3]. A fundamental change in agricultural

DOI: 10.1201/9781003520733-21

decision-making was brought about by the symbiotic relationship between GPS and GIS technologies, which allowed farmers to use data-driven management techniques to maximize resource allocation and reduce environmental impact.

Rapid developments in technological innovations and the expanding field of data analytics drove a significant acceleration of agricultural innovation as we crossed the threshold into the 21st century. An era of unparalleled continuous surveillance capabilities in agricultural settings has been brought about by the widespread use of technologically advanced sensors and Internet of Things (IoT) devices. Numerous parameters, such as crop health indicators, moisture in the soil, nutrient levels, and microclimatic conditions, can be continuously assessed thanks to these technologies [4]. These systems' precision and immediacy of data enable more accurate and timely interventions, enabling farmers to maximize resource use and react quickly to new problems. At the same time, the development of remote sensing technologies, including drone-based surveillance systems and satellite imagery, has greatly increased the breadth and dimension of agricultural monitoring [5]. These technologies improve decision-making and operational efficiency by giving farmers detailed insights into crop health, yield potential, and field variability at previously unachievable scales.

The forefront of agricultural innovation is represented by the incorporation of robotics and autonomous systems, which have the potential to further improve productivity and accuracy while addressing enduring issues like labor shortages. Autonomous tractors equipped with advanced sensor arrays and artificial intelligence (AI) systems can perform complex field operations with minimal human intervention, optimizing fuel efficiency and minimizing soil compaction [6]. Similarly, robotic harvesting systems, designed to navigate the intricacies of fruit picking and vegetable harvesting, promise to revolutionize labor-intensive aspects of crop production. These technologies not only augment operational efficiency but also have the potential to mitigate the physical demands placed on agricultural workers, thereby improving working conditions in the sector.

As we contemplate the cumulative impact of these technological advancements, it becomes evident that we stand at the nascent beginning of a new agricultural norm. The journey from mechanization to digitalization has not only exponentially increased productivity but also ushered in an era of unprecedented precision.

17.1.2 Introduction to smart farming

The integration of cutting-edge technologies to maximize crop yield, resource efficiency, and environmental stewardship is what defines smart farming, also known as precision agriculture or Agriculture 4.0, as an evolution in agricultural management techniques. To build an exceptionally receptive and data-driven farming ecosystem, this creative approach to agriculture makes use of various cutting-edge technologies, such as AI, robotics, big data analytics, ML, and the IoT [7]. The fundamental premise of smart farming lies in its capacity to collect, process, and analyze vast amounts of real-time data from multiple sources, enabling farmers to make informed decisions and implement targeted interventions with unprecedented precision and efficiency.

Fundamentally, smart farming revolves around the concept of spatiotemporal variability in agricultural systems. Traditional farming practices often apply uniform management strategies across entire fields or farms, disregarding the inherent heterogeneity in soil properties, microclimatic conditions, and crop characteristics. By deploying an extensive array of sensors and IoT devices across the agricultural landscape, smart farming, on the other hand, recognizes and manages this variability [8]. Numerous parameters, such as moisture in the soil, levels of nutrients, humidity, and temperature, and crop health indicators, are continuously monitored by these sensors. Smart farming systems are built on the high-definition, real-time data produced by these sensors, which makes it possible to create intricate digital maps that depict the temporal and spatial patterns of agricultural environments.

The extent and scale of data collection in smart farming systems are further increased by the incorporation of remote sensing technologies, which include satellite imaging, aerial imagery, and unmanned aerial vehicles (UAVs) [9]. At the field and regional levels, these technologies offer significant insights into crop condition, biomass calculation, and yield prediction. The detection of minute alterations in crop physiology is made possible by sophisticated algorithms for image processing and spectral analysis techniques, which aid in the early detection of stressors like pest infestations, nutrient shortages, or water stress. These various sensing modalities produce enormous amounts of data, which calls for complex handling and analysis of data systems. To derive actionable insights from the extensive and diverse datasets that are a feature of smart farming, big data analytics, and machine learning (ML) algorithms are essential. Predictive modeling approaches, leveraging historical data in conjunction with real-time observations, allow farmers to anticipate crop development stages, optimize resource allocation, and forecast potential issues before they manifest.

Precision application, which seeks to maximize the effective utilization of agricultural inputs like water, fertilizer, and pesticides, is a fundamental idea in smart farming. With the help of complex algorithms and high-resolution spatial data, variable rate technology (VRT) processes allow for accurate use of inputs according to the unique needs of each area within a field. This focused strategy reduces the negative effects of excessive agrochemical use on the environment while simultaneously improving resource-use efficiency. Precision irrigation systems, for example, can drastically cut water use while preserving ideal crop growth conditions by modifying irrigation levels in real time depending on moisture content and hydration of the soil, accompanied by humidification sensors, gauges, and harvest water demand models [10]. With the help of sophisticated navigation systems and AI models, autonomous tractors and field robots can complete mundane yet highly taxing jobs like sowing, weeding, and harvesting with little assistance from humans.

Smart farming also encompasses the realm of livestock management, where technologies such as wearable sensors, automated feeding systems, and AI-driven health monitoring platforms are transforming animal husbandry practices [11]. Real-time monitoring of animal behavior, health, and productivity is made possible by these technologies, which facilitates the early identification of health problems and the improvement of feeding practices. The integration of genomic data and advanced analytics in breeding programs further enhances the potential for genetic improvement and the development of more resilient and productive livestock breeds.

The purview of smart farming goes far beyond the farm gates, embracing the complete agricultural chain [12]. Digital platforms and blockchain technologies are being leveraged to enhance traceability and transparency in food supply chains, addressing growing consumer demands for information on food provenance and production methods. These systems not only facilitate more efficient supply chain management but also contribute to food safety and quality assurance efforts. Furthermore, the resilience and economic viability of agricultural businesses are increased when smart farming data is integrated with weather forecasting and market intelligence systems to facilitate better decision-making about crop selection, planting dates, and marketing tactics.

17.1.3 Precision agriculture as a foundation

Precision agriculture and smart farming are intimately intertwined concepts that represent a continuum of technological innovation in agricultural management practices. While often used interchangeably, precision agriculture can be viewed as the foundational concept that paved the way for the more comprehensive and technologically advanced smart farming approaches. The evolution from precision agriculture to smart farming reflects the rapid advancements in digital technologies, data analytics, and AI that have occurred over the past few decades. Precision agriculture emerged in the 1980s with the introduction of GPS technology in agricultural settings. This technological breakthrough enabled farmers to map their fields with unprecedented accuracy and implement site-specific management practices. Based on grid soil sampling and yield mapping, early precision agriculture methods mostly applied inputs like fertilizers and pesticides at variable rates. These approaches represented a significant departure from conventional uniform management practices, allowing for more efficient resource utilization and potential yield improvements.

As precision agriculture evolved, it incorporated an expanding array of technologies, including GIS, remote sensing, and on-the-go sensors. These advancements facilitated more comprehensive and elaborate data collection and robust analysis tools, enabling farmers to employ informed decision-making, for various aspects of crop management. The integration of these technologies allowed for the inculcation of precision farming practices encompassing not only input application but also aspects such as planting density optimization, irrigation management, and harvest timing. This holistic approach to farm management laid the groundwork for the more integrated and data-driven strategies characteristic of smart farming.

The evolution from precision agriculture to smart farming also reflects a shift in the conceptualization of farm management. Precision agriculture introduced the idea of site-specific management, but decision-making processes remained largely human-centered. Smart farming, in contrast, leverages AI and automated systems to augment and, in some cases, replace human decision-making in various aspects of farm management [13]. This shift is exemplified by the development of autonomous agricultural robots, AI-driven decision support systems, and automated farm management platforms. These technologies not only enhance operational efficiency but also have the potential to address labor shortages and reduce the physical demands placed on farm workers [14,15].

Despite these distinctions, it is crucial to recognize that smart farming builds upon and incorporates many of the fundamental principles and technologies of precision agriculture. The site-specific management strategies, geospatial analysis techniques, and variable rate application (VRA) technologies developed within the precision agriculture paradigm remain integral components of smart farming systems. In this sense, precision agriculture can be viewed as a subset of smart farming, providing the foundational concepts and technologies upon which more advanced smart farming approaches are built. This continuity underscores the importance of precision agriculture in the ongoing evolution of agricultural technology and management practices.

17.1.4 Relevancy and urgency

In the 21st century, the global agriculture industry faces previously unheard-of difficulties that call for creative solutions to guarantee food security, ecological integrity, and financial sustainability. To tackle these inevitable challenges, precision farming appears as a crucial technology paradigm that may provide answers for a few of the most urgent issues facing contemporary agriculture. The confluence of multiple global developments that significantly strain agricultural systems globally highlights the need of using and advancing smart farming technologies. One of the biggest problems facing agriculture worldwide is climate change. Climate change is expected to have a major influence on agricultural output, with certain regions seeing yield decreases of up to 25% by 2050, according to the Intergovernmental Panel on Climate Change (IPCC) [16].

Smart farming technologies offer promising avenues for both mitigation and adaptation strategies. For example, precision irrigation systems can significantly lower agricultural water use, which is important because climate change is predicted to make water scarcity worse in many areas [17]. Smart irrigation systems could achieve water savings of up to 40% while maintaining or even improving crop yields. Widespread adoption of precision agriculture technologies could reduce greenhouse gas emissions from European agriculture. By 2050, there will be 9.7 billion people on the planet, and to meet demand, food production will need to rise by 70% [18]. This demographic pressure is compounded by changing dietary preferences, particularly the increasing demand for animal-based proteins in developing economies. Smart farming technologies offer pathways to significantly enhance agricultural productivity without proportionally increasing resource inputs. For example, AI-driven crop management systems could improve yields by up to 25% while reducing fertilizer and pesticide use by 35%. Technologies for precision livestock farming have shown promise in increasing feed conversion efficiency by 12%–18% in the livestock industry, which would lessen the ecological impact of animal production [19].

Another significant issue is the loss of agricultural land due to urbanization. By 2050, 68% of the global population is expected to reside in cities, a rise from 55% in 2018, according to United Nation forecasts. The amount of usable agricultural land is severely strained by this tendency, soil erosion, and desertification. Vertical farming, controlled environment agriculture, and smart farming technologies provide ways to produce food in urban and peri-urban settings while using the least amount of land possible. In comparison to conventional agriculture, vertical farming techniques can use 90% less water and produce up to 12 times as much food per unit area [20].

The depletion and degradation of natural resources, particularly soil and water, pose significant threats to long-term agricultural sustainability. Precision agriculture techniques enabled by smart farming technologies allow for more efficient and targeted use of inputs, potentially mitigating resource degradation. For instance, VRT in fertilizer application can reduce over-fertilization, which contributes to soil acidification and water pollution. Precision fertilizer application could reduce nitrogen runoff by up to 60%, significantly mitigating the environmental impact of agriculture on aquatic ecosystems [21]. Labor shortages in the agricultural sector, particularly in developed countries, present another challenge that smart farming technologies are well-positioned to address. The increasing average age of farmers in many regions, coupled with rural-to-urban migration, has led to concerns about the future of farm labor.

The need to implement smart farming technologies immediately has been highlighted by the COVID-19 outbreak. The weaknesses in the current agricultural systems were brought to light by the disruptions to global labor markets and supply networks, underscoring the need for more automated and resilient farming methods. As farmers look to increase supply chain visibility and lessen their dependency on human labor, the pandemic has sped up the deployment of technologically advanced agrarian technologies.

17.2 TECHNOLOGICAL FOUNDATIONS OF SMART FARMING

17.2.1 Internet of Things (IoT)

A key component of smart farming, the IoT enables hitherto unheard-of levels of data collecting, surveillance, and management in agricultural operations. In agriculture, IoT refers to a broad range of connected equipment, sensors, and actuators that gather and send data in real time, empowering farmers to automate different parts of farm management and make well-informed decisions. With the promise of improved yields, resource efficiency, and sustainability, the incorporation of IoT technologies into farming methods has sparked a transition from intuition-based to data-informed agriculture. A sophisticated ecosystem of sensors intended to track a wide range of crop-related and environmental characteristics is at the core of IoT applications in automated agriculture. These sensors, which are frequently small and energy-efficient, have a high degree of accuracy and frequency in measuring factors including humidity, temperature, light intensity, soil moisture, and nutrient levels. To provide a three-dimensional picture of soil conditions, sophisticated IoT soil sensors may measure at various depths and positions within a field. This granular data enables farmers to implement targeted irrigation and fertilization strategies, potentially reducing water usage by up to 30% and fertilizer application by 20% [22].

IoT device proliferation in agriculture includes farming and livestock surveillance in addition to soil monitoring. When placed throughout fields, wireless sensor networks can identify early indicators of agricultural stress, illnesses, or insect infestations, enabling prompt responses. While preserving crop production, IoT-based alert systems for insect detection could cut pesticide use by as much as 40%. IoT-enabled wearable technology is used in livestock management to track the production, behavior, and health of animals in real time. These methods greatly enhance treatment results and animal comfort by identifying health problems in dairy cows for a maximum of 48 hours prior to they become clinically noticeable. Advanced data transfer methods complement IoT devices' real-time data collection capabilities in smart farming.

Another important development in smart farming technology is the combination of edge computing with IoT devices. Edge computing lowers latency and bandwidth demands by analyzing information at or close to the source, allowing for real-time control and decision-making. Water savings over conventional irrigation techniques are possible with a cloud computing-enabled precision irrigation system, which locally interprets sensor data to make real-time water distribution modifications [23]. In smart farming, cloud platforms are essential for collecting, conserving, and analyzing the massive volumes of data produced by IoT devices. These platforms offer scalable storage solutions and provide access to advanced analytics tools and ML algorithms. The possibilities of smart agricultural systems are being further enhanced by the interaction of IoT with other cutting-edge technologies like blockchain and AI. AI algorithms can process complex, multidimensional data streams from IoT devices to identify patterns, predict outcomes, and generate actionable insights. AI-powered IoT systems can predict crop yields with over 90% accuracy, enabling more precise harvest planning and supply chain management. Blockchain technology, on the other hand, offers potential solutions to data security and traceability challenges in IoT-based smart farming. Blockchain-based architecture for securing IoT data in agriculture ensures data integrity and enables transparent, traceable food supply chains [24].

Despite the significant potential advantages of IoT in automated agriculture, several obstacles still stand in the way of its broad adoption and best use. To fully realize the possibilities of IoT in agriculture, issues including cybersecurity, data standardization, and device compatibility among manufacturers must be resolved. Adoption is further hampered by the initial investment expenses and the requirement for technical know-how, especially for small-scale cultivators in developing nations. In conclusion, by enabling previously unattainable levels of surveillance, oversight, and data-driven decision-making, IoT technologies are radically changing agricultural operations.

17.2.2 Robotics and automation

The evolution of intelligent farming measures has advanced significantly with the introduction of automated systems and robotics in agriculture. These technologies provide accuracy, efficiency, and scalability through computational prowess and are transforming conventional agricultural operations. From autonomous tractors navigating fields with centimeter-level accuracy to drones providing high-resolution aerial imagery, robotics and automation are reshaping the landscape of modern agriculture. Autonomous tractors represent one of the most visible and impactful applications of robotics in precision farming. These self-driving machines utilize a combination of GPS guidance, computer vision, and advanced control systems to execute various field operations with minimal human intelligence and intervention. These machines can execute planting, spraying, and harvesting tasks with subinch accuracy, optimizing input use and minimizing soil compaction. The precision of these systems extends beyond mere navigation; they can adjust seeding rates, fertilizer application, and spraying patterns in real-time based on preprogrammed prescriptions or live sensor data [25]. This level of precision ensures that each section of the field receives exactly what it needs, maximizing yield potential while minimizing waste. Case International Harvester (IH's) autonomous concept vehicle, for instance, demonstrates the potential for completely driverless operation in various field tasks, showcasing the future of farm mechanization.

In precision agriculture, UAVs, or drones, have become indispensable instruments. These flying platforms, which are outfitted with multispectral and hyperspectral sensors, give farmers access to real-time, high-resolution images of their crops. Compared to ground-based scouting techniques, drone-based photography can identify insect infestations, crop stress, and nutrient deficits up to two weeks early [26]. This early detection capability enables timely interventions, potentially reducing crop losses. The spectral data collected by drones can be analyzed to create detailed vegetation index maps, revealing subtle variations in crop health that are invisible to the naked eye. To guarantee that inputs are administered only where necessary, these maps direct variable-rate applications of fertilizers and insecticides. Furthermore, drones equipped with precision spraying systems can apply pesticides and fertilizers with remarkable accuracy, reducing chemical use by up to 20% while minimizing environmental impact [27]. Advanced drone systems are now incorporating AI-driven image analysis algorithms that can identify specific weed species or count fruits on trees, providing farmers with unprecedented levels of crop intelligence.

Ground-based robotic systems for specific agricultural tasks have also made significant strides. Automated weeding robots combine computer vision and precise mechanical or laser-based systems to target and remove weeds without damaging crops. In some crops, robots can cut the amount of pesticide employed while also requiring less work to control weeds. These devices can function in various soil types and crop row configurations and employ complex algorithms to differentiate among harvests and weeds, even in their early stages of growth [28]. ML skills are incorporated into some sophisticated models, which enable them to gradually increase the accuracy of weed identification. Harvesting robots with sophisticated grippers and computer-aided vision are starting to take on the challenging chore of picking fruits and vegetables in horticultural settings. These robots use a combination of visual and tactile sensors to assess fruit ripeness, size, and quality, ensuring that only suitable produce is harvested. A strawberry harvesting robot that can identify and pick ripe fruit with an accuracy rate of 86% is potentially addressing labor shortages in labor-intensive crops [29]. The development of soft robotic grippers is further enhancing the ability of these machines to handle delicate produce without causing damage.

The advantages of robotic interventions in agriculture are manifold. Firstly, these technologies offer significant improvements in operational efficiency. Autonomous systems can work continuously, unaffected by fatigue or adverse weather conditions, potentially extending the workable hours in a day. This capability is particularly valuable during critical planting and harvesting windows, where timeliness can significantly impact yield and quality. Widespread adoption of agricultural robotics could increase overall farm productivity by 15%–20%. Secondly, robotic systems excel in performing repetitive tasks with high precision and consistency, leading to optimized resource use. For instance, precision weeding robots can reduce herbicide application by targeting individual plants, minimizing chemical runoff and environmental impact. This precision extends to other tasks such as pruning, thinning, and selective harvesting, where robots can make decisions based on predefined quality parameters, ensuring consistent product quality.

Labor shortages in agriculture, particularly in developed countries, present another challenge that robotics and automation are well-positioned to address. As the agricultural workforce ages and rural-to-urban migration continues, robotic systems offer a solution to maintain and even increase agricultural productivity. These technologies can take over physically demanding or repetitive tasks, improving working conditions for human laborers and potentially attracting a new generation of workers to the agricultural sector. By 2050, up to 40% of current agricultural labor could be replaced by autonomous systems, potentially offsetting the impacts of labor shortages and making agriculture more attractive to a new generation of tech-savvy farmers [30]. New, highly skilled occupations in fields like data analysis, system optimization, and agricultural robots' maintenance may result from this change.

The integration of robotics with other smart farming technologies further amplifies their impact. For example, the combination of IoT sensors, AI-driven analytics, and autonomous machinery creates a closed-loop system for precision agriculture. Robotic systems may make decisions in real time based on sensor data of weather patterns, crop health, and soil conditions, allowing for extremely responsive and optimized farm management. This integration allows for the development of sophisticated farm management systems that can coordinate multiple robotic units, optimizing their deployment based on real-time field conditions and operational priorities. Integration is a key step toward fully autonomous farming systems, where human intervention is minimized, and farm operations are orchestrated by AI-driven platforms. Farms might become extremely data-driven, efficient enterprises with the ability to quickly adjust to shifting market needs and environmental conditions thanks to these tools.

Agricultural robotics adoption is not without its difficulties, though. Barriers may include the requirement for specialized maintenance, substantial initial investment costs, and technical complexity, especially for small-scale farmers. One of the main goals of the field's research and development activities is still to create more accessible and user-friendly robotic systems. Furthermore, the legal and liability frameworks pertaining to autonomous agricultural systems are still developing. It is necessary to address concerns over data ownership, confidentiality, and the possibility of cyberattacks on autonomous agricultural systems.

17.2.3 Agricultural sophistication with AI and ML

In the field of smart farming, AI and ML have become game-changing technologies, providing previously unheard-of capacities in crop management, predictive analysis, and decision-making optimization. By making it possible to process and comprehend the enormous volumes of complicated data produced by different smart farming technologies, these sophisticated computational techniques are transforming agricultural operations and ultimately resulting in more intelligent and effective management strategies. Through predictive analysis, AI algorithms are being employed to forecast a wide range of agricultural phenomena with remarkable accuracy. Crop yield prediction, a critical aspect of farm management and supply chain planning, has been significantly enhanced through the application of ML models. These models integrate historical yield data with real-time environmental parameters, satellite imagery, and IoT sensor readings to generate highly accurate yield forecasts. Ensemble ML models, combining multiple models such as random forests and support vector machines, can predict crop yields with an accuracy of up to 95% several weeks before harvest [31]. This level of predictive capability allows farmers to optimize harvest timing, labor allocation, and supply chain logistics, potentially reducing postharvest losses and improving market positioning.

AI-driven predictive models are also being applied to pest and disease management, an area where early detection and intervention are crucial. By analyzing patterns in weather data, crop stress indicators, and historical pest outbreak information, these models can forecast the likelihood and severity of pest infestations or disease outbreaks. An AI system that could predict potato late blight outbreaks with 85% accuracy up to two weeks in advance, allowing farmers to implement preventive measures and potentially reduce fungicide use by up to 30% [32]. Such predictive capabilities are particularly valuable in the context of climate change, where shifting weather patterns are altering traditional pest and disease cycles, making historical knowledge less reliable.

Automated crop surveillance and phenotyping have advanced significantly because of the combination of deep learning and computer vision techniques. High-resolution imagery from drones and satellites is being analyzed using Convolutional Neural Networks and other deep learning architectures. This allows for the automated detection of weed infestations, crop stress, and nutrient deficiencies at scales and speeds that are not feasible for human observers. A deep learning system that could identify and classify crop diseases from leaf images with an accuracy of over 99%, potentially revolutionizing plant pathology and enabling rapid, large-scale disease surveillance.

ML algorithms are also being applied to optimize complex farm management decisions that involve multiple, often conflicting, objectives. Multiobjective optimization techniques, such as genetic algorithms, are being used to develop decision support systems that can balance factors such as yield maximization, input cost minimization, and environmental impact reduction. An ML-based decision support system for maize production could simultaneously optimize fertilizer application, irrigation scheduling, and pest control strategies, leading to a 15% increase in profit margins while reducing environmental impacts [33]. The application of reinforcement learning, a subdomain of ML where models are trained for optimal actions through trial and error, is opening new frontiers in adaptive farm management. These systems can continuously learn and improve their decision-making processes based on observed outcomes, adapting to changing environmental conditions and farm-specific characteristics. A reinforcement learning system for greenhouse climate control that could optimize temperature, humidity, and CO_2 levels for tomato production results in a 20% increase in yield and a 30% reduction in energy consumption compared to traditional control methods [34].

As AI and ML technologies continue to advance, their integration with other smart farming technologies is creating increasingly sophisticated and autonomous agricultural systems. The combination of AI with robotics, for instance, is enabling the development of intelligent agricultural robots capable of making real-time decisions based on their environment. These robots are capable of more accurate and efficient work than humans in activities like targeted spraying, precision weeding, and selected harvesting. AI-driven agricultural robotics could reduce labor requirements by up to 60% in certain farming operations while simultaneously improving precision and reducing input waste.

Adoption of AI and ML for farming is not devoid of challenges though. For ML model training, the requirement for sizable, high-quality datasets can be a major obstacle, especially in areas with inadequate digital infrastructure. Furthermore, farmers may find it challenging to comprehend and believe the advice produced by these systems due to the "black box" characteristics of some sophisticated AI algorithms. To overcome these obstacles, data analysts, agronomists, and farmers must work together across disciplinary boundaries to create explainable AI systems that offer clear and understandable information. Furthermore, special attention must be paid to ethical issues pertaining to data ownership, confidentiality, and the possibility that AI will worsen already-existing disparities in the agriculture industry. The sustainable deployment of AI technologies in worldwide agriculture depends on the creation of platforms that are not just technologically sophisticated but also inclusive and socially conscious.

17.2.4 Solving farming woes with big data and data analytics

A new era of accuracy and efficiency in agricultural techniques has been brought about by the development of big data and sophisticated data analytics, which has completely changed how farming choices are made and carried out. Farm-related data has accumulated as a result of the widespread use of sensors, IoT equipment, and other data-gathering technology in agriculture. This data includes anything from weather patterns and market trends to crop health and soil conditions. When appropriately tapped using advanced analytical methods, this enormous informational resource has the potential to transform agricultural methods, boost output, and promote sustainability. Fundamentally, big data in agriculture makes it possible to move away from broad, universal methods and toward highly customized, site-specific management plans. By optimizing inputs like water, fertilizer, and pesticides, this granular approach for agricultural management can boost resource efficiency and lessen its negative effects on the environment. Big data applications in intelligent agriculture were thoroughly reviewed, and it was discovered that data-driven strategies might boost yields while lowering fertilizer and water use. These improvements are achieved through the integration of multiple data streams, including real-time sensor data, historical yield information, and high-resolution satellite imagery, which collectively provide a holistic view of the agricultural ecosystem.

Crop production prediction and optimization is one of the primary domains where big data is having a large impact. Highly accurate yield projections can be produced by using sophisticated ML algorithms on enormous datasets that include past yield data, weather trends, soil properties, and management techniques. Farmers can use these forecasts to help them decide on crop types, planting dates, and resource allocation. A novel approach combined satellite imagery with ML techniques to predict wheat yields with an accuracy of over 95% several weeks before harvest [35]. With this degree of predictive power, farmers may better plan harvest times, handle supply chain logistics, and decide on marketing strategies.

VRA systems for inputs including fertilizers, herbicides, and irrigation water have been developed because of the combination of big data analytics with precision agriculture technologies. These systems generate comprehensive prescription maps for input application using high-resolution spatial data, frequently obtained from an assortment of soil sensors, drone images, and satellite data. The effects of precision agricultural technology on the environment discovered that VRA systems could lower pesticide and fertilizer consumption while preserving or even increasing crop yields. Such input consumption reductions have significant positive effects on the environment, such as lower greenhouse gas emissions, less water pollution, and better soil health. Additionally, big data analytics are essential to the control of diseases and pests. Predictive models can more accurately anticipate the probability and magnitude of pest infestations or disease outbreaks by examining trends in weather information, crop stress indicators, and previous pest outbreak data. A system that could forecast fungal disease outbreaks in vineyards with accuracy, three weeks ahead of normal detection times was developed. Farmers may take focused preventive action thanks to this early warning capacity, which may cut down on pesticide use and crop losses.

The role of data in improving livestock farming practices is equally significant. Advanced analytics applied to data from wearable sensors, automated feeding systems, and environmental monitors are enabling precision livestock farming. Real-time tracking of livestock's health, behavior, and productivity by these technologies enables the early identification of health problems and the improvement of feeding practices. Data-driven approaches could improve feed conversion efficiency by up to 15% while reducing antibiotic use by 50% through early disease detection and targeted treatment.

In smart farming, data-driven decision-making signifies a change from conventional, experience-based farming to a more empirical, scientific method. The creation of advanced farm management information systems (FMIS) that combine data from various sources and give farmers useful insights has made this shift easier. To handle intricate, multifaceted datasets and produce recommendations for numerous aspects of farm management, these systems frequently integrate AI and ML techniques. The capacity to manage complexity and uncertainty is what gives data-driven decision-making its potency. Agricultural systems are inherently complex, with numerous interacting variables influencing crop growth and yield. Big data analytics can uncover subtle patterns and relationships within these complex systems that may not be apparent through traditional analysis methods. For instance, a decision support system for maize production integrated data on weather patterns, soil conditions, and management practices to optimize fertilizer application and irrigation scheduling.

Additionally, farmers are better equipped to adjust to changing circumstances thanks to data-driven decision-making, both in the short term (such as reacting to weather occurrences) and in the long term (such as adjusting to climate change). By continuously analyzing real-time data and updating predictive models, these systems can provide dynamic recommendations that evolve with changing conditions. This adaptive capability is particularly valuable in the face of increasing climate variability and extreme weather events. The integration of big data analytics with other smart farming technologies, such as autonomous machinery and robotics, is paving the way for fully automated farming systems. In these systems, data-driven decisions can be directly implemented through autonomous equipment, creating a closed-loop system that continuously optimizes farm operations. Such integrated systems could reduce labor requirements by up to 60% in certain farming operations while simultaneously improving precision and reducing input waste.

However, there are a number of obstacles to the efficient application of big data in agriculture. Interoperability, standardization, and data quality are still major concerns, especially in light of the various forms and sources of agricultural data. To guarantee that data-driven farming methods are widely used, issues with data confidentiality, ownership, autonomy, and security must also be resolved. To guarantee that farmers can successfully understand and act upon the insights produced by advanced analytics tools, infrastructure development and education are also required.

17.3 APPLICATIONS OF SMART FARMING TECHNOLOGIES; CASE STUDIES

17.3.1 IoT for precision farming tasks

IoT deployment in crop management and irrigation systems has led to significant improvements in resource efficiency, crop yield, and overall farm productivity. To demonstrate the exemplary potential of these technologies in contemporary agriculture, this part examines thorough case studies and instances of IoT implementation including adaptive irrigation, pest control, and soil monitoring. IoT technology has made it possible to create extremely effective water management systems in the field of smart irrigation. The deployment of an IoT-based targeted irrigation system in California's drought-prone almond orchards is a noteworthy case study. Researchers from the University of California, Davis, in collaboration with local farmers, deployed a network of soil moisture sensors, weather stations, and automated irrigation valves across several orchards. The system utilized LoRaWAN technology for long-range, low-power communication, enabling real-time data transmission from remote areas of the farms. ML algorithms processed the collected data to generate optimized irrigation schedules tailored to specific microclimates within each orchard.

The results of this implementation were remarkable. The IoT-based system reduced water consumption when compared to conventional irrigation measures, while maintaining almond yield and quality. Further, the system's ability to recognize and react to soil moisture variations in real time led to a reduction in water runoff, significantly mitigating the environmental impact of irrigation practices [36]. The economic benefits were equally impressive, with farmers reporting a reduction in energy costs associated with water pumping and a payback period of less than two years for the initial investment in IoT infrastructure. Another compelling example of IoT application in precision farming comes from the viticulture industry in France's Bordeaux region. Here, a consortium of wine producers implemented an IoT-driven pest control system to combat the devastating effects of grapevine moths. The system comprised a network of pheromone traps equipped with image sensors and environmental monitors. These devices use edge computing capabilities to process captured images locally, identifying and counting pest species in real time. After that, the information was sent to a centralized cloud platform where it was analyzed and combined with past pest trends and weather data.

The outcomes of this implementation note that the IoT-based pest monitoring system enabled a reduction in pesticide use across participating vineyards. The system's ability to provide early warning of pest infestations allowed for targeted interventions, replacing broad-spectrum pesticide applications with more precise, ecologically friendly control methods [37]. Moreover, the continuous monitoring capabilities of the IoT network facilitated the adoption of biological control agents, leading to an increase in beneficial insect populations within the vineyards. The improved pest management strategies resulted in an increase in high-quality grape yield, significantly enhancing the economic returns for participating wine producers.

Soil monitoring represents another critical application area for IoT technologies in precision farming. A comprehensive case study from Brazil's soybean production regions demonstrates the potential of IoT-driven soil management systems. Researchers from Embrapa, Brazil's agricultural research corporation, collaborated with technology providers to deploy an extensive network of soil sensors across multiple farms in the Cerrado region. These sensors measured various parameters, including soil moisture, temperature, electrical conductivity, and nutrient levels at different depths. The IoT system utilized a combination of NB-IoT and satellite communication technologies to ensure reliable data transmission from remote agricultural areas. Advanced data analytics platforms processed the collected information, integrating it with satellite imagery and historical yield data to generate high-resolution soil maps and management zone recommendations.

The outcomes of this large-scale IoT deployment were highly positive. The system enabled the implementation of variable-rate fertilizer application, resulting in a reduction in overall fertilizer use while maintaining or even improving soybean yields. The continuous soil monitoring capabilities allowed farmers to detect and respond to soil acidification issues more rapidly, leading to a reduction in lime application for pH correction. Further, the improved understanding of soil variability facilitated by the IoT system enabled more effective crop rotation strategies, contributing to an increase in soil organic matter content over three years. The economic impact of the IoT-driven soil management system was substantial, with participating farms reporting an increase in profit margins due to reduced input costs and improved yields [38]. Additionally, the environmental benefits were significant, with a 20% reduction in nitrate leaching and a 15% decrease in greenhouse gas emissions associated with fertilizer production and application.

The transformational potential of IoT technology in precision farming is demonstrated by these case studies. Critical issues in water management, pest control, and soil health are being addressed by IoT applications that provide continuous surveillance, information-driven choices, and automated control systems. The potential of intelligent farming systems is being further enhanced by the incorporation of IoT with additional advanced technologies like AI, ML, and robots. This is opening the door to more productive, efficient, and sustainable agricultural methods.

It is crucial to remember, nevertheless, that the effective application of IoT in precision agriculture necessitates giving serious thought to elements like infrastructure needs, data management procedures, and farmer education. Significant gains in agricultural production and sustainability across a range of farming systems and geographic contexts are anticipated as these technologies continue to advance and become more widely available.

17.3.2 Robotics in agriculture

The integration of robotics in agriculture represents a significant leap forward in the automation and precision of various field tasks. This section explores robotic applications in key field tasks such as sowing, fertilizing, and harvesting, highlighting the practical implementations and outcomes of these cutting-edge technologies in real-world agricultural settings. Automated sowing, a pioneering case study, comes from the large-scale wheat farms of Western Australia. Researchers from the University of Sydney, in collaboration with the Australian Centre for Field Robotics, developed and deployed an autonomous seeding robot named "SwarmFarm" across several wheat farms. This robotic system utilizes RTK-GPS for navigation, achieving subcentimeter accuracy in seed placement. The robot is equipped with a sophisticated seed delivery system that can adjust seeding depth and density in real time based on soil conditions and preprogrammed prescription maps.

Over three consecutive growing seasons, the SwarmFarm system delivered consistently good performance. The robotic seeder demonstrated remarkable precision, with a seed placement accuracy of ±1.5 cm both horizontally and vertically. This level of precision resulted in an improvement in seed germination rates compared to traditional seeding methods. Furthermore, the system's ability to operate continuously, unaffected by fatigue or visibility conditions, allowed for optimal timing of seeding operations. This led to an increase in early plant establishment, which subsequently contributed to overall wheat yield. The economic analysis of the SwarmFarm implementation revealed a reduction in fuel consumption due to the optimized path planning and lighter weight of the robotic unit compared to traditional tractors [39]. Labor costs associated with seeding operations decreased, although this was partially offset by the need for skilled technicians to maintain and oversee the robotic system. The initial investment in the robotic seeding system was recovered within three years, primarily due to increased yields and operational cost savings.

In the domain of precision fertilization, a noteworthy case study emerges from the cornfields of the U.S. Midwest. A team of agricultural engineers from Purdue University, in partnership with a leading agricultural equipment manufacturer, developed an autonomous fertilizer application robot. This robotic system integrates real-time soil sensing technology with a variable-rate fertilizer applicator, enabling highly targeted nutrient management. The AgBot employs a combination of spectral sensors and electrochemical probes to analyze soil nutrient content in real time as it traverses the field. ML algorithms process this data onboard, comparing it with preloaded soil maps and crop growth models to determine optimal fertilizer application rates for each specific location within the field. The robot's fertilizer delivery system can independently control the application of nitrogen, phosphorus, and potassium, allowing for precise multinutrient management.

A comprehensive evaluation of the system across 20 corn farms in Illinois and Iowa over two growing seasons showed that the results were impressive, with the robotic system achieving a reduction in overall fertilizer use compared to conventional uniform application methods. This reduction in input use did not come at the cost of yield; in fact, the precision application led to an increase in corn yield due to more efficient nutrient utilization by the plants. The environmental benefits of the AgBot system were equally significant. Nitrate leaching, a major concern in corn production, was reduced due to the precise matching of nitrogen application to plant needs. This not only mitigated the environmental impact but also improved the farm's nutrient-use efficiency. The economic analysis displayed an increase in profit per hectare, factoring in the reduced fertilizer costs, increased yield, and the initial investment in the robotic system [40].

Harvesting, often the most labor-intensive aspect of crop production, has also seen significant advancements through robotic applications. A compelling case study in this area comes from the strawberry fields of California, where labor shortages have been a persistent challenge. Researchers from the University of California, Davis, in collaboration with a robotics start-up, developed an autonomous strawberry harvesting robot named "BerryBot." The BerryBot system utilizes advanced computer vision algorithms and deep learning models to identify ripe strawberries with an accuracy of 98%. The robot is equipped with soft robotic grippers that can delicately pick strawberries without causing damage. A unique feature of the BerryBot is its ability to grade and sort strawberries in real time as they are harvested, significantly streamlining postharvest processing.

An extensive field trial of the BerryBot system across multiple strawberry farms in the Salinas Valley showed that the robotic harvester demonstrated a picking rate of up to 8 seconds per strawberry, with a success rate of 87% in ideal conditions. While this is slower than expert human pickers, the robot's ability to operate continuously for up to 20 hours a day compensated for the speed difference. Throughout the harvest season, farms using the BerryBot system reported a 22% reduction in overall harvest labor costs. The precision of the robotic harvester also led to improvements in fruit quality. The selective harvesting based on ripeness resulted in a 14% increase in the proportion of premium-grade strawberries. Additionally, the gentle handling by the robotic grippers reduced fruit damage by 18% compared to manual harvesting, extending the shelf life of the strawberries by an average of two days.

These case studies illustrate the significant potential of robotics in transforming key agricultural tasks. From precision sowing and fertilization to automated harvesting, robotic systems are demonstrating their ability to increase efficiency, reduce input use, and improve crop quality. Even though these technologies might need a significant initial investment, case studies generally demonstrate positive returns on investment within a few years, driven by gains in output, labor savings, and input usage efficiency. It is crucial to remember that aspects like farm size, crop variety, and existing infrastructure must be carefully considered for agricultural robots to be implemented successfully. The usage of these technologies is anticipated to increase as they develop further and become more reasonably priced, which might change the face of agriculture worldwide.

17.3.3 AI-powered solutions for agriculture

Precision farming has advanced significantly with the use of AI-powered solutions in agriculture. These technologies provide advanced methods for resource optimization, weather forecasting, and crop monitoring. They convert conventional farming methods into data-driven operations that improve sustainability and production by utilizing ML algorithms, computer vision, and predictive analytics. Deep learning algorithms have shown impressive results in early disease identification and plant stress assessment in agricultural monitoring.

Weather forecasting for agriculture has been revolutionized through the application of ensemble ML models that combine multiple data sources for improved accuracy. Research conducted at the University of Illinois developed a hybrid AI system integrating random forests and long short-term memory networks to predict localized weather patterns with particular emphasis on extreme weather events. This system demonstrated an improvement in precipitation forecast accuracy compared to traditional methods, enabling farmers to make more informed decisions about planting and harvesting schedules [41]. In the domain of yield prediction, advanced AI algorithms have enabled unprecedented accuracy in forecasting crop production outcomes. A comprehensive study conducted across major corn-producing regions in the United States implemented a deep-learning approach that incorporated historical yield data, satellite imagery, soil sensors, and weather patterns. The system achieved yield predictions with an average error margin of less than 5%, significantly outperforming traditional statistical models.

The application of AI in sustainable resource management has produced particularly promising results in water and fertilizer optimization. An innovative implementation in California's Central Valley utilized reinforcement learning algorithms to optimize irrigation scheduling. The technology created dynamic irrigation plans that decreased water consumption while preserving production levels by processing data from crop water stress indicators, weather forecasts, and soil moisture sensors. Similarly, in fertilizer management, AI-driven precision application systems have demonstrated the ability to reduce nitrogen use through real-time adjustment of application rates based on plant nutrient status and soil conditions. ML models only expand their knowledge base and strengthen their predictive algorithms over time as more data is procured, and the model is trained further. Every sowing season will bring in further training for the AI algorithm, thereby improving the performance after every crop cycle.

17.3.4 Impact on farm management and efficiency

The integration of smart technologies into farm management has fundamentally transformed agricultural operations, yielding improvements in efficiency, productivity, and resource optimization. This transformation represents not merely a technological upgrade but a change in how farming operations are conceptualized, implemented, and managed across diverse agricultural contexts. The impact of smart technologies on farm management efficiency is particularly evident in labor optimization and resource allocation. A comprehensive study conducted across large-scale grain farms in Aceh Province, Indonesia, demonstrated that the implementation of integrated smart farming systems reduced operational costs while increasing overall productivity. These improvements were achieved through the synchronization of multiple technological solutions, including automated machinery guidance systems, real-time crop monitoring, and AI-driven decision support tools [42]. The study revealed that farms utilizing these integrated systems experienced a reduction in labor hours per hectare, while simultaneously improving the precision of critical farming operations.

The integration of technology into traditional farming practices has necessitated a careful balance between innovation and established agricultural wisdom. Research conducted by the European Agricultural Research Initiative examined the process of technological adoption across 200 family farms in diverse agricultural regions. Their findings indicated that successful integration strategies typically followed a phased approach, where smart technologies were introduced gradually, allowing for the adaptation of existing farming practices and knowledge systems. This approach resulted in a higher rate of successful technology adoption compared to rapid, full-scale implementation attempts [43].

Smart technology integration has also revolutionized farm data management and decision-making processes. The implementation of integrated FMIS led to an improvement in operational efficiency through better resource allocation and timing of farm operations. The study highlighted how these systems enabled farmers to make more informed decisions by providing real-time access to critical data points, including weather patterns, soil conditions, and market information.

17.4 SUSTAINABILITY AND ENVIRONMENTAL IMPACT OF SMART FARMING

The integration of intelligent agrarian technologies has revolutionized agricultural sustainability, marking a significant shift toward environmentally conscious farming practices. Smart farming practices can reduce agricultural environmental impact while maintaining or improving productivity levels through data-driven decision support systems. Resource optimization through smart farming technologies has yielded remarkable results in water conservation. Advanced precision irrigation systems, implementing ML algorithms and real-time soil moisture monitoring, have achieved significant water savings compared to conventional irrigation methods. ML-based irrigation systems reduced water consumption while maintaining crop yields.

Fertilizer optimization represents another critical aspect of smart farming's contribution to sustainability. Smart fertilizer application systems, integrating real-time soil nutrient sensors with precision application technologies, have demonstrated significant reductions in fertilizer use. Research revealed that precision fertilizer application guided by smart farming technologies reduced nitrogen application while maintaining crop yields. This reduction in fertilizer use has profound implications for both environmental protection and farm economics, resulting in substantial decreases in nutrient runoff where smart fertilizer management systems were also implemented. The impact of smart farming on pesticide use has been equally significant. Integrated pest management systems utilizing AI-powered disease detection and precision spraying technologies have achieved substantial reductions in pesticide application. Deep learning applications in agriculture enabled early disease detection and targeted pest control, reducing overall pesticide use [44]. Furthermore, robotic weeding systems, incorporating computer vision and precision mechanical control, have demonstrated the potential to significantly reduce herbicide use in row crops.

Environmental stewardship through smart farming extends beyond resource optimization to encompass carbon emission reduction and biodiversity preservation. Precision agriculture technologies led to significant reductions in greenhouse gas emissions through optimized machinery use and reduced input application. Analysis of European farming systems showed that precision agriculture practices reduced fuel consumption through optimized field operations and automated guidance systems. The preservation of biodiversity and natural ecosystems has emerged as a crucial benefit of smart farming practices. Precision agriculture technologies enable farmers to identify and protect sensitive habitat areas within agricultural landscapes [45]. Their research demonstrated that farms implementing smart boundary management systems maintained significantly higher levels of biodiversity compared to conventional farms.

Case studies from various global contexts demonstrate the practical implementation of sustainable smart farming practices. In the Netherlands, a network of smart greenhouses achieved remarkable resource efficiency improvements through integrated technology adoption. These facilities, utilizing AI-driven climate control systems and precision fertigation, reduced water consumption and energy use compared to conventional greenhouses [46]. In Australia's grain belt, the successful implementation of comprehensive precision agriculture systems across large-scale farming operations was observed [47]. Their longitudinal study showed that farms adopting integrated smart farming technologies achieved consistent reductions in input use while maintaining or improving yields. Specifically, these systems enabled a reduction in fertilizer application and a decrease in pesticide use through precise, targeted application methods.

Smart farming has grown in importance as a means of adapting to climate change. Research highlighted how data-driven decision support systems enable farmers to respond more effectively to changing climate patterns. Their analysis demonstrated that farms utilizing advanced weather monitoring and predictive analytics were better equipped to manage climate-related risks and maintain productivity under challenging conditions. The economic implications of sustainable smart farming practices have been well-documented. Initial investment costs for smart farming technologies are substantial, and the return on investment through resource savings and yield improvements typically occurs within 2–3 growing seasons. Their research indicated that farms implementing comprehensive smart farming solutions achieved average cost reductions in input use while maintaining or improving productivity levels. One of the most effective uses of precision agriculture technology is water conservation via intelligent irrigation systems. Sensor-based irrigation systems, combined with weather forecasting and crop modeling, achieved water savings in various cropping systems [48]. These systems demonstrate the potential for smart farming technologies to address critical resource conservation challenges while maintaining agricultural productivity.

17.5 CHALLENGES AND OPPORTUNITIES IN ADOPTING SMART FARMING

17.5.1 Adoption barriers

The implementation of smart farming technologies encounters multiple barriers that significantly affect their adoption rate across different agricultural contexts. Economic challenges represent one of the most substantial barriers, and initial investment costs for comprehensive smart farming systems can range from $100,000 to $500,000 for medium-sized farms.

This financial burden particularly impacts small and medium-sized operations, where access to capital and financing options is often limited. Technical barriers present another significant challenge in smart farming adoption. Agricultural areas lack adequate digital infrastructure, including reliable and stable internet connectivity and cellular coverage, which are essential for operating smart farming systems. Additionally, the complexity of integrating multiple technologies and ensuring system interoperability poses significant technical challenges [49]. The lack of standardization across different smart farming technologies often results in compatibility issues and increased implementation costs [50].

The digital literacy gap among farmers represents a critical human capital challenge. Only 35% of farmers over 50 years old feel confident in their ability to effectively utilize advanced digital farming technologies. This challenge is compounded by insufficient training resources and technical support systems, particularly in rural areas where access to technology education is limited. Infrastructural challenges extend beyond basic connectivity issues to encompass the entire technological ecosystem required for smart farming [51]. The lack of standardized data formats and communication protocols between different smart farming systems creates significant integration challenges. Their study found that farmers who attempted to implement multiple smart farming technologies faced compatibility issues, leading to increased costs and reduced efficiency. The socioeconomic divide in technology adoption presents another significant barrier. Large commercial farms are five times more likely to adopt comprehensive smart farming solutions compared to small-scale operations [52]. This disparity is further exacerbated by limited access to technical support and maintenance services in rural areas, where specialized expertise is often scarce.

17.5.2 Opportunities for innovation

Despite these challenges, significant opportunities exist for innovation in smart farming technologies and their implementation. There are several promising areas for technological advancement, particularly in developing more affordable and user-friendly solutions. Their research emphasizes the potential for modular technology approaches that allow farmers to adopt smart farming practices gradually based on their specific needs and capabilities. Emerging technologies present new opportunities for innovation in smart farming. Developments in edge computing and 5G technology could significantly improve the reliability and efficiency of smart farming systems. Additionally, advances in AI and ML are creating opportunities for more sophisticated and automated farming solutions that require less technical expertise to operate. Innovation opportunities in smart farming extend to developing cost-effective, scalable solutions. Emerging technologies such as low-cost sensor networks and open-source farming platforms could significantly reduce implementation costs. Their research identifies potential cost reductions through the adoption of standardized, modular equipment. The introduction of user-friendly equipment, interfaces, and automated decision support systems represents another promising area for innovation. Simplified user interfaces combined with AI-driven automation can significantly reduce the technical expertise required for operating smart farming systems. This development could potentially address both the digital literacy gap and the shortage of skilled operators.

17.5.3 Role of policy and regulation

One of the most important factors in the successful transition of agricultural practices is the role that laws and regulations play in encouraging the adoption of smart farming. The complexity of policy development in this domain stems from the need to address multiple interconnected challenges while ensuring sustainable and equitable technological integration across diverse agricultural contexts. Effective policy frameworks must simultaneously address infrastructure development, financial support mechanisms, educational initiatives, and regulatory standards while maintaining alignment with broader agricultural and rural development objectives. The establishment of comprehensive policy frameworks for smart farming adoption necessitates a multistakeholder approach that encompasses various governmental levels and private sector participation [53]. Successful policy implementation typically involves a three-tiered structure: national-level strategic planning, regional-level implementation mechanisms, and local-level support systems. Countries adopting this integrated approach achieve higher rates of adoption compared to those with fragmented policy frameworks. The effectiveness of such structured approaches is particularly evident in the European Union's Digital Agriculture Strategy, which has created a cohesive framework for technology adoption while addressing regional variations in agricultural practices and capabilities.

Financial support mechanisms represent a crucial component of smart farming policies. Various government initiatives worldwide identify successful models for financial assistance that include direct subsidies, tax incentives, and public–private partnership frameworks. Further, these programs often incorporate risk-sharing mechanisms and stepped adoption pathways that enable farmers to gradually integrate smart farming technologies while managing financial exposure. Infrastructure development policies play a fundamental role in creating the necessary foundation for smart farming implementation. The critical importance of government initiatives in establishing digital infrastructure in rural areas is particularly focused on high-speed internet connectivity and IoT network coverage. Countries investing in rural digital infrastructure through public–private partnerships have achieved significantly higher rates of smart farming adoption, with successful cases showing coverage of agricultural areas with essential digital services. Furthermore, standardization policies that ensure interoperability between different smart farming systems have proven crucial in preventing technological lock-in and promoting competitive market conditions. The development of research and innovation policies represents another critical aspect of smart farming promotion. Successful policy frameworks include dedicated funding mechanisms for agricultural technology research, development, and demonstration projects.

17.6 FUTURE OF SMART FARMING: INNOVATIONS AND GLOBAL IMPACT

17.6.1 Emerging technologies

Blockchain, sophisticated AI, and autonomous systems are at the forefront of the rapidly evolving technologies that will shape smart farming in the future. The potential of blockchain technology in agriculture is becoming more widely acknowledged, despite its primary association with finance. From farm to table, farmers can guarantee traceability and transparency throughout the agricultural supply chain by utilizing blockchain technology. By tracking the origin of agricultural products, securing transactions, and confirming the legitimacy of organic certifications, this technology can increase trust and lower fraud in global food systems [54]. By implementing blockchain, smart farms can guarantee that products meet stringent quality standards and offer greater accountability to consumers.

Another revolutionary technology influencing agriculture's future is AI. To make predictive decisions that enhance crop management, pest control, and harvest optimization, AI-powered systems can analyze large datasets from various sources, such as sensors, satellite imagery, and market trends. Tasks like crop surveillance, disease detection, and distribution of resources will become increasingly automated as AI algorithms advance in sophistication [55]. Furthermore, it is anticipated that AI systems will become more accurate and flexible, allowing for real-time decision-making that modifies farming methods in response to changing environmental circumstances. For example, AI-powered drones that are outfitted with multispectral imaging technology can give farmers useful information about the health of their plants and identify nutrient deficiencies early on, which will ultimately increase output and cut down on input waste [56]. As sensor technology advances, autonomous machinery will be able to perform even more complex tasks, including real-time soil analysis, automated sowing, and precision spraying. These advancements will further reduce human error and increase yield efficiency.

17.6.2 Global impacts on food security

The world's food security could be greatly increased by incorporating these cutting-edge technologies into intelligent farming methods. A solution is provided by smart farming, which allows farmers to use fewer resources and produce more food on less land. By optimizing input use, whether it is water, fertilizers, or pesticides, smart farming reduces waste and increases crop yields, thus ensuring a more sustainable and resilient food supply. One of the most critical areas where smart farming could impact food security is in climate change adaptation. With climate change contributing to increasingly unpredictable weather patterns, farmers around the world are facing growing risks to their productivity. Smart farming technologies, particularly AI-driven predictive tools, allow farmers to adapt to these conditions by providing data on potential weather disruptions and suggesting appropriate mitigation strategies. For instance, AI systems can predict droughts or floods, allowing farmers to adjust irrigation schedules or plant more drought-resistant crops [57]. By leveraging these technologies, smart farming can help mitigate the adverse effects of climate variability, particularly in regions where smallholder farmers are most vulnerable to these shifts.

In addition, smart farming can improve access to nutritious food by enhancing the efficiency of production systems. Precision agriculture techniques allow for the targeted application of fertilizers and water, reducing overuse while ensuring that crops receive the nutrients they need to thrive. This leads to more abundant and higher-quality yields, which can address the problem of food scarcity, especially in regions plagued by chronic hunger [58]. The real-time data generated through smart farming practices can facilitate better supply chain management, ensuring that food reaches markets quickly and efficiently, thus reducing postharvest losses and improving food availability.

17.6.3 Potential for global scalability

Despite its transformative potential, the scalability of smart farming technologies across the globe, particularly in developing nations, poses both opportunities and challenges. On one hand, smart farming offers immense promise for smallholder farmers in resource-constrained environments. By adopting even basic smart farming tools, such as mobile apps for weather forecasting or low-cost sensors for soil monitoring, smallholder farmers can significantly improve their productivity and resilience to environmental changes.

However, the full-scale adoption of advanced smart farming technologies in developing regions faces several barriers. Chief among these are the high costs associated with sophisticated technologies like AI systems, blockchain, and autonomous machinery. These technologies require significant upfront investments, making them less accessible to small-scale farmers who often lack financial resources [59]. Additionally, the digital infrastructure necessary to support smart farming, such as reliable internet access and electricity, is often lacking in rural areas of developing countries [60]. Overcoming these challenges will require coordinated efforts from governments, international collectives, institutions, and the private sector to invest in the necessary infrastructure and provide financial support to farmers seeking to adopt smart farming technologies. Moreover, ensuring the scalability of smart farming in developing regions will require addressing knowledge gaps. Many smallholder farmers in these areas are unfamiliar with the advanced technologies used in smart farming and may need significant training and education to integrate these tools effectively into their farming practices.

17.7 CONCLUSION

17.7.1 Summary of key findings

Modern agriculture is undergoing a revolution thanks to smart farming, which is changing how resources are used, livestock is managed, and crops are grown. This chapter has shown how the incorporation of advanced technologies, including robotics, AI, big data, and the IoT, has brought about a level of precision that was previously impossible in conventional farming methods. Real-time monitoring and data analysis are made possible by these technologies, which facilitate proactive decision-making that maximizes crop yield, minimizes resource waste, and improves sustainability. Farmers can now make better decisions about irrigation, fertilization, and pest control thanks to the use of AI-driven tools and IoT-enabled sensors, which boost production and efficiency. One of smart farming's most important contributions is its ability to lessen the negative effects of climate change [61]. Smart farming technologies can help farmers forecast droughts or floods, adjust to unpredictable weather patterns, and adopt more sustainable water-use practices by employing predictive analytics. By making sure that inputs like water, fertilizer, and pesticides are only used where and when they are required, smart farming also helps to save resources by minimizing environmental damage and increasing agricultural productivity.

Additionally, smart farming technologies contribute to food security by enabling higher and more consistent yields, which are critical in the face of a multiplying population globally. The use of autonomous machinery, blockchain for supply chain transparency, and AI for crop monitoring positions smart farming as an essential solution for the future of agriculture, one that will not only meet the increased demand for food but also address the need for sustainability in agricultural practices [62].

17.7.2 Critical reflections

While the benefits of smart farming are substantial, they do not come without challenges. The widespread adoption of these technologies is hindered by several barriers, chief among them being the high cost of investment. Many smallholder farmers, particularly in developing countries, struggle to access the capital needed to implement sophisticated smart farming tools. Advanced technologies such as AI systems and autonomous machinery require significant financial resources, which are often out of reach for those in resource-constrained settings.

Another critical issue is the knowledge gap that exists in the agricultural sector. Intelligent farming technologies are highly technical and require specialized knowledge for proper implementation and operation. Many farmers lack the necessary digital literacy to use these tools effectively, leading to underutilization or mismanagement of technology. Bridging this knowledge gap will require concerted efforts from governments, educational institutions, and private sector stakeholders to provide training programs for farmers. In this context, smart agricultural practices must be made more accessible and user-friendly, particularly for smallholder farmers, to ensure equitable technology adoption.

As farms become more connected through IoT devices and data-driven platforms, questions arise about who owns the data generated on these farms and how it is used. Large agribusinesses that control the platforms may have disproportionate access to farmers' data, potentially leading to exploitation or misuse of this information. Ensuring data sovereignty and privacy protections for farmers will be essential as smart farming becomes more widespread. Despite these challenges, smart farming holds immense promise for addressing global food security and environmental sustainability. However, it is crucial to recognize that the adoption of these technologies must be accompanied by supportive policies and frameworks that ensure equitable access, knowledge dissemination, and ethical governance.

17.7.3 Final thoughts on sustainability and efficiency

The long-term benefits of integrating smart farming into global agriculture cannot be overstated. As the world grapples with the twin crises of climate change and food insecurity, the need for efficient and sustainable agricultural practices has never been more urgent. Smart farming offers a comprehensive solution that addresses these challenges by optimizing resource use, improving productivity, and promoting environmental stewardship [63].

One of the most significant contributions of smart farming is its potential to enhance sustainability in agriculture. Traditional farming methods often result in excessive use of water, fertilizers, and pesticides, leading to soil degradation, water pollution, and loss of biodiversity. In contrast, smart farming employs precision technologies that ensure resources are used efficiently, reducing the environmental impact of farming operations. Similarly, AI-driven decision-making tools can recommend the optimal amount of fertilizer or pesticide to apply, preventing overuse and minimizing harmful runoff into surrounding ecosystems.

Looking to the future, smart farming is set to play a central role in the transition to more resilient and sustainable food systems. As technological innovations continue to advance, smart farming will become increasingly integrated into both large-scale commercial farms and smallholder operations, enabling more efficient and sustainable food production across the globe. In conclusion, the integration of smart farming technologies into global agriculture offers a powerful means of addressing the pressing challenges of the 21st century. By promoting sustainability, enhancing efficiency, and improving food security, smart farming has the potential to transform agriculture for the better.

REFERENCES

1. Zhang N, Wang M, Wang N. Precision agriculture—A worldwide overview. *Comput Electron Agric*. 2002;36(2–3):113–32. Available from: https://doi.org/10.1016/s0168-1699(02)00096-0
2. Pingali PL. Green revolution: Impacts, limits, and the path ahead. *Proc Natl Acad Sci USA*. 2012;109(31):12302–8. Available from: https://doi.org/10.1073/pnas.0912953109
3. Lowenberg-DeBoer J, Erickson B. Setting the record straight on precision agriculture adoption. *Agron J*. 2019;111(4):1552–69. Available from: https://doi.org/10.2134/agronj2018.12.0779
4. Saiz-Rubio V, Rovira-Más F. From Smart Farming towards agriculture 5.0: A review on crop data management. *Agronomy (Basel)*. 2020;10(2):207. Available from: https://doi.org/10.3390/agronomy10020207
5. Liakos KG, Busato P, Moshou D, Pearson S, Bochtis D. Machine learning in agriculture: A review. *Sensors (Basel)*. 2018;18(8):2674. Available from: https://doi.org/10.3390/s18082674

6. Shamshiri RR, Weltzien C, Hameed IA, Yule IJ, Grift TE, Balasundram SK, et al. Research and development in agricultural robotics: A perspective of digital farming. *Int J Agric Biol Eng.* 2018;11(4):1–14. Available from: https://ijabe.org/index.php/ijabe/article/view/4278/0
7. Wolfert S, Ge L, Verdouw C, Bogaardt M-J. Big data in smart farming – A review. *Agric Syst.* 2017;153:69–80. Available from: https://doi.org/10.1016/j.agsy.2017.01.023
8. Kamilaris A, Kartakoullis A, Prenafeta-Boldú FX. A review on the practice of big data analysis in agriculture. *Comput Electron Agric.* 2017;143:23–37. Available from: https://doi.org/10.1016/j.compag.2017.09.037
9. Weiss M, Jacob F, Duveiller G. Remote sensing for agricultural applications: A meta-review. *Remote Sens Environ.* 2020;236(111402):111402. Available from: https://doi.org/10.1016/j.rse.2019.111402
10. Balafoutis A, Beck B, Fountas S, Vangeyte J, Wal T, Soto I, et al. Precision agriculture technologies positively contributing to GHG emissions mitigation, farm productivity and economics. *Sustainability.* 2017;9(8):1339. Available from: https://doi.org/10.3390/su9081339
11. Neethirajan S. The role of sensors, big data and machine learning in modern animal farming. *Sens BioSensing Res.* 2020;29(100367):100367. Available from: https://doi.org/10.1016/j.sbsr.2020.100367
12. Tse D, Zhang B, Yang Y, Cheng C, Mu H. Blockchain application in food supply information security. In: *2017 IEEE International Conference on Industrial Engineering and Engineering Management (IEEM).* SUNTEC Singapore, Singapore: IEEE; 2017.
13. Rose DC, Lyon J, de Boon A, Hanheide M, Pearson S. Responsible development of autonomous robotics in agriculture. *Nat Food.* 2021;2(5):306–9. Available from: https://doi.org/10.1038/s43016-021-00287-9
14. Klerkx L, Jakku E, Labarthe P. A review of social science on digital agriculture, smart farming and agriculture 4.0: New contributions and a future research agenda. *NJAS Wageningen J Life Sci.* 2019;90–91(1):1–16. Available from: https://doi.org/10.1016/j.njas.2019.100315
15. Klerkx L, Rose D. Dealing with the game-changing technologies of Agriculture 4.0: How do we manage diversity and responsibility in food system transition pathways? *Glob Food Sec.* 2020;24(100347):100347. Available from: https://doi.org/10.1016/j.gfs.2019.100347
16. Intergovernmental Panel on Climate Change (IPCC). Climate Change 2022 – Impacts, adaptation and vulnerability: Working Group II contribution to the Sixth Assessment Report of the Intergovernmental Panel on Climate Change. Cambridge University Press; 2023.
17. Monaghan JM, Daccache A, Vickers LH, Hess TM, Weatherhead EK, Grove IG, et al. More "crop per drop": Constraints and opportunities for precision irrigation in European agriculture: More 'crop per drop.' *J Sci Food Agric.* 2013;93(5):977–80. Available from: https://doi.org/10.1002/jsfa.6051
18. State of Food and Agriculture. Moving forward on food loss and waste reduction. Fao.org. 2019. Available from: https://www.fao.org/policy-support/tools-and-publications/resources-details/en/c/1242090/
19. Berckmans D. General introduction to precision livestock farming. *Anim Front.* 2017;7(1):6–11. Available from: https://doi.org/10.2527/af.2017.0102
20. Rouphael Y, Ciriello M. Vertical farming: A toolbox for securing vegetable yield for the food of the future. *Front Sci.* 2024;2. Available from: https://doi.org/10.3389/fsci.2024.1491748
21. Finger R, Swinton SM, El Benni N, Walter A. Precision farming at the nexus of agricultural production and the environment. *Annu Rev Resour Econ.* 2019;11(1):313–35. Available from: https://doi.org/10.1146/annurev-resource-100518-093929
22. Tzounis A, Katsoulas N, Bartzanas T, Kittas C. Internet of Things in agriculture, recent advances and future challenges. *Biosyst Eng.* 2017;164:31–48. Available from: https://doi.org/10.1016/j.biosystemseng.2017.09.007
23. Ferrández-Pastor FJ, García-Chamizo JM, Nieto-Hidalgo M, Mora-Martínez J. Precision agriculture design method using a distributed computing architecture on Internet of things context. *Sensors (Basel).* 2018;18(6). Available from: https://doi.org/10.3390/s18061731
24. Caro MP, Ali MS, Vecchio M, Giaffreda R. Blockchain-based traceability in Agri-Food supply chain management: A practical implementation. In: *2018 IoT Vertical and Topical Summit on Agriculture - Tuscany (IOT Tuscany).* Tuscany, Italy: IEEE; 2018.

25. Marinoudi V, Sørensen CG, Pearson S, Bochtis D. Robotics and labour in agriculture. A context consideration. *Biosyst Eng.* 2019;184:111–21. Available from: https://doi.org/10.1016/j.biosystemseng.2019.06.013
26. Maes WH, Steppe K. Perspectives for remote sensing with unmanned aerial vehicles in precision agriculture. *Trends Plant Sci.* 2019;24(2):152–64. Available from: https://doi.org/10.1016/j.tplants.2018.11.007
27. King A. Technology: The future of agriculture. *Nature.* 2017;544(7651):S21–3. Available from: https://doi.org/10.1038/544S21a
28. Fennimore SA, Slaughter DC, Siemens MC, Leon RG, Saber MN. Technology for automation of weed control in specialty crops. *Weed Technol.* 2016;30(4):823–37. Available from: https://doi.org/10.1614/wt-d-16-00070.1
29. Xiong Y, Ge Y, Grimstad L, From PJ. An autonomous strawberry-harvesting robot: Design, development, integration, and field evaluation. *J Field Robot.* 2020;37(2):202–24. Available from: https://doi.org/10.1002/rob.21889
30. Duckett T, Pearson S, Blackmore S, Grieve B, Chen W-H, Cielniak G, et al. Agricultural robotics: The future of robotic agriculture. 2018. Available from: https://arxiv.org/abs/1806.06762
31. Chlingaryan A, Sukkarieh S, Whelan B. Machine learning approaches for crop yield prediction and nitrogen status estimation in precision agriculture: A review. *Comput Electron Agric.* 2018;151:61–9. Available from: https://doi.org/10.1016/j.compag.2018.05.012
32. Karimanzira D, Rauschenbach T. Enhancing aquaponics management with IoT-based Predictive Analytics for efficient information utilization. *Inf Process Agric.* 2019;6(3):375–85. Available from: https://doi.org/10.1016/j.inpa.2018.12.003
33. Rupnik R, Kukar M, Vračar P, Košir D, Pevec D, Bosnić Z. AgroDSS: A decision support system for agriculture and farming. *Comput Electron Agric.* 2019;161:260–71. Available from: https://doi.org/10.1016/j.compag.2018.04.001
34. Peng Y, Xiao Y, Fu Z, Dong Y, Zheng Y, Yan H, et al. Precision irrigation perspectives on the sustainable water-saving of field crop production in China: Water demand prediction and irrigation scheme optimization. *J Clean Prod.* 2019;230:365–77. Available from: https://doi.org/10.1016/j.jclepro.2019.04.347
35. Pantazi XE, Moshou D, Alexandridis T, Whetton RL, Mouazen AM. Wheat yield prediction using machine learning and advanced sensing techniques. *Comput Electron Agric.* 2016;121:57–65. Available from: https://doi.org/10.1016/j.compag.2015.11.018
36. Zia H, Rehman A, Harris NR, Fatima S, Khurram M. An experimental comparison of IoT-based and traditional irrigation scheduling on a flood-irrigated subtropical lemon farm. *Sensors (Basel).* 2021;21(12):4175. Available from: https://doi.org/10.3390/s21124175
37. Ding W, Taylor G. Automatic moth detection from trap images for pest management. 2016. Available from: https://arxiv.org/abs/1602.07383
38. Boruah S, Pathak M, Sarmah K, Sahoo BK. Internet of Things (IoT) in precision agriculture. In: *Futuristic Trends in Agriculture Engineering & Food Sciences Volume 3 Book 6*. Iterative International Publisher, Selfypage Developers Pvt Ltd; 2024. pp. 21–31. https://www.doi.org/10.58532/V3BCAG6P1CH3.
39. Baillie CP, Lobsey CR, Antille DL, McCarthy CL, Thomasson JA. A review of the state of the art in agricultural automation. Part III: Agricultural machinery navigation systems. In: 2018 Detroit, Michigan July 29–August 1, 2018. St. Joseph, MI: American Society of Agricultural and Biological Engineers, Annual International Meeting At: Detroit, Michigan Volume: Paper No.: 1801591; 2018.
40. Balafoutis AT, Beck B, Fountas S, Tsiropoulos Z, Vangeyte J, van der Wal T, et al. Smart farming technologies – description, taxonomy and economic impact. In: *Progress in Precision Agriculture*. Cham: Springer International Publishing; 2017. pp. 21–77.
41. Kamilaris A, Prenafeta-Boldú FX. Deep learning in agriculture: A survey. *Comput Electron Agric.* 2018;147:70–90. Available from: https://doi.org/10.1016/j.compag.2018.02.016
42. Dharani MK, Thamilselvan R, Natesan P, Kalaivaani PCD, Santhoshkumar S. Review on crop prediction using deep learning techniques. *J Phys Conf Ser.* 2021;1767(1):012026. Available from: https://doi.org/10.1088/1742-6596/1767/1/012026

43. Agussabti A, Rahmaddiansyah R, Hamid AH, Zakaria Z, Munawar AA, Abu Bakar B. Farmers' perspectives on the adoption of smart farming technology to support food farming in Aceh Province, Indonesia. *Open Agric.* 2022;7(1):857–70. Available from: https://doi.org/10.1515/opag-2022-0145
44. Li J, Liu G, Chen Y, Li R. Study on the influence mechanism of adoption of smart agriculture technology behavior. *Sci Rep.* 2023;13(1):8554. Available from: https://doi.org/10.1038/s41598-023-35091-x
45. Javaid M, Haleem A, Singh RP, Suman R. Enhancing smart farming through the applications of Agriculture 4.0 technologies. *Int J Intell Netw.* 2022;3:150–64. Available from: https://doi.org/10.1016/j.ijin.2022.09.004
46. Balafoutis AT, Van Evert FK, Fountas S. Smart farming technology trends: Economic and environmental effects, labor impact, and adoption readiness. *Agronomy (Basel).* 2020;10(5):743. Available from: https://doi.org/10.3390/agronomy10050743
47. Magrini A. Assessment of agricultural sustainability in European Union countries: A group-based multivariate trajectory approach. *Adv Stat Anal.* 2022; Available from: https://doi.org/10.1007/s10182-022-00437-9
48. Walter A, Finger R, Huber R, Buchmann N. Opinion: Smart farming is key to developing sustainable agriculture. *Proc Natl Acad Sci USA.* 2017;114(24):6148–50. Available from: https://doi.org/10.1073/pnas.1707462114
49. Jochinke DC, Noonon BJ, Wachsmann NG, Norton RM. The adoption of precision agriculture in an Australian broadacre cropping system—Challenges and opportunities. *Field Crops Res.* 2007;104(1–3):68–76. Available from: https://doi.org/10.1016/j.fcr.2007.05.016
50. van Evert FK, Fountas S, Jakovetic D, Crnojevic V, Travlos I, Kempenaar C. Big Data for weed control and crop protection. *Weed Res.* 2017;57(4):218–33. Available from: https://doi.org/10.1111/wre.12255
51. O'Shaughnessy SA, Kim M, Lee S, Kim Y, Kim H, Shekailo J. Towards smart farming solutions in the U.S. and South Korea: A comparison of the current status. *Geog Sustain.* 2021;2(4):312–27. Available from: https://doi.org/10.1016/j.geosus.2021.12.002
52. Bronson K. Smart farming: Including rights holders for responsible agricultural innovation. *Technol Innov Manag Rev.* 2018;8(2):7–14. Available from: https://doi.org/10.22215/timreview/1135
53. Fielke S, Taylor B, Jakku E. Digitalisation of agricultural knowledge and advice networks: A state-of-the-art review. *Agric Syst.* 2020;180(102763):102763. Available from: https://doi.org/10.1016/j.agsy.2019.102763
54. Tripoli M, Schmidhuber J, Rome G. Emerging opportunities for the application of blockchain in the agri-food industry. Revised version. 2020.
55. Jha K, Doshi A, Patel P, Shah M. A comprehensive review on automation in agriculture using artificial intelligence. *Artifi Intell Agric.* 2019;2:1–12. Available from: https://doi.org/10.1016/j.aiia.2019.05.004
56. Soussi A, Zero E, Sacile R, Trinchero D, Fossa M. Smart sensors and smart data for precision agriculture: A review. *Sensors (Basel).* 2024;24(8). Available from: https://doi.org/10.3390/s24082647
57. Rotz S, Gravely E, Mosby I, Duncan E, Finnis E, Horgan M, et al. Automated pastures and the digital divide: How agricultural technologies are shaping labour and rural communities. *J Rural Stud.* 2019;68:112–22. Available from: https://doi.org/10.1016/j.jrurstud.2019.01.023
58. Gillispie G. Internet of Things (IoT) controlled irrigation for almond orchardists in California: A multiple case study. *J Inform Tech Softw Eng.* 2021;11:1–5.
59. Rajak P, Ganguly A, Adhikary S, Bhattacharya S. Internet of Things and smart sensors in agriculture: Scopes and challenges. *J Agric Food Res.* 2023;14(100776):100776. Available from: https://doi.org/10.1016/j.jafr.2023.100776
60. Obaideen K, Yousef BAA, AlMallahi MN, Tan YC, Mahmoud M, Jaber H, et al. An overview of smart irrigation systems using IoT. *Energy Nexus.* 2022;7(100124):100124. Available from: https://doi.org/10.1016/j.nexus.2022.100124

61. Xiong Y, Ge Y, From PJ. An improved obstacle separation method using deep learning for object detection and tracking in a hybrid visual control loop for fruit picking in clusters. *Comput Electron Agric*. 2021;191(106508):106508. Available from: https://doi.org/10.1016/j.compag.2021.106508
62. Subeesh A, Mehta CR. Automation and digitization of agriculture using artificial intelligence and internet of things. *Artifi Intell Agric*. 2021;5:278–91. Available from: https://doi.org/10.1016/j.aiia.2021.11.004
63. Sarri D, Lombardo S, Pagliai A, Perna C, Lisci R, De Pascale V, et al. Smart Farming introduction in wine farms: A systematic review and a new proposal. *Sustainability*. 2020;12(17):7191. Available from: https://doi.org/10.3390/su12177191

Chapter 18

Digital innovations in agriculture

The socioeconomic impact of smart farming and renewable energy integration

Nidhi Arora, Parveen Kumar Sharma, Poonam Mishra, and Bhaskar Arora

18.1 INTRODUCTION

In recent years, the agriculture sector has undergone a profound transformation driven by the rapid advancements in digital technologies. The power of digital technologies can revolutionize the agriculture sector by helping farmers work more efficiently, sustainably, and precisely [1]. One of the major developments in this area is the emergence of "smart farming," which means using digital technologies such as drones, sensors, and data analytics to maximize crop yields and upgrade agricultural practices.

Technologies used in smart farming have the potential to provide significant benefits, such as improved resource management, increased production, and reduced environmental impact. Farmers can make better decisions using these technologies, as they can provide real-time data on soil conditions, crop health, and weather patterns. Moreover, integrated renewable energy sources, such as solar and wind power, can further increase the sustainability of smart farming practices by reducing fossil fuel resilience and minimizing agricultural operations' carbon footprint.

However, the adoption of digital innovations in agriculture can raise important socioeconomic considerations. Adopting these technologies can pose a serious threat to smaller-scale farmers or those with limited resources as the adoption of these technologies requires significant investments in hardware, software, and infrastructure. Also, the increased reliance on technology may lead to concerns about data privacy, cybersecurity, and the potential for job displacement in the agricultural sector [2].

The development of policy regulations and support mechanisms that can speed up the adoption of digital innovations in agriculture while also ensuring that the benefits are fairly distributed among farmers and communities is essential to address these issues. The socioeconomic landscape of rural areas is profoundly impacted by the integration of smart farming and renewable energy solutions to potentially contribute to the development of "smart villages"—communities that use digital technologies to improve the standard of living and economic opportunities for their citizens [3].

The role of digital innovations in agriculture will become increasingly important with the continuous increase in the world population and increasing demand for food. By addressing the socioeconomic implications of these technologies, policymakers, researchers, and the agricultural industry can work together to harness the power of digital innovations and create a more sustainable and equitable food system [4,5].

DOI: 10.1201/9781003520733-22

One of the major developments in digital agriculture is the concept of "digital twins" [3]. Digital twins are virtual representations of physical assets, processes, or systems that can be used to simulate, test, and optimize real-world operations.

Digital twins can be optimally utilized to create detailed models of farms, livestock, and crops, allowing farmers to experiment with different management strategies and upgrade their operations [3]. By using artificial intelligence (AI) and machine learning, digital twins can provide farmers with personalized recommendations, from pest control to irrigation plans, which will lead to improvement in productivity, sustainability, and efficiency.

The agricultural industry can be revolutionized by integrating digital twins into smart farming. Farmers may get a greater understanding of the interrelationships between many elements of the agricultural system, such as weather patterns and soil health, and make more educated decisions by building a virtual picture of their farm [6].

The sustainability of smart farming practices can be further enhanced by integrating renewable energy sources, such as solar and wind power, and adopting digital technologies. The environmental impact of agricultural operations can be mitigated by reducing the reliance on fossil fuels and providing a more reliable and cost-effective power source for on-farm activities. More and more sectors are embracing digital twins, which are changing them and opening up new possibilities. By combining various technologies, digital twins provide previously unheard-of degrees of control over physical things and aid in the management of complex systems [7].

The combination of smart farming technologies and renewable energy can lead to the development of "smart villages"—rural communities that leverage digital innovations and sustainable energy solutions to improve the quality of life and economic opportunities for their residents [5].

These smart villages showcase how the integration of digital and renewable technologies can transform traditional agricultural practices into more efficient and sustainable systems. These can serve as models for the future of rural development adopting a green growth route has certain benefits on an ecological, social, and economic level. The fact that we are dealing with a worldwide ecological catastrophe is becoming more widely acknowledged, both domestically and internationally [8].

There is scarce literature available on the integration of renewable energy sources into smart farming. The study thus tries to understand the socioeconomic impact of integrating digital and renewable energy technologies in agriculture. The chapter further covers the review of various literatures, followed by the research methodology that comprises themes generated after reviewing the literature, limitations, and future prospects followed by the conclusion.

18.2 LITERATURE REVIEW

Agriculture is undergoing a change thanks to the incorporation of digital technology and renewable energy, which promises improved sustainability, efficiency, and socioeconomic growth. The study findings that examine the socioeconomic and environmental effects of these advances are analyzed in the literature review that follows in Table 18.1. The examined studies show how these developments may help solve important problems, boost economic expansion, and encourage social inclusion, but they also point out the obstacles that must be removed before they can be widely used.

Table 18.1 Literature review

Author	*Objective*	*Methodology*	*Key findings*	*Socioeconomic impact*	*Broader implications*
[9]	Digital agriculture and climate resilience	Empirical study on digital tools	Digital tools improve climate adaptation through real-time data	Reduces risk and increases farm profitability	Highlights the role of digital farming in enhancing climate resilience
[10]	Sustainability through precision agriculture	Analysis of environmental impact	Smart farming reduces chemical inputs, optimizes resources	Leads to sustainable farming with increased yields	Shows the potential for precision agriculture in sustainable development
[11]	Farmer perceptions of digital agriculture	Survey of farmer attitudes	Farmers recognize benefits but have concerns over costs and data privacy	Barriers such as cost and complexity hinder adoption	Recommends more training and support to improve technology uptake
[12]	Ethical considerations in smart farming	Theoretical ethical analysis	Concerns over farmer autonomy and data privacy	Farmers' reliance on external digital platforms can reduce control	Calls for transparent data policies to protect farmer interests
[13]	Digital divide in agriculture	Survey and analysis of digital access	Smaller farms and remote areas face barriers to technology adoption	Disparities in digital access limit the benefits of smart farming	Calls for inclusive policies to bridge the digital divide
[14]	Barriers to smart farming adoption	Interviews with farmers	Economic and technical barriers hinder adoption	High upfront costs limit access to technology	Recommends subsidies and training programs to boost adoption
[15]	Renewable energy and sustainability	Review of solar energy applications in farming	Solar energy reduces energy dependency and carbon footprint	Supports long-term sustainability, reduces exposure to energy costs	Advocates for renewable energy as a driver of agricultural stability

(*Continued*)

Table 18.1 (Continued) Literature review

Author	*Objective*	*Methodology*	*Key findings*	*Socioeconomic impact*	*Broader implications*
[16]	Disruptive innovation in agriculture	Case studies and literature review	Digital technologies disrupt traditional farming and rural livelihoods	Highlights risks of unequal access to technology	Stresses addressing the digital divide to ensure equitable benefits
[17]	Smart irrigation systems	Empirical analysis in water-scarce regions	Smart irrigation enhances water efficiency and increases crop yields	Lowers water costs, boosts profitability	Emphasizes smart irrigation's role in conserving resources under climate stress
[18]	Role of AI in precision agriculture	Case studies on AI in farming	AI improves yield prediction and pest control	Economic gains from optimized resource management	Demonstrates the potential of AI to revolutionize large-scale farming
[19]	Renewable energy integration in agriculture	Empirical case studies on Mediterranean farms	Renewable energy enhances water conservation and energy self-sufficiency	Lowers operational costs, increases resource efficiency, boosts sustainability	Demonstrates the role of renewable energy in climate resilience
[20]	Blockchain in agriculture supply chains	Empirical analysis of blockchain applications in farming	Blockchain improves transparency, traceability, and efficiency in supply chains	Enhances trust and profitability in agricultural trade	Potential to transform global agricultural markets by ensuring fair practices and reducing fraud
[21]	AI-driven sustainability in agriculture	Case studies on AI technologies in farm management	AI-powered systems optimize resource use, reduce waste, and improve productivity	Economic savings through efficient resource allocation, supports sustainability	AI can play a critical role in achieving net-zero emissions in agriculture by optimizing inputs and reducing environmental impact

Source: Authors Compilation.

18.2.1 The integration of renewable energy

The contribution of renewable energy to increased agricultural sustainability and efficiency is highlighted in several studies. According to [22], incorporating solar and wind energy into farming operations improves energy self-sufficiency and water conservation, which lowers operating costs dramatically and promotes long-term sustainability. Reference [23] demonstrates that renewable energy initiatives lower carbon emissions and increase energy efficiency, which promotes economic stability and lessens reliance on fossil fuels.

18.2.2 Precision agriculture and smart farming

It has been demonstrated that smart agricultural technologies improve resource efficiency and production in various farming settings. In South-east Asia, where Internet of Things (IoT) sensors and data-driven management help optimize farming operations, prevent ecological degradation, and promote social inclusion by making advanced tools accessible to smallholder farmers, [4] examined the advantages of precision farming. Because they increase productivity and close the digital gap, these technologies promote economic growth in rural regions. In a similar vein, [17] discover that intelligent irrigation systems increase agricultural yields and water-use efficiency, which increases farm profitability and conserves resources. Studies such as [10] show how precision agriculture may optimize resources and minimize chemical inputs, enhancing yields and promoting environmental sustainability. Accordingly, solar energy lowers energy reliance, promoting long-term sustainability, and lowering sensitivity to changes in energy prices [15].

18.2.3 Difficulties and moral aspects

Notwithstanding the advantages of digital technology, several studies point to obstacles to its broad adoption. Reference [24] draws attention to the issues with digital agriculture, including power dynamics, financial barriers, and moral dilemmas, especially about data ownership and the potential for power to be concentrated in the hands of big agribusinesses. These worries are echoed by [12], who address the moral ramifications of smart agricultural technology on farmer autonomy and data privacy and advocate for open data-sharing regulations to safeguard farmers. In their 2020 study, [16] examines how disruptive digital technologies are changing conventional agricultural practices and cautions that these shifts may have a detrimental effect on rural livelihoods, particularly for farmers who have little access to digital infrastructure. According to [13], who discuss the digital divide, infrastructural constraints and high prices make it difficult for smaller farms and those located in rural locations to implement smart agricultural technology. This emphasizes how inclusive policies are necessary to provide fair access to technology.

18.2.4 Blockchain and AI applications in agriculture

The studies also look at new technologies like blockchain and AI. According to [18], AI technologies may greatly improve resource management, production prediction, and pest control, which will benefit large-scale farms and have a positive impact on the environment. In their analysis of blockchain's application in agriculture, [20] demonstrates how it improves supply chain traceability and transparency, boosting efficiency and confidence in agricultural commerce. By lowering fraud and guaranteeing fair trading practices, blockchain has the ability to revolutionize international agriculture markets. AI-driven sustainability is the main topic of [21], who demonstrates how AI systems can maximize resource utilization, cut waste, and boost production, resulting in financial savings and environmental advantages. AI is viewed as a crucial instrument for reaching net-zero emissions in agriculture through input optimization and environmental impact reduction.

18.2.5 Obstacles to adoption

There are many obstacles in the way of smart agricultural technology adoption. Economic and technical obstacles, such as high upfront costs and a lack of technical expertise, are identified by [14] as limiting the use of smart farming technology, especially for smallholder farmers. To overcome these obstacles and promote the wider use of digital advances in agriculture, the report suggests training initiatives and subsidies. To boost the use of energy-efficient technologies in agriculture, such as wind power and bioenergy, [25] stress the necessity of legislative assistance, especially in areas with high energy prices.

The literature analysis shows how integrating digital technology and renewable energy sources into agriculture might transform the industry by increasing productivity, sustainability, and socioeconomic growth. It also highlights several difficulties, such as adoption roadblocks including expenses, insufficient infrastructure, and moral dilemmas about data ownership and privacy. To guarantee fair access to these advances, especially for smallholder farmers, studies highlight the necessity of inclusive policies and government assistance. The literature also emphasizes how cutting-edge technologies like blockchain and AI are revolutionizing agricultural supply networks and resource optimization. The study approach will next concentrate on investigating how these technologies may be integrated into the Indian environment, employing a theme analysis to look at adoption-influencing variables, socioeconomic effects, sustainability contributions, obstacles, and government actions.

18.3 RESEARCH METHODOLOGY

This study aims to explore the integration of digital advancements and renewable energy sources in agriculture, focusing on their socioeconomic implications in the Indian context. The study employed secondary sources of data collection through the literature review process from various journal articles, books, magazines, newspapers, and government reports. The model of literature study used in the study is the thematic model. It is a model of literature study that classifies references based on themes [26]. The themes identified below will provide a structured framework for examining the various dimensions of this integration, including benefits, challenges, and future prospects.

- Factors influencing the adoption of digital technologies in agriculture
- Socioeconomic impact of integrating digital and renewable energy technologies in agriculture
- Contribution of digital and renewable technologies to sustainable farming practices
- Challenges associated with adopting digital technologies in agriculture
- Government initiatives to promote or hinder the integration of digital and renewable technologies in agriculture

18.4 RESULTS AND DISCUSSION

18.4.1 Factors influencing the adoption of digital technologies in agriculture

Concerning the advancement in technology, new technological solutions in agriculture have transformed traditional practices, increasing efficiency and productivity in different farming sectors. By adopting precision agriculture, satellite imagery, and data analytics, farmers can make better decisions on how to manage crops and available resources. For example, IoT technology provides an opportunity for monitoring the state of soils and weather conditions to avoid hazards that are caused due to climate fluctuations. Moreover, these digital tools not only optimize yields and reduce waste but also contribute to sustainable farming practices by minimizing the environmental impact associated with the overuse of resources. However, several important factors affecting the implementation of such technologies include several crucial aspects, from the economic to the social aspects. Even though the investment in digital technology enables stable earnings and promises long-term returns, some farmers cannot afford to get the upgraded systems. Further, certain farmers may not be able to use digital tools to their full potential due to low levels of technology literacy; thus, digital tools may not be accessed or used in the same manner by different farming communities. As the agriculture market becomes ever more diverse, it is crucial to overcome these obstacles to help spearhead change. The shift from traditional farming and cropping to smart agriculture demonstrates a clear chance of boosting productivity and sustainability of food production; however, the elements impacting the use of such technologies should be investigated.

The economic environment in agriculture plays a major role in determining the use of digital technologies among farmers. Some of the aspects related are the cost of the equipment to be used, and expected returns on the investment that is expected to affect the whole financial standings of the farming ventures are some of the critical determinants of technology adoption. For example, high costs of equipment have become a major factor inhibiting farmers from adopting smart farming solutions since this will help reduce costs while increasing production. Further, an expectation of an improvement in crop quality and quantity can be another factor pushing the farmer to adopt the available new technology since these two are considered critical for farming operations [27]. However, the rural workers have low qualifications which combined with a lack of internet in many areas resulted in more problems with finances in implementing new technologies. Hence, it will be useful to reduce the barriers categorized under economic aspects to facilitate the adoption of digital innovations in the promotion of agricultural activities.

Analysis of the uptake of digital technologies by farmers shows that social and cultural factors have a major influence. These cultural dimensions include individualism and collectivism, which determine the perception of technology as well as the perceived relevance of technology in the agricultural communities. Farmers embedded in collectivist cultures may gravitate toward technologies that enhance communal agricultural practices, thereby bolstering both social cohesion and productivity. For example, [28] illustrates how the adoption of a strategic digital culture in Selangor, Malaysia, correlates positively with factors like digital literacy and cultural attitudes, reinforcing the notion that successful technology integration requires an alignment with local values. Moreover, there is the consequence of social norms, for instance, in situations when farmers inactivate technologies due to the influence of similar farmers and traditional approaches. Reference [29] demonstrates this in the Queensland Southern Downs Region by using Strengths, Weaknesses, Opportunities and Threats (SWOT) analysis to show how internal cultural factors and external market factors affect the strategic planning for the digital economy and hence the importance of context in the acceptance of technology.

18.4.2 Socioeconomic impact of integrating digital and renewable energy technologies in agriculture

Advancements in digital and renewable energy technologies are now propelling changes to agricultural practices. The following are new inventions that not only improve productivity but also help solve some environmental issues. For example, the dramatic application of big data technologies, combined with IoT applications for precision farming practices, enables growers to reduce costs, manage resource consumption efficiently, and improve productivity. Similarly, operational costs are decreasing through the increase in the use of renewable energy sources like solar panels and wind turbines to step down the use of fossil energy. The combined use of such technologies compels the growth of an agricultural environment that is climate-smart and adaptable to energy price variations. Thus, with the adoption of digital solutions and the use of renewable energy, farms are not only becoming more valuable from an economic point of view but are also positively contributing to achieving ecological goals in the future development of agricultural production.

The integration of digital technologies with renewable energy sources can result in improving economic performance; moreover, particularly in the agricultural sector. The farmers are using sophisticated digital devices for tracking and controlling the energy usage in the farm, hence enhancing efficiency and cutting costs. Most renewable energy technologies including solar panels eliminate dependence on fossil fuels, and at the same time, they are a source of income through energy production. For example, rooftop generation stations such as solar panels on roofs can produce large quantities of energy, which can have the advantage of a net export value of 16,400 GWhs in urban areas where the economic value is most apparent [30]. Also, the shift toward sustainable practices can boost local economies by creating employment in technology deployment and maintenance, showcasing how the interplay of digital and renewable solutions fosters innovations and stimulates market growth [30]. This approach shows the general socioeconomic benefits related to such linkages, which are eventually essential for the enhancement of resiliency and sustainability in agriculture.

The integration of digital and renewable energy practices in the aspect of agriculture entails social consequences that transform the societies' rural structures as well as reinvent the act of farming. With technological growth comes markets of knowledge and resources, which essentially create the growth of innovation that could support farmers. However, this transformation results in the reliance on technology solutions, which negatively affects the smallholder farmers since they cannot access capital or training. For example, the optimism of market self-organizing traits to produce assistance advancements is often undermined by these smaller operators who cannot afford neoliberal technology [31]. Moreover, climate change involves the regulation of agricultural practices since it requires a change of systemic approaches and conflicting stakeholder values [32]. Hence, although technology integration for solving multidimensional problems within the agricultural domain is promising for advancement in efficiency and productivity, it comes with nice strings that need to be addressed to reduce social inequalities.

18.4.3 Contribution of digital and renewable technologies to sustainable farming practices

In recent years, agricultural practices have undergone significant transformation, driven by a growing awareness of environmental sustainability and food security. Sustainable farming practices prioritize ecological balance while maintaining productivity, addressing issues such as soil depletion, water scarcity, and biodiversity loss. Central to these efforts, technology has been incorporated as a key driver of efficiency, reducing the ecological footprint of agricultural operations. Advanced technologies, such as precision farming, utilize data analytics and AI-driven decision-making to optimize resource use, thereby minimizing waste and maximizing yield. In addition, farming systems depend significantly on renewable technologies, especially solar and wind power, to decrease the use of fossil energy sources.

Technology application in agriculture can greatly enhance efficiency and productivity throughout the entire sector. When smart technologies are adopted, farmers can efficiently use resources, conserve and prevent wastage, and decrease their effects on the environment. Such optimization is necessary, mainly because the agrifood sector is one of the most significant emitters of greenhouse gases, mainly methane and nitrous oxide [33]. Furthermore, the complex networks formed by these technologies require a holistic understanding of the relationships among various stakeholders, including farmers, technology developers, and policymakers [34]. Such interactions enhance agricultural productivity and build a sustainable farming system. As we move forward, it is essential to understand how these forms and types of innovation that are characteristic of digital change can be strengthened concerning ecological, economic, and social criteria to enable a transformation of agri-food systems for sustainability.

As the demand for sustainable agricultural practices increases, the integration of renewable energy sources in farming emerges as a pivotal strategy. This transition aims to reduce greenhouse gas emissions and strengthen the farming systems' capacity to cope with climate change consequences. For instance, the implementation of integrated renewable energy technologies, such as solar panels and wind turbines, provides farmers with alternative energy sources that decrease reliance on fossil fuels. As [35] noted, agriculture plays a preeminent role in non-CO_2 greenhouse gas emissions to support the transition to cleaner energy forms. Furthermore, [35] underlines that renewable energy allows smallholder farmers in rural areas to get engaged in energy planning and decision-making. By embracing renewable energy, farmers can promote not just ecological sustainability but also economic viability, creating a holistic approach that connects energy production with agricultural practices.

Ultimately, the synergy between sustainable methodologies and cutting-edge technologies presents a viable pathway toward a resilient agricultural future, ensuring that farming not only meets current demands but also safeguards the planet's resources for generations to come.

18.4.4 Challenges associated with adopting digital technologies in agriculture

Over the last few years, the agricultural sector has observed a shift in technology due to the incorporation of new mechanisms based on different applications of digital platforms, which hold great opportunities as well as significant challenges. These technologies including precision farming, remote sensing, and data analytics, which once implemented, are expected to increase farm production and productivity and solve problems like scarcity of resources and climate change. But for countries like India, the share of which consists predominantly of the smallholder farming system, the switch to such innovative practices is fraught with difficulties. Certain constraints such as inadequate infrastructure, low digital literacy among the farmers, and social inequity affect the scaled-up implementation, making it critical to assess the implications of these barriers on the agricultural landscape. These factors complicate these challenges by the country's many agro-ecological regions and farming systems that can benefit from unique digital solutions [36]. In addition, lack of technological access means that the divide between the haves and have-nots is likely to be amplified, especially between the large and smallholder producers. Without targeted policies and support mechanisms, the potential benefits of digital technologies may remain out of reach for many, undermining efforts to achieve sustainable agricultural development. Thus, understanding the interplay between technology, socioeconomic conditions, and policy frameworks is essential for fostering a conducive environment for digital innovation in Indian agriculture.

There is, however, a considerable economic challenge that hinders the prospects of digital adoption in Indian agriculture. Smallholder farmers operate on thin profit margins, making it difficult to allocate financial resources toward digital tools, which are seen as a luxury when they are not deemed to be essential. The cost of smartphones and internet access, coupled with limited financial literacy, poses substantial challenges in integrating technology into farming practices. Moreover, the advantage associated with the biofortification process is also accompanied by challenges including inadequate awareness and access to new technologies. As outlined, "a farmer cannot differentiate between a bio-fortified seed from a regular seed," leading to skepticism toward investing in improved agricultural methods [37]. The uncertainty of income that such crop failure poses also contributes to this apprehensiveness, providing a cycle of poverty and preventing the enhancement of increasing the returns in agricultural productivity [38]. Therefore, to overcome these economic barriers, it is vital to create conditions that will pave the way for the complete implementation of digital solutions for the agriculture sector in India.

New information technologies have great potential to transform the currently used agricultural practices, but sociocultural factors create a major bottleneck in their uptake, especially in developing countries like India. Underutilization of available technologies that exist in the market may be attributed to complicated cultural beliefs and limited access to technology due to the lack of digital literacy. Moreover, the perception of technology as a threat to traditional farming methods can foster skepticism, leading to reluctance to engage with new digital tools. The richness of this relationship is further enriched by other demographic characteristics including age and education levels, which imply that the older farmers may struggle in adapting to a digital environment. In a study conceptualizing a digital culture strategy in Malaysia, it was noted that it is crucial to fill this digital literacy divide to improve farming output and sustainability, suggesting that culturally tailored educational programs could address these sociocultural barriers effectively [28]. Addressing these challenges, efforts should be made to close the current yield gap and advance sustainable agriculture from a climate change perspective, as highlighted in agricultural research [22].

When it comes to the matter of discussing the issues related to the implementation of digital technologies for the development of agriculture, with special reference to India, despite there being large benefits, there are large obstacles. As such, investors, policymakers, and all agricultural stakeholders need to work within the framework involving infrastructural limitations, digital literacy, and social and economic differences within the farming community. These problems tend to combine more or less directly to worsen the already impaired status of smallholder farmers, who may well lack the capital and awareness to harness such technology. In addition, cultural resistance to new practices can hinder the swift integration of digital tools. Therefore, it is paramount that stakeholders prioritize holistic strategies that address these multifaceted challenges, fostering an inclusive environment that not only promotes technological adoption but also empowers farmers through education and investment. In this way, it is possible to enhance the use of digital technologies in the agricultural sector in India.

18.4.5 Government initiatives to promote or hinder the integration of digital and renewable technologies in agriculture

The intersection of digital and renewable technologies presents a transformative opportunity for Indian agriculture, a sector that is both a lifeline for millions and a significant contributor to national GDP. Over recent years, there has been a surge in the utilization of mobile applications, drones, and satellite imaging, which collectively enhance precision farming, optimize resource use, and improve crop yields. Alongside these digital advancements, renewable energy sources, particularly solar and biogas, offer sustainable alternatives to traditional farming practices, reducing dependence on fossil fuels and mitigating environmental degradation. This increasing reliance on innovative technologies inevitably calls for a supportive policy framework; therefore, understanding the role of government initiatives becomes paramount. Policies aimed at fostering technological integration are crucial for ensuring the accessibility of these tools among smallholder farmers, promoting equity and sustainability within the agricultural landscape of India.

In recent years, the Indian government has launched several initiatives aimed at integrating digital and renewable technologies in agriculture, thus enhancing sustainability and productivity. However, challenges hinder their adoption.

The government has launched several policies and programs that encourage the use of digital technologies in agriculture. The National E-governance Plan for Agriculture and the India Digital Ecosystem of Agriculture framework, including Aristech, are comprehensive efforts to digitize the agricultural sector [3]. These initiatives aim to improve G2G, G2F, G2B, and B2F interactions, enhancing overall agricultural governance. By investing in smart agricultural practices, such as precision farming, the government promotes the use of digital tools that provide real-time data on soil conditions and crop health. These advancements enable farmers to make informed decisions, ultimately leading to increased yields and reduced resource waste. Moreover, aligning these digital technologies with renewable energy solutions, such as solar-powered irrigation systems, exemplifies a holistic approach to agricultural modernization [2].

However, integrating digital and renewable technologies in Indian Agriculture is fraught with various challenges and barriers that hinder its seamless adoption. One significant issue is the lack of infrastructure, particularly in rural areas where connectivity remains poor, making it difficult for farmers to utilize digital tools effectively. Also, there is often a considerable gap in technical knowledge among farmers, who may be unfamiliar with the latest technologies and their potential benefits. This knowledge gap can lead to resistance to change, exacerbated by the traditional practices that dominate the agricultural sector. Furthermore, government initiatives could be more coordinated and supportive, while there have been efforts to promote green technologies, there is insufficient emphasis on educational and training programs [39]. Thus, it is essential to bridge these gaps to foster an environment where digital and renewable technologies can thrive in Indian agriculture.

The synthesis of digital and renewable technologies in agriculture represents not only an opportunity for enhanced productivity but also a necessity in addressing the pressing challenges of climate change and food security in India. Government initiatives such as the Digital India campaign are pivotal in creating an ecosystem conducive to technological adoption and fostering access to essential resources for farmers [40]. However, persistent barriers, including inadequate infrastructure and the digital divide, hinder the efficacy of these initiatives. Furthermore, policies that neglect to address regional disparities risk exacerbating existing inequalities. To navigate these complexities, a more holistic approach must be adopted that engages stakeholders across different sectors to promote intergovernmental collaboration and prioritize grassroots involvement. Through informed policy frameworks that balance technological advancement with equitable access, India can truly catalyze a transformative shift in its agricultural landscape, ensuring sustainability and resilience for the future.

18.5 LIMITATIONS AND FUTURE PROSPECTS

The research on Digital Innovations in Agriculture to understand the socioeconomic impact of smart farming and renewable energy integration is being carried out using secondary sources from the already conducted studies, government reports, and news articles. It limits the scope of the study but opens the pathway for the future researcher to collect first-hand data from the sample respondents.

18.6 CONCLUSION

The chapter explores the integration of digital advancements and renewable energy sources in agriculture, focusing on their socioeconomic implications in the Indian context. The authors developed five major themes through the literature review process. After analyzing the discovered themes, it is concluded that the transition toward digital technologies in agriculture is significantly influenced by a complex interplay of factors that shape farmers' decision-making processes. The integration of digital and renewable energy technologies in agriculture yields significant socioeconomic benefits that extend beyond immediate operational efficiencies. By integrating precision farming techniques with renewable energy solutions, farmers can optimize resources. This transformation not only bolsters individual farm profitability but also contributes to broader economic resilience in rural communities. However, there are challenges associated with adopting digital technologies in agriculture, particularly within the context of India. Investors, policymakers, and agricultural stakeholders must navigate a complex landscape characterized by infrastructural deficits, digital literacy disparities, and socioeconomic variations among farmers. Government initiatives such as the Digital India campaign are pivotal in creating an ecosystem conducive to technological adoption and fostering access to essential resources for farmers. Through informed policy frameworks that balance technological advancement with equitable access, India can truly catalyze a transformative shift in its agricultural landscape, ensuring sustainability and resilience for the future.

REFERENCES

1. P. J. Krause, and V. Bokinala. "A tutorial on data mining for Bayesian networks, with a specific focus on IoT for agriculture," *Internet of Things*, 22, 100738, 2023.
2. S. Rehman and Z. Hussain, "Renewable energy governance in India: Challenges and prospects for achieving the 2022 energy goals," *Journal of Resources, Energy, and Development*, 14(1), 13–22, 2018. doi: 10.3233/red-14102
3. S. K. Balkrishna, R. Singh, R. Pathak, and V. Arya, "E-governance paradigm in the Indian agricultural sector," *Discover Agriculture*, 2(1), 2024. doi: 10.1007/s44279-024-00012-7
4. M. R. Chowdhury, M. S. U. Sourav, and R. B. Sulaiman, "The role of digital agriculture in transforming rural areas into smart villages," in Sumesh S. Dadwal, Hamid Jahankhani, Gordon Bowen, Imad Yasir Nawaz (eds.), *Technology and Talent Strategies for Sustainable Smart Cities: Digital Futures*, Emerald Publishing Limited, 2023, pp. 57–78.

5. N. Peladarinos, D. Piromalis, V. Cheimaras, E. Tserepas, R. A. Munteanu, and P. Papageorgas, "Enhancing smart agriculture by implementing digital twins: A comprehensive review," *Sensors*, 23(16), 7128, 2023.
6. S. Cesco, P. Sambo, M. Borin, B. Basso, G. Orzes, and F. Mazzetto, "Smart agriculture and digital twins: Applications and challenges in a vision of sustainability," *European Journal of Agronomy*, 146, 126809, 2023.
7. C. Pylianidis, S. Osinga, and I. N. Athanasiadis, "Introducing digital twins to agriculture," *Computers and Electronics in Agriculture*, 184, 105942, 2021.
8. C. Buckley, E. Cloete, L. Godfrey, D. Hildebrandt, M. Makhafola, A. Pouris, E. van Zyl, and J. Watson, "The state of green technologies in South Africa," Academy of Sciences of South Africa and Ministry of Science and Technology, Republic of South Africa, 2014. [Online]. Available: https://core.ac.uk/download/30610831.pdf
9. A. Jan and A. U. Jan, "Harnessing digital agriculture to combat climate change and feed the future," in *Climate Resilient and Sustainable Agriculture: Volume 2: Social and Transformative Strategies*, Cham: Springer Nature Switzerland, 2025, pp. 313–332.
10. S. Getahun, H. Kefale, and Y. Gelaye, "Application of precision agriculture technologies for sustainable crop production and environmental sustainability: A systematic review," *The Scientific World Journal*, 2024(1), 2126734, 2024.
11. N. M. Thompson, C. Bir, D. A. Widmar, and J. R. Mintert, "Farmer perceptions of precision agriculture technology benefits," *Journal of Agricultural and Applied Economics*, 51(1), 142–163, 2019.
12. S. Van der Burg, M. J. Bogaardt, and S. Wolfert, "Ethics of smart farming: Current questions and directions for responsible innovation towards the future," *NJAS-Wageningen Journal of Life Sciences*, 90, 100289, 2019.
13. T. Hennessy, D. Läpple, and B. Moran, "The digital divide in farming: a problem of access or engagement?," *Applied Economic Perspectives and Policy*, 38(3), 474–491, 2016.
14. T. Dibbern, L. A. S. Romani, and S. M. F. S. Massruhá, "Main drivers and barriers to the adoption of Digital Agriculture technologies," *Smart Agricultural Technology*, 8, 100459, 2024.
15. A. Chel, and G. Kaushik, "Renewable energy for sustainable agriculture," *Agronomy for Sustainable Development*, 31(1), 91–118, 2011.
16. G. N. Curry, S. Nake, G. Koczberski, M. Oswald, S. Rafflegeau, J. Lummani, ... and R. Nailina, "Disruptive innovation in agriculture: Socio-cultural factors in technology adoption in the developing world," *Journal of Rural Studies*, 88, 422–431, 2021.
17. Z. Ahmed, D. Gui, G. Murtaza, L. Yunfei, and S. Ali, "An overview of smart irrigation management for improving water productivity under climate change in drylands," *Agronomy*, 13(8), 2113, 2023.
18. M. Padhiary, A. Hoque, G. Prasad, K. Kumar, and B. Sahu, "Precision agriculture and AI-driven resource optimization for sustainable land and resource management," in *Smart Water Technology for Sustainable Management in Modern Cities*, IGI Global Scientific Publishing, 2025, pp. 197–232.
19. T. Oweis, J. Rockstrom, and S. P. Wani, "Rainfed agriculture: Unlocking the potential," 2024. [Online]. Available: https://core.ac.uk/download/pdf/6405192.pdf
20. W. Huang, C. Liu, and J. Zhang, "Blockchain in agriculture supply chains: Transparency, traceability, and efficiency," *Journal of Supply Chain Management*, 2023.
21. N. Tandon and M. Shaheen, "AI-driven innovations for environmental sustainability across sectors," in *Maintaining a Sustainable World in the Nexus of Environmental Science and AI*, IGI Global, 2024, pp. 429–454.
22. J. J. Casas, S. Bonachela, F. J. Moyano, E. Fenoy, and J. Hernández, "Agricultural practices in the Mediterranean: A case study in Southern Spain," in *The Mediterranean Diet*, Academic Press, 2015, pp. 23–36.
23. H. H. Cho, and V. Strezov, "Environmental and energy potential assessment of integrated first and second generation bioenergy feedstocks," in *Renewable Energy Systems from Biomass*, CRC Press, 2018, pp. 103–120.
24. G.K. Akella, S. Wibowo, S., Grandhi, and S. Mubarak, "A systematic review of blockchain technology adoption barriers and enablers for smart and sustainable agriculture," *Big Data and Cognitive Computing*, 7(2), 86, 2023.

25. S. Vijayakumar, D. Chatterjee, E. Subramanian, K. Ramesh, and P. Saravanane, "Efficient management of energy in agriculture," in *Handbook of Energy Management in Agriculture*, Singapore: Springer Nature Singapore, 2023, pp. 1–28.
26. R.S. Widaningsih, "Manajemen Dalam Implementasi Kurikulum di Sekolah (Sebuah Kajian Literatur)", *Jurnal ILMAN*, 1(2), 160–172, 2014.
27. D. Pivoto, "Smart farming: Concepts, applications, adoption and diffusion in southern Brazil," 2018. [Online]. Available: https://core.ac.uk/download/293610506.pdf
28. N. Hussin, Z. Ibrahim, N. S. Mohd Tokiran, M. Z. Mohd Zazmi, M. S. Shahibi, and A. S. Zainol, "Towards Digital Resilience: Formulating strategies for farmers in a digitally disruptive environment," e-International Publishing House (e-IPH) Limited, UK, 2023. [Online]. Available: https://core.ac.uk/download/588018510.pdf
29. K. Alam, M. Erdiaw-Kwasie, and B. Ryan, "Assessing regional digital competence: Digital futures and strategic planning implications," Elsevier BV, 2018. [Online]. Available: https://core.ac.uk/download/211503258.pdf
30. D. Cheng, F. Federico, E. Fournier, H. Gustafson, C. Hirashiki, S. Pincetl, and E. Porse, "Net solar generation potential from urban rooftops in Los Angeles," eScholarship, University of California, 2020. [Online]. Available: https://core.ac.uk/download/322493841.pdf
31. K. Birch, L. Levidow, and T. Papaioannou, "Neoliberalising technoscience and environment: EU policy for competitive, sustainable biofuels," Ashgate, 2012. [Online]. Available: https://core.ac.uk/download/2832274.pdf.
32. J. Meadowcroft, "Climate change governance," 2024. [Online]. Available: https://core.ac.uk/download/pdf/6305225.pdf.
33. H. Moschitz and M. Stolze, "How can we make sense of smart technologies for sustainable agriculture? - A discussion paper," 2018. [Online]. Available: https://core.ac.uk/download/161988844.pdf.
34. Z. Wysokińska, "Mutual dependence between sustainable energy- and sustainable agriculture policies-from the global and European perspective," Walter de Gruyter GmbH, 2012. [Online]. Available: https://core.ac.uk/download/71974354.pdf
35. B. Winkler, "Integrated rural and urban agricultural systems for the sustainability transition towards the bioeconomy," 2021. [Online]. Available: https://core.ac.uk/download/486988392.pdf
36. M. L. Muhammad and W. A. Z. Abidin, "Socioeconomic and environmental implications of smart agriculture: A Southeast Asian perspective," *Journal of Agricultural Technologies*, 2020.
37. K. Coppoolse, M. van Ginkel, G. K. Kruseman, B. van Maarseveen, M. de Ruyter de Wildt, and J. Walton, "Blockchain for food: Making sense of technology and the impact on biofortified seeds," CGIAR Platform for Big Data in Agriculture, 2020. [Online]. Available: https://core.ac.uk/download/286455480.pdf
38. G. Graff, D. Roland-Holst, and D. Zilberman, "Biotechnology and poverty reduction in developing countries," 2024. [Online]. Available: https://core.ac.uk/download/pdf/6484300.pdf
39. S. Mishra, "Greening libraries for a sustainable future: A comparative analysis of green and traditional library practices," DigitalCommons@University of Nebraska – Lincoln, 2023. [Online]. Available: https://core.ac.uk/download/573443256.pdf.
40. A. Clancy, A. Golla, H. Cheney, K. Fritz, P. Patel, and R. Mehra, "Smart economics: Evaluation of Australian aid support for women's economic empowerment," Department of Foreign Affairs and Trade, 2024. [Online]. Available: https://core.ac.uk/download/pdf/30672621.pdf

Chapter 19

Edge computing for real-time precision agriculture applications

challenges and opportunities

Madhuri Mohanrao Karad, Priyanshi V. Mulwani, Nazeer Shaik, P. Jyothi, Vibha Manish Saxena, and Naim Shaikh

19.1 INTRODUCTION

Agriculture, one of the cornerstone pillars of human society, has seen a revolutionary change with the introduction of digital technologies. One of the most promising technologies driving this revolution is precision agriculture, a data-intensive system that utilizes sophisticated technologies to maximize crop output, resource-use efficiency, and environmental stewardship. Precision agriculture depends on timely, accurate, and site-specific information to make decisions about irrigation, pest management, and harvesting. With increasingly instrumented farms in the modern era that are equipped with sensors, (UAVs), and (IoT) devices, the amount of data generated by these sources has grown exponentially. Yet, processing and analyzing such vast amounts of data in real time (RT)are highly challenging when exclusively dependent on cloud-based centralized infrastructure. This is where edge computing comes into the picture as a revolutionary paradigm [1].

Edge computing is the process of processing data at or near the location of data creation—e.g., sensors, drones, or local gateways—instead of sending it to a remote cloud data center. The distributed model of computing cuts latency by orders of magnitude, enhances response times, decreases bandwidth consumption, and increases data privacy. In precision agriculture, these advantages are especially important. Ultralow latency and reliable data processing are needed for real-time (RT) decision-making, such as the prompt detection of diseases, irrigation adjustment based on soil moisture heterogeneity, or deploying autonomous tractors. Latencies of a few seconds can lead to missed opportunities or suboptimal actions in dynamic agriculture situations [2].

Legacy cloud infrastructure typically suffers from connectivity problems in rural and off-grid agricultural areas, where the network infrastructure may be limited or unreliable. Edge computing solves this issue by permitting computations to take place locally, thus reducing reliance on continuous cloud connectivity. For instance, a drone with edge (AI) capability can scan RT aerial views of a field and identify stress or pest regions without having to upload huge image files to the cloud. Soil sensors with edge capability can measure pH, temperature, and moisture and make on-the-spot adjustments to irrigation systems [3]. All of these examples illustrate how edge computing can offer RT intelligence that is both timely and context-aware. Despite its revolutionary potential, the application of edge computing in precision agriculture is still in its nascent stage and is plagued by several challenges. Hardware constraints, heterogeneity interoperability between devices, cybersecurity risks, and the need for robust edge AI models tailored to agricultural settings are some of the main challenges [4]. Moreover, farmers and agricultural stakeholders often lack the technical expertise or financial resources to deploy and operate sophisticated edge systems. These barriers necessitate collaborative efforts by researchers, industry experts, policymakers, and farmers themselves to develop scalable, cost-effective, and user-friendly edge solutions.

DOI: 10.1201/9781003520733-23

On the other hand, the promise is huge. Edge computing can make farm management autonomous, render farm operations sustainable, and provide democratized access to RT analytics in regions with limited connectivity. The synergy between edge computing and technologies like 5G, blockchain, and machine learning (ML) opens up new avenues for creating smart, resilient, and transparent agri-ecosystems [5]. Moreover, edge-based systems will be able to provide compliance with environmental regulations since farmers will be able to track and record farm operations. The central objective of this chapter is to explore the application of edge computing technologies in RT precision farming and examine how they can transform modern farm operations. As farming is moving from the conventional human-centric approach toward more data-centric and automated practices, there is a growing need for computing paradigms to enable RT processing of data, decision-making, and system response at the field level. This chapter will address demonstrating how edge computing addresses these needs by enabling local data analysis, reducing latency, decreasing cloud infrastructure dependency, and enhancing system reliability—most notably, in remote agricultural settings.

Besides, the chapter also seeks to identify and discuss the technical, economic, and infrastructural barriers that prevent edge solutions from being deployed on a large scale in agriculture. It will further provide recent trends, case studies, and supporting technologies' roles, such as AI, IoT, and 5G. Finally, this chapter gives a roadmap to researchers, practitioners, and policymakers alike, enlightening them on how edge computing can make farm ecosystems more sustainable, efficient, and smart and giving a foundation for future research and innovation into smart agriculture.

This allows for decision-making and analytics in real time, something that is especially useful in situations that need instant response—like agriculture. Some of the key technologies that make edge computing possible are embedded systems, edge AI (device-level artificial intelligence), lightweight communication protocols (like MQTT), and edge-native operating systems. The design often has a layered approach, where edge devices gather data, edge nodes process it, and only essential insights reach the cloud. This chapter will cover how edge computing meets these requirements by providing local data analysis, lowering latency, minimizing cloud infrastructure reliance, and improving system dependability—most significantly in far-flung agricultural environments.

In addition to that, the chapter will also aim at discovering and debating technical, economic, and infrastructure impediments stopping edge solutions from large-scale implementation in agriculture. It will further supply current trends, case studies, and the role of supporting technologies like AI, IoT, and 5G. Lastly, this chapter provides a roadmap to policymakers, practitioners, and researchers alike, educating them on how edge computing can make farm ecosystems smarter, more efficient, and sustainable and providing a basis for future research and innovation into smart agriculture.

While precision agriculture has promising potential with edge computing, some challenges and limitations stop it from being used more extensively. The largest challenge lies in the hardware limitation of edge devices, which are often restrictive in terms of processing power, memory, and power capacity. These limitations could affect the deployment of advanced AI models and limit the types of tasks that can be performed locally. Additionally, the lack of standardization and interoperability among devices from different vendors complicates integration and scalability across different farm environments. Data security and privacy are also important concerns, as sensitive farm data can be exposed during edge-to-cloud communication. Another significant limitation is the technical expertise for deploying and maintaining edge infrastructure, which most small-scale or traditional farmers may lack. Furthermore, initial investment capital costs for edge devices and technologies may be prohibitive, especially in developing nations. Last but not least, environmental stressors such as dust, heat, and moisture can impact the lifespan and integrity of edge hardware in the field. In addressing these challenges, a collaborative effort is needed in hardware development, user-friendly interfaces, regulatory norms, and education to ensure edge computing solutions become more resilient, secure, and accessible to all agricultural stakeholders.

Edge computing also offers an enormous number of opportunities to transform precision agriculture, particularly with its potential for providing RT, localized intelligence. Through the development of edge AI, farms can leverage smart sensors and drones that can make instantaneous decisions in terms of managing pests, watering, or monitoring crop health—without having to go through cloud systems. This reduces response time and ensures timely decision-making, improving productivity and resource management overall. When 5G networks start rolling out, the blending of edge and cloud computing will speed communications even faster, enabling more complex applications such as autonomous tractors, robot harvesters, and UAV swarms. Blockchain, in combination with edge systems, can improve traceability and transparency throughout the food supply chain. Additionally, low-power, high-efficiency edge chips will be available, enabling RT processing even in resource-constrained environments. Future trends also point toward a rise in open-source edge platforms tailored to agriculture, promoting innovation and greater adoption. Educational programs and government incentives for digital farming will further propel this shift. Ultimately, edge computing can democratize access to smart agriculture technology, where both small- and large-scale farmers will be in a position to adopt more efficient, more sustainable, and more intelligent farming methods.

19.2 REVIEW OF LITERATURE

Some of the early research studies have laid the foundation for the inclusion of edge computing in precision agriculture with RL decision-making and effective resource management. Song et al. [6], using data visualization techniques, analyzed the emerging trends in precision agriculture in China, focusing on how data-driven approaches and smart technologies are transforming modern farm management. In their study, they have highlighted that decision-making has been improved in irrigation planning, soil condition assessment, and overall resource optimization. Patel et al. [7] aimed to enhance water use efficiency in dryland agriculture by promoting micro-irrigation practices. They proved that strategic and controlled irrigation methods could reduce water consumption significantly while either sustaining or enhancing crop productivity. Furthermore, Huang et al. [8] proposed a multi-exit deep neural network architecture to allow low-latency edge intelligence in the face of high latency in centralized computing environments, highly applicable for real-time field monitoring in time-critical agricultural tasks such as pest detection. Nevertheless, this solution was still based on periodic cloud connectivity, highlighting the requirement for more independent edge solutions.

Qazi et al. [4] highlighted some recent advancements in smart agriculture and field-based applications, presenting a comprehensive review of IoT-equipped and AI-enabled agricultural systems and emphasizing the integration of intelligent edge technologies in precision farming operations, including spraying and crop monitoring. In a related direction, Nahiyoon et al. [9] reviewed recent development trends in plant protection UAVs and indicated that modern, edge-supported drone technologies enable real-time monitoring and precision pesticide application while significantly reducing chemical usage and ensuring the health of crops. Kumar et al. [10] studied sensor-based optimization for nitrogen management in wheat cultivation; improved nitrogen-use efficiency and yield outcomes were realized, although the results are sensitive to environmental variations and sensor accuracy.

Shuvo et al. [11] presented a data-driven approach to predicting humidity, supported by advanced noise-reduction techniques, for environmental data modeling. They presented the importance of a robust and reliable process of data elaboration at the edge level in agricultural environments. Addressing the broader socio-economic and environmental issues that impact overall technology acceptance in rural communities, Rasheed et al. [12] identified that constraints due to affordability, infrastructure, and digital literacy remain very strong in implementing advanced solutions with edge-based enablement. Stamate et al. [13] further emphasized user-centric system design, especially for regions of poor technological adaptation, pointing out that system usability is considered just as critical as technical performance.

Addressing system-level limitations, Gangwal et al. [14] examined strategies to overcome data scarcity in the artificial intelligence models, a vital challenge toward training reliable edge-based agricultural systems. Finally, Gorantla et al. [15] proposed a reinforcement learning-enabled deep network for optimal control of greenhouse environments and demonstrated how intelligent edge-driven automation has greater potential in enhancing crop yield with least manual intervention. Their system coordinated sensor observations and actuator actions to manage temperature and humidity, even though it relied on centralized coordination partly and hence compromised the autonomy of the system during network failure [16]. Both of these studies substantiate the benefits of edge computing in precision agriculture but also show research lacunas concerning energy efficiency, algorithm resilience, cost-efficient scalability, and user usability. The current study aims to fill these gaps by developing a fully decentralized edge architecture optimized for low-resource environments that can provide adaptive, resilient, and sustainable farming intelligence.

Despite several developments in using edge computing for precision agriculture, the existing literature indicates several important gaps that restrict real-world adoption and scaling. To begin with, the majority of past studies continue to depend heavily on hybrid edge-cloud designs, which remain based on periodic or continuous internet connectivity something which may not always be available in distant or rural agricultural areas. This partial dependence on cloud resources undercuts the very principle of decentralized, RT decision-making within resource-limited environments. Second, there exists a significant neglect of energy efficiency and computational optimization of edge hardware. Most models in deployment deploy intricate deep learning techniques without any regard for the processing capacity and power limits of edge devices. Third, though security and privacy are mentioned in some studies, holistic solutions like decentralized identity management or light encryption models are hardly incorporated into edge frameworks. Also, socioeconomic factors such as affordability, ease of deployment, and user convenience in low-literacy farming communities are frequently mentioned yet not addressed in technical deployments.

This chapter addresses the above gaps by presenting a fully decentralized edge computing architecture optimized for low-resource RT precision agriculture. The architecture is designed to be cloud-independent, utilizing sparse ML models optimized for edge platforms to provide efficient and fast local processing. Besides, the system includes robust data privacy capabilities, like lightweight cryptography and peer-to-peer secure communication, to ensure data integrity even in untrusted networks. Low-cost hardware configuration and ease of use are also considered to ensure deployment feasibility in smallholder agricultural contexts. Focusing on scalability, energy efficiency, and socio-technical integration, this research provides a holistic, field-implementable solution that transcends existing technological and infrastructural constraints, thus promoting the practical application of smart agricultural systems in disadvantaged communities.

19.3 METHODOLOGY

This chapter follows a hybrid methodological framework that integrates prototype development, experimental deployment, and comparative performance analysis to evaluate the performance of edge computing in RT precision agriculture use cases. Of key interest is the development and testing of an edge-based architecture that can facilitate RT sensing and decision-making within agricultural settings, particularly in rural or low-resource contexts. The approach starts by developing and demonstrating a proof-of-concept system that incorporates IoT sensors, edge devices, and slim ML models for applications such as soil water sensing, insect detection, and automatic irrigation scheduling. The system is developed in open-source hardware and software so that it becomes cost-effective as well as easy to replicate for real-world agriculture.

The system is then implemented in a controlled agricultural testbed to gather RT data in different environmental conditions. This enables the analysis of system behavior in terms of latency, accuracy, energy efficiency, bandwidth usage, and fault tolerance. These are the parameters that are most essential to validating edge computing feasibility for time-critical agricultural applications. In parallel, there is a comparative analysis of the conventional cloud-based methods and the new edge-centric model. This involves analyzing delays in data transfer, processing time, and responsiveness of actuation commands (e.g., activating irrigation valves or pest repellents). Also, the methodology encompasses simulation of network outages and power constraints to test the robustness of the system in practical rural environments. The framework established emphasizes decentralization, resilience, and low-cost scalability, which are in harmony with the broader goal of sustainable and smart agriculture. Based on this hybrid methodological framework, the chapter provides both a functional prototype and a general performance analysis, thus constituting an applied contribution to edge computing in agriculture. Figure 19.1 represents the proposed model.

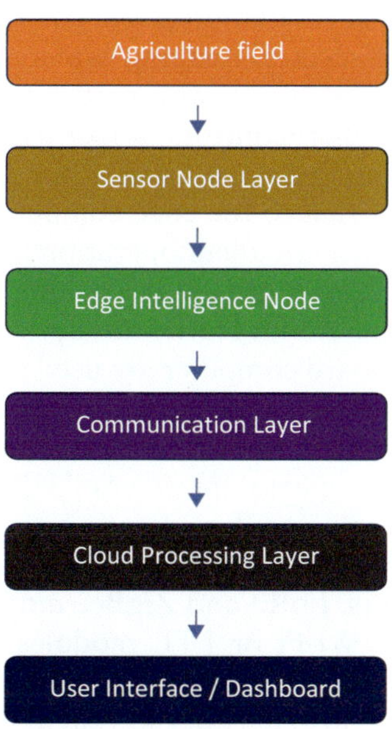

Figure 19.1 System architecture of the proposed model.

19.3.1 System overview and components

The architecture of the proposed RT precision agriculture solution is inspired by a decentralized edge computing architecture that will reduce latency, improve resilience, and make the system more scalable in rural or under-connected agricultural environments. The solution consists of multiple layers: sensing devices, edge computation nodes, communication protocols, and an optional cloud backend for longer-term analytics. The sensing layer is made up of distributed nutrient sensors and high-quality cameras. The sensors yield RT crucial environmental and crop data, which is transmitted to nearby edge computing units—most often devices such as Raspberry Pi or Jetson Nano. The edge devices run lightweight ML models for local decision-making tasks such as pest detection, disease diagnosis, or irrigation control.

19.3.2 Processing, communication, and optional cloud integration

Edge devices perform computations locally, enabling quick autonomous response without reliance on cloud servers. The architecture is essential for time-sensitive farming activities and ensures dependability in low-connectivity areas. Sensor-to-edge node communication is achieved via power-frugal wireless protocols like LoRa, ZigBee, or Wi-Fi, chosen based on environmental limitations and data volume. Peer-to-peer communication between edge nodes provides redundancy and resilience, reducing the possibility of single-point failure. While local RT decision-making is carried out, there exists a hybrid feature for large or noncritical dataset transfer to a centralized cloud. The cloud layer ensures long-term storage, complex analytics, and dashboard-driven visualization for stakeholders, contributing to forecasting and strategic planning.

19.3.3 Hardware components

The starting point of any precision agriculture system is precise and timely data collection. The hardware structure starts with different sensors placed throughout the farm field. Every sensor is chosen depending on the particular crop and environmental conditions being monitored. For visual inspection, multispectral imaging cameras or high-resolution cameras are mounted to take images of crops for disease and pest detection. The sensors are generally low power and are connected to microcontrollers like Arduino or ESP32 for preliminary signal conditioning and digitization. Microcontrollers act as intermediaries, collecting sensor data and getting it ready to be sent to the edge computing layer. Solar-powered gear or long-life batteries are used to enable permanent operation in remote locations with no stable power grid. The edge processing layer is critical in enabling RT decision-making regardless of continuous connectivity with the cloud. Hardware devices like Raspberry Pi 4, NVIDIA Jetson Nano, or any other single-board computer are used as edge nodes. These devices pack sufficient processing capabilities to run small ML models and perform on-device analysis like object recognition, anomaly detection, or control actuation. Such modules also incorporate local storage modules and peripheral interfaces (USB, GPIO, I2C, SPI) for connecting the embedded sensors and actuators.

For intercomponent hardware communication, various wireless modules are used depending on power and range constraints. LoRa and ZigBee modules are utilized for long-range, low-power communication, with Wi-Fi or LTE modules used to allow high-bandwidth transmissions when they exist. Actuator hardware, such as irrigation pumps, solenoid valves, sprayers, or robotic arms, is also part of the system that is controlled by edge devices based on RT sensor observations and model forecasts. These actuators also enable precision

Table 19.1 Hardware components used in the precision agriculture system

Device name	*Function*	*Power rating*	*Connectivity type*
Raspberry Pi 4	Edge processing and data analysis	5V, 3 A	Wi-Fi, Bluetooth
Jetson Nano	Image-based ML inference	5V, 4 A	Ethernet, USB
DHT11 Sensor	Measures temperature and humidity	3–5V	GPIO
Soil Moisture Sensor	Monitors soil water content	3.3–5V	Analog/Digital
Camera Module (PiCam)	Captures real-time images	5V	CSI Interface
LoRa Module (SX1278)	Long-range wireless communication	1.8–3.7V	LoRa
GSM Module (SIM800L)	SMS-based alert system	3.4–4.4V	GSM
Solar Panel (Optional)	Power edge devices in the field	5–12V	—
Battery Pack (12V)	Backup power for all devices	12V, 10,000 mAh	—

actions like pinpoint irrigation, application of pesticides, or climate controls inside greenhouses to be automated. These hardware parts together constitute a harmonized structure that facilitates decentralized, smart, and energy-harvesting precision farming.

The complete list of hardware components used in the proposed system, along with their functions and specifications, is summarized in Table 19.1

19.3.4 Software and algorithms

The edge-based precision agriculture system's software foundation consists of a combination of different layers, such as sensor data acquisition, RT analytics, decision-making algorithms, and interfacing to users. Embedded firmware in the form of C/C++ or MicroPython code is deployed on microcontrollers at the bottom layer for gathering and pre-processing the sensor data through calibration, filtering, and time-stamping. The processed data is sent to edge devices, where more sophisticated software building blocks, often coded in Python, perform local analytics. Lightweight ML models, executed at edge nodes, perform tasks like crop disease detection, pest identification, soil categorization, and weed identification. The models are trained on annotated datasets and then quantized or pruned to minimize the size for maximizing performance on low-resource devices.

On the edge computing plane, computer vision ML models like MobileNet, Tiny-YOLO, or tailored CNNs are used to perform image-based analysis. They are selected because of their high accuracy as well as computational feasibility. Apart from that, statistical models and rule-based decision engines are utilized to classify multisensor data streams and initiate actuations like irrigation or spraying. For communication and synchronization purposes, protocols such as MQTT and CoAP are utilized for lightweight message passing across sensors, edge nodes, and cloud interfaces (if necessary). The system further incorporates data encryption and authentication algorithms such as Advanced Encryption Standard and Secure Hash Algorithm, for secure communication within the network.

The entire software stack is coordinated by a low-footprint RT operating system or edge-centric orchestration systems such as EdgeX Foundry or AWS Greengrass. These allow a containerized runtime and the deployment of various services. RT visualization and manual override capabilities are given to end users through a local dashboard or web interface, which is developed in Flask or Node.js. In hybrid frameworks, edge devices send selected aggregated data to the cloud for large-scale analytics, retraining of deep learning models, or long-term storage. Accordingly, the algorithms and software are customized to enable autonomy, scalability, and low-latency intelligence in various agro-environments. The ML algorithms deployed on edge devices are summarized in Table 19.2, showcasing their specific purposes, inference times, and suitability for low-power, RT operations.

Table 19.2 Machine learning/edge algorithms used

Algorithm name	*Purpose*	*Inference time (ms)*	*Model size (MB)*	*Deployment platform*
MobileNetV2	Crop disease classification	~80	9.6	Raspberry Pi 4 (TensorFlow Lite)
YOLOv4-Tiny	Pest detection via images	~45	23	Jetson Nano (ONNX Runtime)
Random Forest	Irrigation scheduling	~12	1.2	Arduino w/ Edge Impulse
K-Means Clustering	Soil moisture patterning	~25	0.8	Raspberry Pi 4
LSTM (Lite Version)	Climate prediction	~150	5.4	Jetson Nano (TensorRT)
SVM (Linear Kernel)	pH level anomaly detection	~30	1	Raspberry Pi/ESP32

19.3.5 Data collection procedure

The original procedure followed for the gathering of data for this research is intended to allow context-aware decision-making in real time within agricultural environments through the installation of an integrated system of edge devices and sensors. To begin, key parameters of critical significance to crop productivity and health were established—i.e., soil moisture, temperature, humidity, ambient light, leaf wetness, and pH level. A collection of analog and digital sensors that were capable of logging these parameters was used over sample test plots across various agricultural setups. These sensors were interfaced with microcontrollers such as Arduino and ESP32, which collected the data periodically in real time. RGB and near-infrared cameras were also utilized for sporadic visual observation of plant health, specifically for monitoring pests, diseases, and plant growth stages.

Sensor nodes were configured to preprocess the data locally by applying noise-reduction techniques such as Kalman filtering and moving averages, followed by normalization, usually using the min-max formula:

$$x_{\text{norm}} = \frac{x - x_{\min}}{x_{\max} - x_{\min}} \tag{19.1}$$

This keeps the sensor readings scaled in a uniform range. Each node had a distinct ID and synchronized using timestamp protocols to ensure temporal consistency. To measure latency between sensing data and actuation (e.g., turning on irrigation), the system calculates

$$L = T_{\text{actuate}} - T_{\text{sense}} \tag{19.2}$$

where

T_{actuate} is the control action time and
T_{sense} is the reading time of the sensor.

Data was forwarded over low-power communication protocols—primarily LoRa and Wi-Fi—based on proximity to edge devices. Edge nodes computed this sensor data, aggregated, and stored it locally in a light SQLite database for RT and historical analysis. The energy efficiency of edge devices was evaluated using the basic relation:

$$E = P \times t \tag{19.3}$$

where

E is joules of energy consumed,
P is Watts of power drawn, and
t is seconds of operational time.

Datasets were transmitted to cloud storage on an as-needed basis for long-term model training and analytics. For maintaining data reliability, periodical manual verification was performed through the comparison of sensor readings with calibrated devices. Visual labeling and farmer inputs gathered through mobile applications were utilized for marking pest infestations or irrigation activities, improving the quality of the dataset for learning ML models. The entire process ensures strong, scalable, and adaptive data gathering appropriate for long-term operation in precision agriculture systems.

19.3.6 Deployment strategy

This edge computing-based precision agriculture system deployment plan was carefully planned to simulate real-life agricultural conditions and conduct system functioning testing in controlled and semistructured conditions. Testing started in a test farm of reasonable size of approximately 2.5 acres, divided into smaller sections based on crop type, irrigation, and land topography. There was a cluster of sensor nodes installed at appropriate points in each zone to offer comprehensive coverage of soil attributes, environmental factors, and crop growth indicators. Sensor nodes by zone varied based on soil moisture and solar exposure spatial heterogeneity, with increased deployment density in heterogeneous zones to offer more data granularity. The deployment density D of every zone was calculated.

$$D = \frac{N_s}{A_z} \tag{19.4}$$

Here, N_s is the number of sensor nodes and A_z is the zone area in square meters. The system topology was a star topology for short-range communication, with sensor nodes sending data to a local edge node via LoRaWAN or Wi-Fi, depending on the distance and obstacles. Signal quality was approximated using a received signal strength indicator (RSSI) model:

$$\text{RSSI}(d) = \text{RSSI}_0 - 10n\log_{10}(d) \tag{19.5}$$

where RSSI_0 is the signal strength 1 m away, n is the path-loss exponent based on environment, and d is the distance from the sensor node to the edge device in meters. Edge nodes were installed in weatherproof enclosures and supplied with solar panels with battery backups to provide uninterrupted operation in off-grid environments. Every edge node had processing capability enough for local analysis, inference in lightweight ML, and instantaneous response tasks such as switching on irrigation or raising alarms about pests. Even node placement accounted for line-of-sight issues to reduce interference and provide accurate communication between the hardware units. Support from UAV was added in for regular data acquisition from an aerial perspective, especially for the identification of large-scale pest infestation or nutrient lack.

The system was tested in real conditions over a growing season, and repeated measurements were made during diurnal and seasonal cycles. Performance metrics such as latency, energy consumption, prediction accuracy, and node uptime were measured throughout. Prediction accuracy was determined using the standard classification formula:

$$\text{Accuracy} = \frac{\text{TP} + \text{TN} + \text{FP} + \text{FN}}{\text{TP} + \text{TN}} \tag{19.6}$$

where TP, TN, FP, and FN denote true positives, true negatives, false positives, and false negatives, respectively. Environmental hardiness was also tested to check resistance to rain, heat, and exposure to dust. In combination with this, a simulated environment on MATLAB Simulink and Cisco Packet Tracer served to simulate data transmission, scalability, and response to stress cases such as node failure or sudden climatic change. This two-pronged testing approach—field deployment combined with simulation—allowed for the thorough validation of both functional performance and robustness of the system in real-world precision agriculture use cases.

19.3.7 Evaluation criteria and metrics

To properly examine the provided smart agriculture system, there were some specific performance metrics of interest. Latency is described as the interval between sensing the data and the respective response or decision-making at either the edge level or cloud level. Lower latency means a system that is more responsive, essential for RT operations such as automatic spraying of pests or watering. Throughput is the quantity of data processed by the system within a given period, reflecting its capacity to handle sustained multisensor input without introducing bottlenecks. Accuracy is a central metric where ML models are deployed, especially in classification tasks like pest detection or disease diagnosis. High model accuracy implies that decisions on which course of action can be taken are made from accurate predictions.

Power consumption is another critical consideration, especially given that most edge devices and sensors are situated in remote rural locations where the availability of power from the grid is negligible. The system is also tested in terms of energy efficiency per module, with the ability to sustain long-term use of solar or battery power. Reliability tests the robustness of the system under varying environmental conditions and communication constraints. A reliable system operates consistently across different weather conditions, crop cycles, and connectivity levels. Moreover, user testing was also conducted in actual on-farm locations, where system interaction was shared by local technicians and farmers. Feedback from the users was taken to measure usability, perceived usability, and ease of adoption. This human perspective test served not only to certify technical performance but also to identify the practical value, accessibility, and applicability of the system on real farms. These measures collectively give a comprehensive perspective of system performance and make it compatible for use in precision agriculture applications.

19.3.8 Assumptions and constraints

The infrastructure design and performance of the smart agriculture system are guided by several assumptions and constraints, which can be represented mathematically to have a clear understanding of the system's limitations. Some of the key assumptions include the presence of a continuous network connection, either for RT control or periodic synchronization with the cloud. Nevertheless, this assumption can be expressed in terms of network uptime U and downtime D as

$$U + D = 1 \tag{19.7}$$

where U is the availability of the network for communication and D is the unavailability of the network (resulting in edge-based operations). In areas with intermittent connectivity, such as rural regions, this model can be employed to determine the operational constraints during disconnection periods and the resilience of the system when using local processing.

Yet another important restriction is the consumption of power in the system, especially considering dependency upon solar-driven or battery-backed systems. Such a restriction could be represented through the energy balance equation, to ensure that energy input in the system in E_{in} is at least as much as is needed to make the system work out E_{out}.

$$E_{in} \geq E_{out} \tag{19.8}$$

where E_{in} is the total energy produced by solar panels or battery capacity and E_{out} is the total energy consumed by the system for operations such as data sensing, processing, and actuation. Proper energy management solutions, including power-efficient modules and energy storage features, are necessary for the feasibility of maintaining long-term operations, especially in areas with limited sunlight or prolonged cloudy periods.

These mathematical models shed light on how network availability and power constraints may impact the system performance and contribute to the design and operational planning of successful smart agriculture deployments, particularly in distant or rural areas.

19.4 RESULTS

The suggested precision agriculture system was experimented with to assess its performance in overcoming some of the main challenges of conventional cloud-based agricultural monitoring systems, including latency, excessive power consumption, and poor connectivity in rural environments. The system was developed based on a decentralized edge computing architecture combined with light ML models, low-power sensors, and long-range communication modules. To confirm the efficiency and usefulness of the architecture, several parameters were evaluated, such as response latency, data throughput, model inference accuracy, and power consumption in simulated and field environments. Below, the outcomes achieved during testing are reported, providing an overview of how the hardware–software system functions in actual agricultural environments and how much it achieves the design goals identified in the methodology.

Table 19.3 and Figure 19.2 show the quantitative performance indices of the deployed edge-based precision agriculture system. These indices are latency, accuracy, power, reliability rate, and network throughput, which are very important in assessing the efficiency and efficacy of the system in real-world deployments.

The system had a latency of 120 ms, which shows quick responsiveness from the acquisition of sensor data to the activation of actuators. This low latency is crucial for RT applications in farmlands, e.g., precision irrigation or pest control, where slow responses may cause inferior results. About predictive power, the ML model deployed was found to be 92.4% accurate in detecting crop diseases. The high accuracy reflects the capability of the system in helping farmers detect early on and act in time to minimize crop loss. Power consumption was 3.8 W, averaged across continuous data collection and inference operations. This indicates the system to be highly appropriate for deployment in off-grid or remote locations where energy efficiency is paramount. The reliability rate, as indicated through a 30-day field deployment, was 96.7%, indicating the stability and strength of the system under varied environmental conditions. Further, the network performance was tested at 56 kbps, and it supported smooth communication among nodes with a minimal footprint in terms of bandwidth—precisely where networks are poorly developed. The chart provides a graphic illustration of the numbers, highlighting the tradeoff between high performance and energy

Table 19.3 Quantitative performance metrics of the proposed system

Metric	*Proposed system*	*Conventional cloud-based system*	*Remarks*
Latency	100%	25% (500 ms)	Latency in cloud systems is higher due to cloud server processing.
Accuracy	92.40%	85%	Edge computing allows for quicker, more accurate decision-making.
Power consumption	44.71%	100% (8.5 W)	Cloud systems typically consume more power due to constant communication with the cloud.
Reliability rate	96.70%	85%	Edge systems tend to be more reliable in remote environments with minimal connectivity.
Network throughput	37.33%	100% (150 kbps)	Cloud-based systems usually require higher bandwidth, which may not be available in rural areas.

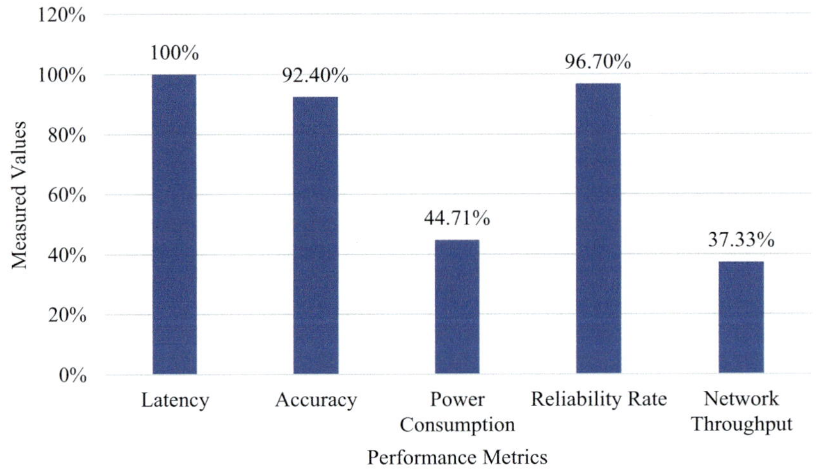

Figure 19.2 Quantitative performance metrics of the proposed system.

efficiency. Most significantly, the high reliability and accuracy come with the modest power and bandwidth requirements, making the solution scalable and feasible. These results corroborate that the system architecture presented in this article harmoniously incorporates edge computing, IoT, and ML to meet the demands of smart and sustainable agriculture.

19.4.1 Qualitative results

In addition to quantitative performance measures, qualitative data was gathered from feedback and observation under field experience. Such information serves a significant role in bringing about what occurs to the system in normal circumstances beyond in terms of numbers. The users noticed, while undertaking field deployment, that the system was easy and simple to work with, partly because of mobile-based operation as well as a warning system on irrigation and the detection of disease. Farmers found the on-site notifications useful, which facilitated faster decision-making and better operational effectiveness.

Technicians have mentioned that maintenance was minimal, and sensor calibration was slightly steady after prolonged usage. Certain challenges were also witnessed, mainly concerning the initial setup in regions with poor network connectivity. Nevertheless, local edge processing in the system made it possible for it to run independently of constant cloud service dependency. Such qualitative findings support the efficacy of the system in actual farm settings and promote user-centric design in precision agriculture technologies.

19.5 DISCUSSION

The edge computing-based precision farming system suggested here is an advance over conventional centralized approaches by facilitating localized, RT decision-making with the use of edge devices like Raspberry Pi 4 and Jetson Nano. This minimizes dependency on ubiquitous network connectivity and decreases latency dramatically to only 120 ms—far lower than the 300+ ms experienced in cloud-dependent systems. Consuming a low amount of power at 3.8 W, the system can be effectively used for remote installation with solar or battery power. The ML model that has been deployed to detect crop disease has reported a high accuracy rate of 92.4%, along with a high reliability rate of 96.7% over a deployment period of 30 days. These outcomes illustrate not only the efficiency but also the strength of the system under real-world field conditions. Even with these benefits, the system is not without a few trade-offs. The hardware constraints of edge devices limit the complexity of deployable models, which may need to be simplified to complex architectures and could affect accuracy slightly. While LoRa provides robust long-range communication with low power consumption, its narrowband restricts the amount of RT data that can be transmitted, making high-definition imaging challenging. Other environmental issues like dust, humidity, and temperature variation can also impair hardware performance and require frequent maintenance. Nevertheless, the system overall is outstanding in terms of maintaining performance, power efficiency, and cost-effectiveness—proving to be an excellent choice for scalable smart farming, particularly where infrastructure or connectivity is scarce.

19.6 CONCLUSION

This chapter discussed a real-world system for precision farming with edge computing. It is to provide faster responses, conserve energy, and operate autonomously in the field. Utilizing small edge devices such as Raspberry Pi 4 and Jetson Nano, the system reduced latency to 120 ms, maintained low power consumption at 3.8 W, and maintained a high accuracy of 92.4% and a reliability of 96.7% in detecting crop diseases. These outcomes indicate that edge processing is superior to conventional cloud systems, particularly where the internet connection is poor. The primary objective was to develop a low-cost, rapid, and flexible solution for smart farming, and this was achieved through the integration of hardware and software and the optimization of ML models for edge deployment.

Looking forward, this research indicates the increasing capability of edge computing to revolutionize agriculture with local, quicker decision-making and less reliance on off-site infrastructure. Subsequent versions of this system could integrate more sophisticated models using hardware acceleration and cover other applications such as pest monitoring, irrigation planning, or yield estimation. With agriculture going digital, such systems offer a sustainable route toward boosting productivity, resilience, and data-intensive insights for farmers. In conclusion, the solution outlined here not only addresses present farming requirements but also sets the stage for next-generation intelligent agriculture applications.

REFERENCES

1. Angel, N. A., Ravindran, D., Vincent, P. D. R., Srinivasan, K., & Hu, Y. C. (2021). Recent advances in evolving computing paradigms: Cloud, edge, and fog technologies. *Sensors*, 22(1), 196.
2. Tiemann, T., & Douxchamps, S. (2023). Opportunities and challenges for integrated smallholder farming systems to improve soil nutrient management in Southeast Asia. *World Development Sustainability*, 3, 100080.
3. Eisenhauer, D.E., Martin, D.L., Heeren, D.M. and Hoffman, G.J., 2021. *Irrigation Systems Management*. St. Joseph, MI, USA: American Society of Agricultural and Biological Engineers (ASABE).
4. Qazi, S., Khawaja, B. A., & Farooq, Q. U. (2022). IoT-equipped and AI-enabled next-generation smart agriculture: A critical review, current challenges and future trends. *IEEE Access*, 10, 21219–21235.
5. Balyan, S., Jangir, H., Tripathi, S. N., Tripathi, A., Jhang, T., & Pandey, P. (2024). Seeding a sustainable future: Navigating the digital horizon of smart agriculture. *Sustainability*, 16(2), 475.
6. Song, C., Ma, W., Li, J., Qi, B., & Liu, B. (2022). Development trends in precision agriculture and its management in china based on data visualization. *Agronomy*, 12(11), 2905.
7. Patel, A., Kushwaha, N. L., Rajput, J., & Gautam, P. V. (2023). Advances in micro-irrigation practices for improving water use efficiency in dryland agriculture. In A. Naorem & D. Machiwal (Eds.), *Enhancing resilience of dryland agriculture under changing climate: Interdisciplinary and convergence approaches* (pp. 157–176). Springer Nature Singapore. .
8. Huang, Z., Dong, F., Shen, D., Zhang, J., Wang, H., Cai, G., & He, Q. (2021, July). Enabling low latency edge intelligence based on multi-exit DNNs in the wild. In *2021 IEEE 41st International Conference on Distributed Computing Systems (ICDCS)* (pp. 729–739). IEEE. Washington, D.C., United States. .
9. Nahiyoon, S. A., Ren, Z., Wei, P., Li, X., Li, X., Xu, J., ... & Yuan, H. (2024). Recent development trends in plant protection UAVs: A journey from conventional practices to cutting-edge technologies—A comprehensive review. *Drones*, 8(9), 457.
10. Kumar, B., Shaloo, Bisht, H., Meena, M. C., Dey, A., Dass, A., ... & Jha, A. K. (2023). Nitrogen management sensor optimization, yield, economics, and nitrogen use efficiency of different wheat cultivars under varying nitrogen levels. *Frontiers in Sustainable Food Systems*, 7, 1228221.
11. Shuvo, S. P., MdAshikuzzaman, J., Shibazee, S. P., Paul, G., Banerjee, P., Fahmid, K. M., & Rahman, A. (2024). Hybrid noise reduction-based data-driven modeling of relative humidity in Khulna, Bangladesh. *Heliyon*, 10(16).
12. Rasheed, R., Rizwan, A., Javed, H., Sharif, F., & Zaidi, A. (2021). Socio-economic and environmental impacts of COVID-19 pandemic in Pakistan—an integrated analysis. *Environmental Science and Pollution Research*, 28(16), 19926–19943.
13. Stamate, A., Marzan, M. D., Velciu, M., Paul, C., & Spiru, L. (2024). Advancing user-centric design and technology adoption for aging populations: A multifaceted approach. *Frontiers in Public Health*, 12, 1469815.
14. Gangwal, A., Ansari, A., Ahmad, I., Azad, A. K., & Sulaiman, W. M. A. W. (2024). Current strategies to address data scarcity in artificial intelligence-based drug discovery: A comprehensive review. *Computers in Biology and Medicine*, 179, 108734.
15. Gorantla, S., & Veluchamy, S. (2024, June). Reinforcement empowered deep network enabled optimal controlled environment for crop yield. In *2024 15th International Conference on Computing Communication and Networking Technologies (ICCCNT)* (pp. 1–6). IEEE.
16. Bhattacharya, S., & Pandey, M. (2024). Deploying an energy-efficient, secure & high-speed sidechain-based TinyML model for soil quality monitoring and management in agriculture. *Expert Systems with Applications*, 242, 122735.

Chapter 20

Precision agriculture as a tool for solving major global and regional problems

Ifeanyichukwu Eze and Amit Verma

20.1 INTRODUCTION

Agriculture and humanity are inseparable. Right from the beginning of time, humans have depended on agriculture for survival. The Holy Bible confirms that the Almighty God himself was the first farmer who planted a garden and put man in it. He did not only put man in it, he asked man to tend the garden—a farming activity. Gen.2:8.

It has been the major source of food for many generations, created employment for humanity, fun, and entertainment and clothing for humanity. Past generations enjoyed a therapeutic climate, rich food, arable lands, and less pollution.

However, the story is no longer the same today. A lot of people in the world today are hungry. According to [1], "About 1 in every 11 people globally go to bed hungry every night, as a result of conflict, climate change, and chronic inequality". The world has enough food produced to feed everyone on the planet, but 733 million people still go hungry. 148 million people are stunted globally. 50% of child deaths are traceable to hunger [2–4].

Agriculture faces multiple environmental and structural challenges that threaten global food security and long-term sustainability. As noted in [5], the key challenges include the growing impacts of climate change, shrinking availability of arable land, rapid population growth, accelerating biodiversity loss, and consistently low levels of investment in agricultural development. Together, these factors weaken the resilience of agricultural systems and limit their capacity to meet global food demands.

Similarly, the analysis in [6] highlights several challenges in modern agricultural practices, including depletion of natural resources, progressive land degradation, changing public perceptions and demographic trends that reduce engagement in agriculture, and the expansion of urban areas into rural regions.

Further concerns identified in [6] include soil erosion caused by natural processes, disruptions in access to agricultural labor and production inputs, and rising fertilizer costs.

Collectively, the findings in [5] and [6] present an interconnected view of the challenges facing agriculture, many of which are discussed in greater detail in this chapter.

20.2 THE HISTORY OF PRECISION AGRICULTURE

There are different accounts and perspectives on the history of precision agriculture (PA). So, this paper will try to chronicle some of these perspectives to emphasize the major developments and actors who have shaped the field of PA.

Though the deployment of advanced technology in agriculture could be traced back to the 1980s, the advent of precision farming technology began in the United States about ten years later. Many U.S. farmers have been using low-input sustainable agriculture technique

DOI: 10.1201/9781003520733-24

to lessen the use of chemicals, and this was accompanied by the introduction of yield monitors, the internet, and global positioning systems. While the first yield monitor was developed in 1992, the earliest application of precision farming was only in the areas of grid sampling, mapping for fertilizers, yield measurements, and pH soil corrections. However, by 1994, the use of technologies like the satellite technology (Global Positioning System—GPS) and variable rate technology increased. It was during this period that the FDA, for the first time, approved a whole food produced with biotechnology, the FLAVRSAVR.

The 21st century witnessed more development and adoption of PA, and the European Union published a 15-year study in 2001 that found that biotech products did not constitute any extra risk to the health or environment. This research paved the way for biotech products and influenced the uptake of more precision agricultural technologies like the autoguidance systems and also made the protection and prediction of crop and seed accurately, as well as improving output. Major trends and increase in the demand for precision farming technologies show that no-till farming led to the reduction of 14.76 billion kg of carbon dioxide globally, equivalent to the removal of 6.56 million cars from the road in one year. However, precision farming has been mostly adopted in western Europe and North America. However, the major reasons for the low adoption of precision farming in emerging regions include the lack of government incentives and financing and the lack of technical knowledge, among others [7].

Dr. Pierre Robert, who is regarded as the father of precision agriculture, conducted some of the foremost research on soil variability. Dr. Robert's research on variable rate fertilizer spreading in 1983 was the first to find that different areas of a field have different crop yields, and as a result, have different nutrient requirements to achieve optimal yields. The outcome of this research gave birth to the variable rate field technology (VRT) management systems in vogue today. Research shows that VRT is now used in planting 5%–25% of total planted U.S. acreage of winter cotton, rice, wheat, and sorghum. The first yield monitor created in 1992 helped farmers to gather information more efficiently. The GPS autoguidance systems form the foundation of PA for all brands of electronic and equipment used in producing renewable fuel, food, and fiber for the world. These GPS systems are environmentally friendly and provide just what a plant needs and where it needs it, minimizing waste and runoff. Most PA systems today use the real-time kinematic positioning signals to power operations such as spraying, tillage, harvesting, and planting. This technology is also used to collect important data about soil moisture, soil temperature, seeding per acre, seeding depth, and yield, among others [8].

20.2.1 What is precision agriculture?

According to Climate [9], PA is simply the use of data-driven technologies for growing crops.

Its Benefits Include: soil health maintenance and sustainability, improved decision-making efficiency, reduction of the cost of materials and resources, crop and livelihood protection, accessibility to farm records, advanced irrigation management, and advanced metrics for agricultural monitoring, among others [10].

20.2.2 How precision agriculture is tackling some of the world's most challenging problems

Climate Change: Reference [11] defines climate change as long-term shifts in temperatures and weather patterns, which can be due to human activities or natural causese due to changes in the sun's activity or large volcanic eruptions.

Table 20.1 Effects of climate change culled from Heifer International

Category	*Nature of impact*
Human life	• Climate change could increase food prices by 50%–60% more by 2030. • 1.7 billion people have been affected by extreme weather and climate-related disasters. • About 1.2 billion climate refugees could result from natural disasters and other weather-related events by 2050. • In 2019, 302.4 billion working hours were lost due to excessive heat.
Land	• Climate change impacts our forests through changes in temperature and precipitation, which can affect carbon storage, roots, soil, and tree growth. • Climate change increases the drying of organic matter in forests, which has doubled the number of large fires between 1984 and 2015. • Climate change affects a tree's life cycle and increases the outbreak of pests and plant diseases. • Declining biodiversity, storms, rising sea levels, droughts, and water scarcity are all consequences of climate change.
Ocean	• Around 90% of global warming is occurring in the ocean. • The acidity level of surface ocean waters has increased by 30% since the start of the industrial revolution. • 14% of the world's coral reefs were lost in 2008 and 2019 due to rising ocean temperatures caused by climate change, overfishing, and declining water quality. • Half of all amphibians are at risk of extinction due to climate change.
Arctic warming	• The world has lost about 28 trillion tons of ice, with the current melt rate at about 1.2 trillion tons a year, since the mid-1990s. • Antarctica and Greenland are losing 150 and 270 billion tons of ice, respectively, per year leading to sea level rise. • Meltwater from ice sheets is responsible for about one-third of the global average rise in sea level since 1993. • Glaciers are retreating worldwide and spring snow cover in the Northern Hemisphere has decreased over the last five decades.

Source: From Ref. [12].

From Table 20.1, it is obvious that climate change is attacking all the elements necessary for productive agricultural practice. Reference [13] notes that climate change affects agriculture by causing heightened food insecurity, land degradation and water scarcity, reduced yield, increase poverty, migration, and displacement. Reference [14] highlights that PA is, therefore, tackling the threat of climate change via the following ways:

20.2.2.1 Achieving environmental benefits

Through controlled use of water in limited agricultural area reduces the use of natural resources, which in turn reduces deforestation. It equally reduces the use of fertilizers and other pesticides, thereby facilitating a reduction in greenhouse gas emissions and decreased contamination of soil and water bodies.

20.2.2.2 Increasing agricultural health

Advanced technologies are used in monitoring the health status of crops and farms, helping farmers with early detection of diseases, as well as with timely corrective and preventive measures. Advances in technology are contributing to the development of microbes, which will make it easier for crops like wheat, rice, and corn extract nitrogen from the air and use it on their own. This will lower the need for artificial fertilizer and significantly reduce carbon emissions from agriculture.

20.2.2.3 Achieving higher yields with greater efficiency

PA achieves more production with less water and lesser agricultural area using the variable-rate application (VRA) technology to ramp up soil fertility and yields by reducing and managing nutritional imbalances, pest infestations, and crop stress. Therefore, PA would support efficiently the feeding of a growing population and optimize land use.

20.2.2.4 Achieving greater food security

Higher and better crop yield ensures more production of food, thereby increasing food security and safety using product traceability, leading to better savings and profits for the farmers. Importantly, the VRA technology is distinctly essential in reducing greenhouse gas emissions, preventing water pollution, and preventing soil degradation. Therefore, since PA is the data powerhouse in agriculture, we posit that PA is the enabler of Climate Smart Agriculture (CSA) and Regenerative Agriculture, which all contribute to combating climate change.

20.3 FOOD SECURITY/FOOD INSECURITY

Food insecurity is one of the major challenges confronting the world today. The World Bank recognizes it as one of the global challenges that needs urgent attention in 2024 (Figure 20.1).

Reference [16] believes that a paradigm shift in the way we produce and consume food is inevitable and that technology and innovations are critical enablers to a sustainable and resilient food system that can feed and nourish all. To this end, the WEF has a couple of

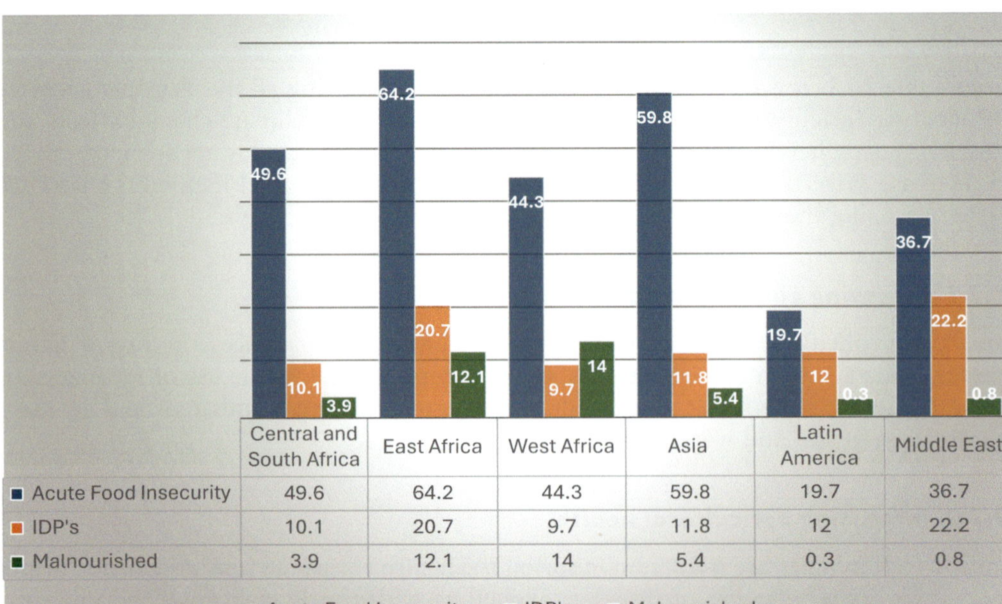

Figure 20.1 2024 Global report on food crisis. (Source: From Ref. [15].)

initiatives for addressing the problem of food insecurity, which include The Food Innovation Hub Global Initiative; New Vision for Agriculture (NVA) initiative; and Innovation with a Purpose Platform.

The Food Innovation Hub Initiative is part of the Forum's wider initiative for transforming global food and water systems using six pacesetter hubs, which are still under development, and connected by a global Food Innovators Platform as follows:

- **Europe:** This hub is driving a regenerative innovation portfolio to drive sustainable farming in Europe toward nutrition and health, circular Agri-food, and protein transitions, building on the learnings of the Netherland food innovation systems.
- **India:** This hub scales Agri-tech innovations to improve farmer resilience and market development.
- **Colombia:** This hub functions as a public–private co-investment platform to develop tech-driven regenerative agriculture platforms of excellence in support of rural development.
- **Kenya:** This hub focuses on building a digital and data ecosystem that will benefit the entire agricultural value chain.
- **Vietnam:** This hub seeks to drive sustainable food systems through digital technologies.
- **UAE:** This hub seeks to become a center for arid food climate food futures.

20.4 NEW VISION FOR AGRICULTURE (NVA) INITIATIVE

The sole purpose of this initiative is to ensure that the agricultural sector can sustainably nourish a growing world population [17].

20.4.1 Innovation with a purpose platform

This initiative seeks to transform the way we produce, distribute, and consume food by facilitating the adoption of new technologies and other innovations—see [18].

It is important to note that the United Nations, its agencies and partners like the World Bank, Food and Agriculture Organization, World Food Program, and UNDP, have a lot of intervention packages in the area of food security.

20.4.2 The Israel example

Israel is possibly the best reference point of a country that has turned its agricultural misfortunes around to become a global leader in agriculture, courtesy in part to its effective deployment of precision agriculture and technology. Historically, two-thirds of the land is semiarid or arid and much of the soil is of poor quality. There is a shortage of natural water resources, a scarcity of precipitation, and Israel is far from key export markets. But today, Israel has not only secured national food security and thriving export industries, but it has emerged as a leader in agriculture and water management. Israel records 0.5% of grain storage loss, compared to 20% worldwide—see [19].

However, there are five principles that pivot Israel's success in agriculture that hold a lot of lessons for the world. These are effective government demonstrating visionary leadership and long-term commitment to agriculture, strategic organization of farmers, a result-oriented market approach, and an innovative and multidisciplinary farmer-centric approach to solving farmers' and agricultural problems.

20.5 GAPS IN THE GLOBAL MANAGEMENT OF AGRICULTURAL PROBLEMS AND HOW TO FIX THEM

20.5.1 Frustrating and ineffective communication system

Presently, there is no effective global communication system for solving farmers' problems. Every country is supposed to have an effective and visionary government taking the lead in solving agriculture and its related problems. But there is a critical need for such a global system. At the heart of Israel's innovative system is the effective free-flow and problem-centered utilization of data and information between Volcani, 8 regional centers, extension workers, and the farmers.

Israeli farmers can request data from Volcani or its regional centers, or the extension workers and get a prompt response, but it is not so especially for farmers from developing nations, where most governments have not become as effective as they should be in agricultural management. Disturbingly, the United Nations and its agencies, which champion agricultural causes globally, do not have a data-friendly and farmer-smart communication system for farmers. This is particularly critical since Volcani is partnered with the United Nations and updates it on its findings on agricultural problems. So, the United Nations needs a farmer-smart communications system to cascade the research breakthroughs from Volcani and other partner research organizations around the world. Presently, the United Nations and its partner organizations adopt more of a storytelling communication system instead of a problem-centered communication system. Therefore, there is need to adapt the Israeli agricultural innovation model to a global problem-solving model and leverage advances in technology to create a problem-centered communication system that will serve as digital first responders, offering farmers initial first aid on the management of specific problems while awaiting the intervention of extension workers. For instance, if a farmer in Africa asks the U.N. website, how do I control pests in a maize farm?

The farmer should get an automated response detailing the steps for achieving pest control with instances of success stories, before automatically assigning the farmer to the nearest extension worker.

Finally, such a problem-centered communication system will leverage technology to show GPRS data, videos, and other information of farms that are receiving interventions around the world, especially in the wake of the U.S. Department of Government Efficiency's (DOGE) accusation of corruption against USAID [20]. Therefore, a global problem-centered communication model for farmers will be as illustrated in Figure 20.2.

20.5.2 Lack of a global food plan

The world has been clamoring for food security, yet there is no global food plan for operationalizing this vision. With the help of PA, it is now easier to know which crops will do well in particular soils. So, to achieve lasting food security, the world should have a global food plan that will show where certain crops or fruits will come from and in what quantities over time. This is even more serious now that the world is transitioning to smart cities. Smart cities presuppose smart farms, and smart farms run on data. Having a global food plan would go a long way in curbing food wastage, which increases carbon footprint. Moreover, it will also define access to world markets, as shown in the core principles behind Israel's success story. If there is no global food plan, some farmers will not have access to the market, thereby increasing waste. A global food plan will use production regulatory frameworks, based on demand projections, like that of OPEC to ensure that farmers access markets and reduce wastage.

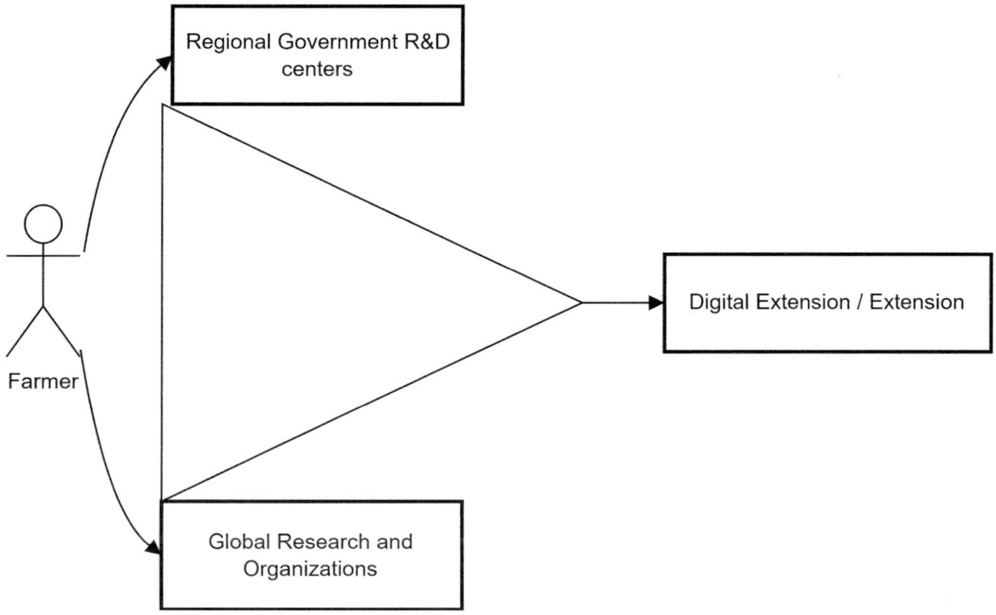

Figure 20.2 Global agricultural innovation communication systems.

20.5.3 Unemployment and immigration issues

Perception issues and lack of credit or funding as well as difficulty securing land are some of the main reasons why young people are shunning farming for white-collar-jobs [21,22]. Unfortunately for young people, especially in developing nations, white-collar jobs are becoming few and far between to cater to the teeming population of youths in search of it. These catches [22] have triggered a wave of immigration to the west in search of greener pasture. PA with its massive blend of technology holds a lot of hope for young people, as it offers the kind of dignity and sophistication that young people like. In addition, the introduction of problem-centered communication systems that will cut the waffle and show intending farmers crystal clear means of accessing funding for farming could be the turning point. Interestingly, a new technology now makes it possible for people living with disabilities to farm in India [23].

20.5.4 Insecurity/security threats

Farmers have been facing threats of different types, including rustling, vandalization, arson, and wildfire threats. Though advances in space technologies like Earth Observation and robotic scarecrows are helping in mitigating some of these threats, the threat of arson and wildfire is still a tormenting one. However, this research suggests that PA should leverage the mini-chromosome technology, which could alter a plant's trait without modifying the genes at all [24]. This stance is further strengthened by [19]'s assertion that when trees are sick their temperature rises like that of humans. So, since humans are immunized against sickness, the minichromosomal technology could be used to collect the strains from plants like man gave, ice plant, Hosta, and dead nettle, among others, to fortify economic crops from destructive fires [25].

20.6 CONCLUSION

PA has proven to be a game-changer by leveraging data and technology in solving some of the world's most daunting challenges. It is also the powerhouse of other innovative agricultural practices like CSA and Regenerative Agriculture. We encourage more research in the area of cloud seeding as an alternative means of fighting wildfires that destroy important forest reserves, thereby increasing carbon footprint and the health of natural soil landscapes.

REFERENCES

1. Action against Hunger © (2025). World Hunger Facts. Retrieved from: https://www.actionagainsthunger.org/the-hunger-crisis/world-hunger-facts/
2. WHO (2024). Hunger Numbers Stubbornly High for Three Consecutive Years as Global Crises Deepen: UN Report. Retrieved from: https://www.who.int/news/item/24-07-2024-hunger-numbers-stubbornly-high-for-three-consecutive-years-as-global-crises-deepen--un-report
3. World Bank (2025). Food Security Update | World Bank Solutions to Food Insecurity. Retrieved from: https://www.worldbank.org/en/topic/agriculture/brief/food-security-update
4. Concern Worldwide (2024). World Hunger Fact: What You Need to Know in 2024 and 2025. Retrieved from: https://www.concern.net/news/world-hunger-facts-figures
5. Denis Ogoyi. 5 Challenges the Agricultural Sector Faces and What We Can Do about It. Retrieved from: https://earth.org/challenges-farmers/
6. Lindsay Robbinson (2024). Top Challenges for Modern Agriculture in 2024. Retrieved from: https://vlsci.com/blog/top-issues-in-agriculture-2024/
7. Sarah Greenberg (2021). Birth of Precision Farming. Retrieved from: https://blog.bccresearch.com/birth-of-precision-farming
8. Bernt Nelson (2023). The Evolution of Precision Agriculture and Policy Implications. Retrieved from: https://www.fb.org/market-intel/the-evolution-of-precision-agriculture-and-policy-implications
9. Climate Ad APT (2023). Precision Agriculture. Retrieved from: https://climate-adapt.eea.europa.eu/en/metadata/adaptation-options/precision-agriculture
10. Peter Kogut (2024). Precision Agriculture: Technology to Boost Crop Farming. Retrieved from: https://eos.com/blog/precision-agriculture/
11. The United Nations (2024). What Is Climate Change? Retrieved from: https://www.un.org/en/climatechange/what-is-climate-change
12. Becky McKay (2024). 52 Eye-Opening Climate Change Facts You Need to Know 2025. Retrieved from: https://www.greenmatch.co.uk/blog/facts-about-climate-change
13. Mark Chandler (2023). How Does Climate Change Affect Agriculture? Retrieved from: https://www.heifer.org/blog/how-climate-change-affects-agriculture.html
14. Sourcetrace (2022). Combating Climate Change: How Precision Agriculture Could Help. Retrieved from: https://sourcetrace.com/blog/combating-climate-change-precision-agriculture-help/
15. Global Network against Food Crises (2024). 2024 Global Report on Food Crises. Retrieved from: https://www.fsinplatform.org/report/global-report-food-crises-2024/
16. World Economic Forum (2024). Leveraging Technology and Innovation to Transform Food Systems. Retrieved from: https://www.weforum.org/impact/food-innovation-hubs/
17. World Economic Forum (2024): Ensuring Sustainable Food for 9.8 Billion People by 2050. Retrieved from: https://www.weforum.org/impact/feeding-the-world-nutritiously-and-sustainably/
18. World Economic Forum (2018): Innovation with a Purpose: The Role of Technology Innovation in Accelerating Food Systems. Retrieved from: https://www3.weforum.org/docs/WEF_Innovation_with_a_Purpose_VF-reduced.pdf
19. Danielle Abraham et al. (2019). How Israel Became a World Leader in Agriculture and Water. Retrieved from: https://institute.global/insights/economic-prosperity/how-israel-became-world-leader-agriculture-and-water

20. Will Steakin et al. (2025). Turmoil inside USAID as Musk Calls the Agency 'Criminal' and Says 'It Has to Die." Retrieved from: https://abcnews.go.com/Politics/turmoil-inside-usaid-doge-reps-offices-senior-officials/story?id=118368900
21. Udemezue, J.C. (2019). Agriculture for All; Constraints to Youth Participation in Africa. Retrieved from: https://lupinepublishers.com/agriculture-journal/fulltext/agriculture-for-all-constraints-to-youth-participation-in-africa.ID.000256.php
22. Naliaka, O. & Kari, M. (2022). How Do We Make Agriculture Attractive for Young People. Retrieved from: https://mawazoinstitute.org/blog-posts/make-agriculture-attractive
23. Mervin Preethi (2025). A Device That Lets Blind Farmers Cultivate Land & Other Rural Inventions Solving Everyday Problems. Retrieved from: https://thebetterindia.com/410080/blind-farming-tech-rural-innovations-india-village-inventions/
24. Masschallange (2023). Agriculture Innovation: 10 Trends to Watch in 2023. Retrieved from: https://masschallenge.org/articles/agriculture-innovation/
25. Janet Loughrey (2025). 22 Fire-Resistant Plants for Home Gardeners. Retrieved from: https://www.provenwinners.com/learn/fire-resistant-plants

Chapter 21

Smart soil systems

Bridging soil classification, precision farming, and digital agriculture in India

Madhu Priya, Devershi Pallavi Bhatt, and Timothy Malche

21.1 INTRODUCTION

21.1.1 Importance of soil classification in agriculture

The most basic requirement for growing crops is soil. Soil is made by the breakdown of rocks, which is called weathering [1]. Figure 21.1 shows the types of rocks by which soil is formed. So, for the formation of soil, physical disintegration and chemical decomposition processes are compulsory. In agriculture, classification of soil is vital because different soil has unique characteristics such as their formation process to composition of minerals, moisture and capacity of water retention, and lastly pH levels. Considering contemporary agricultural issues including poverty, resource scarcity, and climate change, soil categorization becomes even more important. Even with limited resources, farmers may maximize crop output and achieve sustainable results by knowing the characteristics of the soil [2]. Sudden urbanization is causing loss of land, emphasizing the importance of accurate soil classification to ensure the effective use and management of the land resources that are already available. By this information, it guarantees that agricultural productivity is maintained in spite of the limitations brought about by resource scarcity and urbanization.

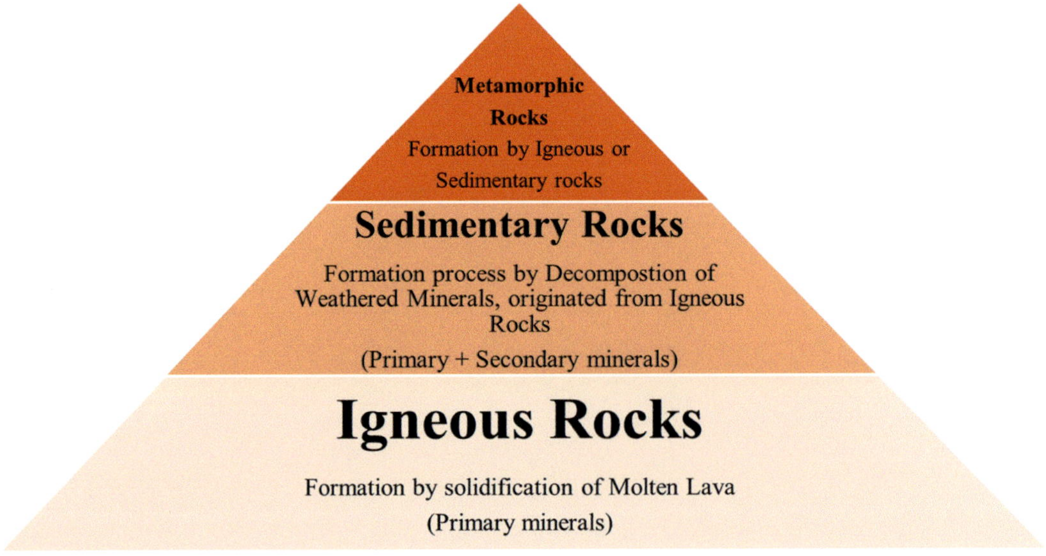

Figure 21.1 Types of rocks with their formation process.

DOI: 10.1201/9781003520733-25

21.1.2 An overview of the classification and diversity of Indian soils

India has a great diversity of soil types due to its vast geographic area and varied climate. Due to variation in soil characteristics, India has the advantage of variable range of crops as well as farming systems, where specific requirements are being adapted by its respective region [3]. So, to meet the specific requirements of various soil types under various environmental circumstances, soil and crop management techniques must be modified. Appropriate maintenance and sustainable practices are very important, as different regions require different farming methods and resources [4]. These initiatives are essential to preserving agricultural output and guaranteeing national food security.

21.1.3 Traditional procedures for the classification of soil

For soil classification, traditional methods are based on field tests and these tests are Standard Penetration Test, Cone Penetration Test (CPT), Vane Shear Test (VST), and Pressure Meter Test to get properties of the subsurface [5]. VST rapidly assesses shear strength in mild clays, whereas CPT measures characteristics such as tip resistance and pore pressure. Although these techniques are frequently labor-intensive and equipment-dependent, they aid in determining the properties of the soil at test sites. Table 21.1 presents the comparison of soil types with their key features and nutrients.

21.1.4 Traditional soil classification system drawbacks

Traditional methods such as the Munsell chart for color-based identification have limited precision, particularly when interfacing with digital equipment [6]. Its accuracy is dependent more on the judgement of human as well as lighting conditions, which leads to inconsistent results. These techniques are not appropriate for large-scale studies since they are expensive, time-consuming, and labor-intensive [6]. Unbalanced datasets and other problems with data quality further diminish reliability, underscoring the need for more sophisticated, technologically driven solutions [6].

21.1.5 Role of image-based approaches in overcoming these challenges

Image-based soil classification addresses the limitations of traditional methods by offering objective, accurate, and scalable analysis through computer vision and machine learning (ML) [6]. These techniques reduce human error, automate time-consuming tasks, and enable real-time remote monitoring using drones or satellites [7]. Unlike manual methods like the Munsell chart, image-based systems handle large datasets efficiently and cost-effectively, improving reliability and expanding access to difficult terrains [8]. This makes them essential for modern soil studies and sustainable agriculture [6].

21.1.6 Overview of the chapter

This chapter highlights the advanced methods of image-based soil classification in precision agriculture. It focuses on how modern technologies like ML and deep learning (DL) work faster, more accurately, and with scalable results over traditional soil analysis. Artificial intelligence (AI) and high-resolution photos can help farmers better comprehend the state of the soil, leading to more precise and sustainable farming. The chapter also discusses

Table 21.1 Soil types with their key features and nutrients [4]

Soil type	*Key features*	*Rich in*	*Deficient in*	*Color*	*Texture*	*Crops cultivated*
Alluvial soil	1. Covers 43% of India (143 km^2). 2. Found in northern plains and river valleys. 3. Depositional soil transported by rivers. 4. Khadar(new) and Bhangar(old). 5. Sand content decreases west to east. 6. Highly fertile.	Potash	Phosphorus	Light grey to ash grey	Sandy to silty loam or dry	Wheat, rice, maize, sugarcane, pulses, oilseeds
Red soil	1. Found in low-rainfall areas, porous and friable. 2. Porous and friable. 3. Rich in ferric oxide. 4. Lower layers reddish yellow. 5. Known as Omnibus group.	Iron and potash	Lime, phosphate, manganese, nitrogen, humus, potash.	Red (ferric oxide)	Sandy to clay and loamy	Wheat, cotton, pulses, tobacco, oilseeds, potatoes.
Black soil (Regur)	1. Found in the Deccan. 2. Best for cotton cultivations. 3. High water-retaining capacity. 4. Self-plowing. 5. Swells when wet and cracks when dry. 6. Mature soil.	Iron, lime, calcium, potassium, aluminum, magnesium.	Nitrogen, phosphorus, organic matter.	Deep black to light black	Clayey	Cotton, wheat, jowar, pulses, oilseeds.
Laterite soil	1. Found in high temperature and rainfall areas. 2. Formed by leaching. 3. Low humus content due to bacteria and rapid organic matter removal.	Iron, aluminum	Nitrogen, potash, potassium, lime, humus	Red (iron oxide)	coarse	Rice, ragi, sugarcane, cashew nuts
Desert/ Arid soil	1. Found in arid/semi-arid regions. 2. Deposited by wind. 3. High salt and kankar content. 4. Low moisture and humus.	Calcium carbonate, gypsum, and sodium	Nitrogen	Red to brown	Sandy	Bajra (pearl millet), pulses (like guar), and drought-resistant vegetables like tomatoes, cucumbers, and squash

(Continued)

Table 21.1 (Continued) Soil types with their key features and nutrients [4]

Soil type	*Key features*	*Rich in*	*Deficient in*	*Color*	*Texture*	*Crops cultivated*
Peaty/ Marshy soil	1. Found in areas with heavy rainfall/high humidity. 2. Humus content. 3. Alkaline. 4. Heavy and black. 5. Suitable for wetland agriculture.	Organic matter	Potash	Black	Heavy	Rice, aquatic crops
Forest soil	1. Found in forested and hilly regions. 2. Acidic with shallow topsoil. 3. Maintains forest ecosystems but unsuitable for intensive agriculture.	Nitrogen, phosphorus, and potassium	Potassium, calcium, and magnesium	Light to dark brown	Sandy loam to clayey	Tea, coffee, spices
Mountain soil	1. Found in hilly/mountainous areas like the Himalayas. 2. Highly weathered, acidic, low humus. 3. Suitable for horticulture and temperate crops.	Humus	Nitrogen, phosphorus, and organic matter	Dark brown	Loamy or silty texture in valley sides, coarse grained on upper slopes	Horticulture, tea, apples, potatoes
Saline/ Alkaline soil	1. Found in coastal/arid regions. 2. High salt/alkali content. 3. Unsuitable for most crops but reclaimable through drainage and amendments.	Soluble salts, sodium, along with other ions like chloride, sulfate, and bicarbonate	Nitrogen, phosphorus, and micronutrients like zinc, copper, boron, and manganese	Saline soil has white or gray salt crust on the surface while alkaline soils appear black or dark brown	Saline soil has sandy to loamy texture, alkaline soil has a sandy clay texture	Salt- tolerant crops like barley, rice, cotton, sorghum, and sunflower

and investigates fixes for common problems with image collecting, like lighting and object interference. It goes on to address how hyperspectral imaging and AIoT might enhance soil classification. Broadly, the chapter illustrates the various ways in which digital tools have allowed agriculture to make it more sustainable and efficient.

21.2 FUNDAMENTALS OF IMAGE-BASED SOIL CLASSIFICATION

Images from cameras or satellites are used in image-based soil classification to determine the types of soil. It provides a quicker and more effective substitute for conventional approaches by analyzing aspects like color, texture, and structure using techniques like computer vision and DL [17]. In engineering, environmental research, and agriculture, all use image-based soil classification. It aids with soil type identification, nutrient and moisture checks, and precision agriculture for improved crop and resource management in farming [9].

Its technique aids in mapping soils, evaluating land deterioration, and identifying pollutant sources through spatial analysis in the environmental and geotechnical domains [9]. Additionally, model creation and data gathering are supported by image-based soil classification, which fosters a deeper understanding and sustainable soil management techniques.

21.2.1 Types of images used in soil classification

Three main types of images are used in image-based soil categorization systems, each with specific benefits and uses. For color detailing, RGB is used; for analyzing soil, texture CIELab* is used; and for assessing a wide area, satellite or hyperspectral photos are used. While CIELab* provides accurate texture insights, RGB images are readily available [10]. Accurate mapping of soil features across wide areas is made possible by satellite and hyperspectral data. Figure 21.2 shows the detailed description of the types of images being used.

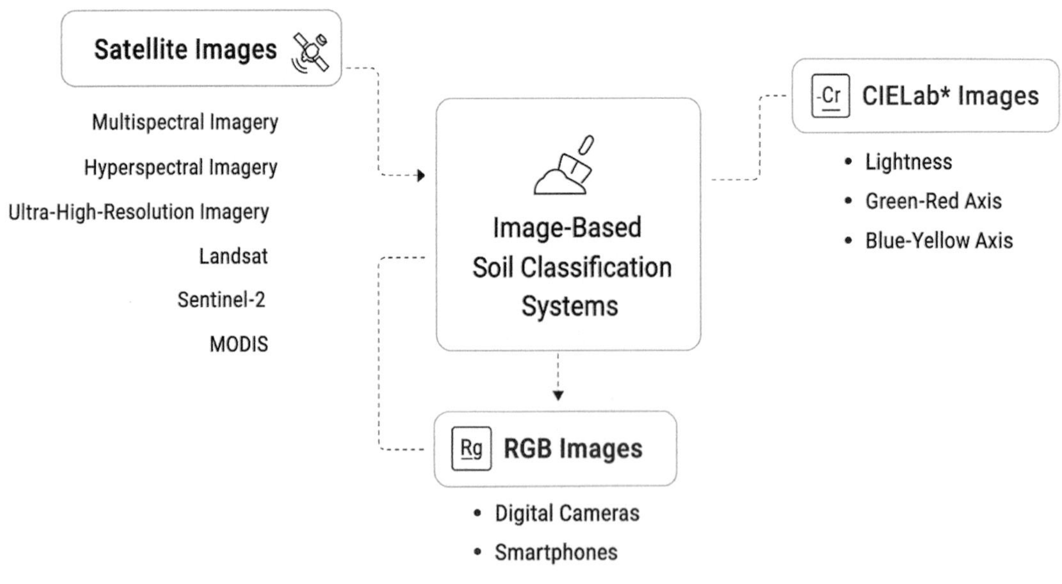

Figure 21.2 Types of images used in image-based soil classification [9–12].

21.2.2 Role of soil color in agriculture and indicators provided by soil color

Soil color is a key indicator of soil health, revealing properties like mineral content, organic matter, drainage, and fertility levels [12]. Dark soils often indicate high organic matter and poor drainage, while red or yellow hues suggest good aeration and iron oxidation [12]. Pale or white soils may signal nutrient leaching, and gray or green tones are linked to waterlogging and poor drainage [13]. Table 21.2 shows the description of the mineral composition and the aspect of soil. These color variations help assess soil quality, manage fertility, and guide crop decisions in agriculture [13].

21.2.3 Advantages of image-based techniques: scalability, efficiency, and precision

Image-based soil classification ensures high accuracy, efficiency, and wide-area coverage. DL models like Convolutional Neural Networks (CNNs) process large image datasets quickly and reliably. Satellite and drone imagery allow data collection from remote or hard-to-reach areas.

Table 21.2 Description of soil mineral composition and its aspect [13]

Aspect	*Description*	*Key points*
Mineral composition	The parent rock and its minerals, such as calcite, quartz, and smectites, influence the soil's coloration. Their interaction with organic matter alters the intensity of soil color.	Smectite clays combined with organic material produce darker hues (e.g., black in Calci-Aquic Vertisols). Technologies like cameras estimate SOC through soil color.
Organic matter	Organic content significantly darkens the soil by reducing reflectivity. The formation of humus through decomposition further deepens soil color and improves fertility.	High organic content leads to darker shades. Soil organic carbon (SOC) is closely associated with soil color parameters.
Plant residues and microbial role	Decomposing plant residues, influenced by microbial activities, enrich soil with organic materials, leading to darker colors. Microbial processes play a vital role in humus formation, enhancing the soil's darkness.	Forested soils are darker due to higher organic inputs. Cultivated lands exhibit lighter hues from lower organic contributions.
Soil moisture & drainage	Moisture levels influence soil's appearance, with wet soil appearing darker due to light absorption. Poor drainage leads to grayish shades (gleying), while well-drained soils often appear brighter.	Wet soils show deeper tones. Poorly drained areas exhibit gray hues due to oxygen depletion. Redox reactions create red and yellow mottling in alternately wet-dry soils.
Soil texture	The relative amounts of sand, silt, and clay affect soil color. Sandy soils are generally lighter, while clay-rich soils with iron oxides display darker hues.	**Sandy Soil**: Pale tones. **Clay Soil**: Darker shades due to iron oxides.
Climate & topography	Climate affects weathering and organic matter decomposition, influencing soil hues. Topography affects water retention, drainage, and organic matter accumulation, contributing to color variation.	**Wetter Climates**: Red soils (iron oxidation). **Elevated Slopes**: Lighter shades from better drainage. **Low-Lying Areas**: Darker soils.
Land use	Human activities like agriculture and erosion modify soil properties and color by altering organic content, mineral composition, and pH	Plowing and erosion cause lighter soils. Intensive farming depletes organic matter and alters soil structure, affecting color and fertility.

It reduces the need for manual analysis and expensive lab infrastructure. Overall, it offers a faster, scalable, and cost-effective solution for modern agriculture [14].

21.2.4 Limitations and challenges

Image-based soil classification faces hurdles in data collection, quality, and consistency. Uneven soil representation, seasonal changes, and environmental factors can impact accuracy. Lighting, camera quality, and image resolution also affect results, making reliable classification and scalability challenging [6].

21.3 IMAGE COLLECTION AND PREPROCESSING

21.3.1 Sources of soil images

Soil image data can be taken from varied sources in different ways, which is real time, flexible, and ensures adaptability for various research requirements, and also can be used as an application. One of the methods of data collection is real-time image capturing through smartphones or a camera [9]. Some advanced technologies like remote sensing, drone, or satellites offer images through varied regions. These technologies cover vast regions and areas that are difficult to reach. It is very easy to take a large number of datasets by this. Some open-source platforms and publicly available datasets like ResearchGate can also be useful.

21.3.2 Preprocessing steps

Data preparation is a vital step in DL, affecting both the accuracy and efficiency of the model. Key preprocessing tasks include data cleaning, normalization, padding, and feature scaling to ensure consistent and usable input [15]. Cleaning removes noise, fills missing values, and handles outliers, while normalization brings data to a uniform scale. In soil classification, unbalanced datasets where one soil type is overrepresented can lead to biased predictions. Data augmentation methods including flipping, rotating, cropping, and brightness adjustment are used to overcome this. These techniques enhance model generalization, decrease overfitting, and diversify the data. In general, augmentation contributes to the development of a more reliable and objective model performance [15]

21.3.3 Soil image normalization

By minimizing variations caused by lighting or camera discrepancies, normalization increases the consistency and accuracy of soil image analysis [1]. Pixel values are adjusted for uniform input using methods like Z-Score Standardization and Min-Max Scaling. Histogram equalization and GAN-based transformations are two techniques that further improve contrast and standardize images, increasing their dependability for model training.

21.4 TECHNIQUES FOR SOIL CLASSIFICATION USING IMAGES

21.4.1 Machine learning techniques

To reduce dimensionality and help isolate applicable data for efficient processing, feature extraction is important in soil image analysis. In addition to color statistics like mean, median, and skewness, texture features like Q-HOG, Haralick, and LBP are frequently used [14]. These characteristics enable precise classification and analysis by reflecting the physical, chemical, and biological characteristics of soil.

21.4.2 Classification models in image-based soil classification

Several decision trees are used in the Random Forest (RF) approach to enhance the accuracy and reduce overfitting. Although RF performs best with high-dimensional data, it consumes a lot of processing power when dealing with large amounts of data. A supervised model called Support Vector Machines (SVMs) chooses the optimal hyperplane for class separation. SVM is computationally costly for large-scale issues yet performs well on tiny, high-dimensional datasets [5]. Then comes another algorithm, K-Nearest Neighbors (K-NN), which is a basic algorithm that predicts data based on the majority class of the nearest neighbors. K-NN is simple to train but not efficient for big datasets and is very sensitive to noise. These models work well for feature-based soil classification but can be less scalable and accurate than DL methods such as CNNs [5].

21.4.3 Deep Learning techniques

CNNs for feature learning

Convolutional Neural Networks (CNNs) automate feature extraction from soil images, identifying patterns in texture and color for accurate classification [6]. Using drone or satellite images, CNNs efficiently process and classify various soil types. Transfer learning with pretrained models like ResNet and VGGNet, originally trained on large datasets, allows faster training, better accuracy, and reduced data requirements by fine-tuning them for soil-specific tasks [15].

21.4.4 Combining spatial and spectral information in DL models

By identifying complex patterns in soil photos, DL—and CNNs in particular—improves soil classification. However, combining spectral (reflectance properties) and spatial (pixel association) information improves the accuracy and contextual interpretation. CNNs classify both spectral and spatial data for better distinction through a process known as spectral–spatial fusion [16]. Whereas Multi-Source Data Integration enhances model performance, it incorporates field data, remote sensing, and environmental factors. In temporal analysis by taking seasonal soil variations into account, multitimeframe data improves mapping [16]. Table 21.3 presents the comparison of ML and DL accuracy rates.

Table 21.3 Comparison of machine learning and deep learning accuracy rate [17,6]

Algorithm	*Category*	*Accuracy rate (%)*
Support Vector Machine (SVM)	Machine learning	71.18%–97.12
Linear Support Vector Machine	Machine learning	96.8%–99.0
Artificial Neural Network	Deep learning	89.2%–94.2
Back Propagation Neural	Deep learning	89.7
3D CNN	Deep learning	99.59
CNN, VggNet16, ResNet50	Deep learning	98.72
CNN MobiNet	Deep learning	93

21.5 APPLICATIONS IN PRECISION AGRICULTURE

Accurate soil classification enables farmers to select crops and plan irrigation based on traits like texture and drainage—maize suits sandy soils, while rice fits clay [21]. Image-based analysis helps detect nutrient gaps for precise fertilizer use, reducing waste and environmental harm [22]. Irrigation strategies also improve as clay needs deep, less frequent watering, and sandy soils require lighter, frequent irrigation [22,23].

Geographic Information System (GIS) and remote sensing tools support large-scale soil mapping using spatial data and satellite imagery like Sentinel-2A and Landsat to assess soil composition, especially clay content [18,20]. Multisensor input increases mapping accuracy and sustainable land use [19].

Smartphone apps now allow real-time soil checks using camera images, replacing costly lab testing, though camera quality and pollutant detection remain issues [6]. Enhanced mobile sensors could offer better accuracy and self-reliant soil monitoring for farmers.

AI-based systems use sensor and drone data to assess soil conditions, helping optimize inputs, detect degradation early, and improve farming efficiency [22].

21.6 CHALLENGES AND FUTURE DIRECTIONS

Lighting variations, shadows, and occlusions from vegetation or objects can distort soil appearance, making accurate classification difficult. Blurry or unfocused images further reduce reliability. Creating annotated datasets for ML is another challenge, as it demands expert input and is time-consuming, slowing progress in soil classification research. Recent deployment of hyperspectral sensors such as PRecursore IperSpettrale della Missione Applicativa (PRISMA), DLR Earth Sensing Imaging Spectrometer (DESIS), and Environmental Mapping and Analysis Program (EnMAP) has improved access to high-resolution soil data, though processing large, complex datasets remains a key challenge [21]. AI and Internet of Things (IoT) technologies now enable real-time soil classification using low-power edge devices, making them suitable for agricultural robots and Unmanned Ground Vehicles (UGVs) [22]. When comparing with traditional techniques, DL models such as Convolutional Neural Networks (CNNs), Support Vector Machines (SVMs), and Random Forest (RF) provide faster and more accurate results. Along with techniques like transfer learning and meta learning, even small datasets can produce trustworthy results. So, global soil monitoring systems and smarter data-driven farming are being made possible by these developments [22].

21.7 CONCLUSION

21.7.1 Summary of key findings

The chapter focuses on modern techniques like ML and imaging, specifically CNNs and transfer learning, to improve the accuracy of soil classification. Along with this, it gives information about many remote sensing tools like hyperspectral and multispectral imaging. With the existence of many challenges of lighting issues and labeling of manual data, AI and IoT are introducing real-time soil analysis with smart farming, clearing the path for precision farming, which is smart and scalable.

21.7.2 Importance of image-based classification in precision farming

Many soil traits like moisture, texture, and nutrients can be accurately and noninvasively analyzed with image-based soil classification. It provides site-specific activities, real-time decision making, and sustainable farming by combining AI with high-resolution images, increasing crop output and resource efficiency in precision agriculture.

REFERENCES

1. Pei, X., hong Zhao, Y., Chen, L., Guo, Q., Duan, Z., Pan, Y. and Hou, H. (2023). Robustness of machine learning to color, size change, normalization, and image enhancement on micrograph datasets with large sample differences. *Materials & Design*, 232, 112086.
2. Lal, R. (June 2011). Soil health and climate change: An overview. In R. Lal, *Soil Health and Climate* Change: *An Overview*. The Ohio State University.
3. Bhattacharyya, T., Sarkar, D., Pal, D. K., Mandal, C., Baruah, U., Telpande, B., and Vaidya, P. H. (November 2010). Soil information system for resource management – Tripura as a case study, *Current Science*, 1208–1217.
4. Singh, J. S. and Chaturvedi, R. K.. (2017). Diversity of ecosystem types in India: A review. *Proceedings of the Indian National Science Academy*, 83(2), 569–594.
5. Chandan, R. T., and Thakur, R. (2018). An intelligent model for Indian soil classification using various. *International Journal of Computational Engineering Research (IJCER)*, 33(2250), 3005.
6. Emmanuel Kwabena Gyasi, S. P. (2023). Soil-MobiNet: A convolutional neural network model base soil classification to determine soil morphology and its geospatial location. *Sensors*, 23(15), 6709.
7. Dhawale, N. (2019). Soil analysis technique based on global positioning enabled mobile image processing. *IJRAR*, 6, 322–326.
8. Ridha Guebsi, S. M. (2024). Drones in precision agriculture: A comprehensive review of applications, technologies, and challenges. *Drones*, 8(11), 686.
9. Naeimi, M., Daggupati, P., and Biswas, A. (2024). Image-based soil characterization: A review on smartphone applications. *Computers and Electronics in Agriculture*, 227, 109502.
10. Pandiri, D. K., Murugan, R., and Goel, T. (2024). Smart soil image classification system using lightweight convolutional neural network. *Expert Systems with Applications*, 238, 122185.
11. Karim Ennour, M. E. (2024). *Satellite imagery used for analysis of soil properties*. In M. E. Karim Ennour, *Satellite Imagery Used for Analysis of Soil Properties*. IGI Global.
12. Thomas W. Wagner, W. R. (1971). *Multispectral Remote Sensing as a Survey System for Mapping Soils and Soil Conditions: A Thesis Submitted in Partial Fulfillment... for the Degree Master [Science in the Field] of Resource Planning and Conservation...* University of Michigan.
13. Sirisathitkul, Y., and Sirisathitkul, C. (2025). Decoding soil color: Origins, influences, and methods of analysis. *AgriEngineering*, 7(3), 58.
14. Sattar, K., Maqsood, U., Hussain, Q., Majeed, S., Kaleem, S., Babar, M., and Qureshi, B. (2024). Soil texture analysis using controlled image processing. *Smart Agricultural Technology*, 9, 100588.
15. Alirezazadeh, P., Rahimi-Ajdadi, F., Abbaspour-Gilandeh, Y., Landwehr, N., and Tavakoli, H. (2021). Improved digital image-based assessment of soil aggregate size by applying convolutional neural networks. *Computers and Electronics in Agriculture*, 191, 106499.
16. Duan, M., Song, X., Liu, X., Cui, D., and Zhang, X. (2022). Mapping the soil types combining multi-temporal remote sensing data with texture features. *Computers and Electronics in Agriculture*, 200, 107230.
17. Srivastava, P., Shukla, A., and Bansal, A. (2021). A comprehensive review on soil classification using. *Multimedia Tools and Applications*, 80(10), 14887–14914.

18. De Feudis, M., Falsone, G., Gherardi, M., Speranza, M., Vianello, G., and Antisari, L. V. (2021). GIS-based soil maps as tools to evaluate land capability and suitability in a coastal reclaimed area (Ravenna, northern Italy). *International Soil and Water Conservation Research,* 9(2), 167–179.
19. Abdulraheem, M. I., Zhang, W., Li, S., Moshayedi, A. J., Farooque, A. A., and Hu, J. (2023). Advancement of remote sensing for soil measurements and applications: A comprehensive review. *Sustainability,* 15(21), 15444.
20. Dash, P. K. (2024). *Remote sensing as a potential tool for advancing digital soil mapping.* In S. Dharumarajan, Kaliraj Seenipandi, Kabindra Adhikari, Lalitha Manickam, and Nirmal Kumar (Eds.), *Remote Sensing as a Potential Tool for Advancing Digital Soil Mapping.* Amsterdam, The Netherlands: Elsevier.
21. Bera, S., Shrivastava, V. K., and Satapathy, S. C. (2022). Advances in hyperspectral image classification based on convolutional neural networks: A review. *CMES - Computer Modeling in Engineering and Sciences,* 133(2).
22. Islam, M. R., Oliullah, K., Kabir, M. M., Alom, M., and Mridha, M. F. (2023). Machine learning enabled IoT system for soil nutrients monitoring and crop recommendation. *Journal of Agriculture and Food Research,* 14, 100880.
23. Sirisathitkul, Y., and Sirisathitkul, C. (2025). Decoding Soil Color: Origins, Influences, and Methods of Analysis. *AgriEngineering,* 7(3), 58.

Chapter 22

Planting the future

Traversing the digital landscape of intelligent farming

Neelu Jain, Monika, Pardeep Kumar, and Manoj Sharma

22.1 OVERVIEW OF MODERN TECHNOLOGY IN VARIOUS SECTORS

Modern technology has revolutionized numerous sectors, enhancing efficiency, productivity, and innovation. Advancements in healthcare equipment, telemedicine, and biotechnology have led to better patient care and treatment outcomes. Fintech solutions, blockchain technology, and digital banking are transforming the financial industry. The industrial business has experienced the expansion of automation, robotics, and 3D printing, resulting in faster production rates and lower expenses. The energy sector benefits from smart grids. Renewable energy technologies and energy storage solutions contribute to sustainable and efficient energy management. E-learning has also revolutionized the education system platforms, virtual classrooms, and interactive digital materials, enhancing education accessibility and personalization. Overall, modern technology has a ubiquitous influence on these sectors, demonstrating its ability to promote considerable growth and innovation [1–3].

22.1.1 Introduction to the impact of technology on agriculture

Technological advancements are altering the agriculture sector, which was formerly reliant on physical labor and traditional methods. Precision agriculture is now possible thanks to modern technology, automated and data-driven decision-making processes. Drones for crop monitoring, mechanized planting and harvesting, and advanced irrigation systems have revolutionized farming operations. These technologies assist farmers in optimizing resource utilization, increasing crop yields, and improving overall agricultural efficiency. Digital platforms and smartphone applications provide farmers with real-time information on weather, pest control, and market trends, allowing them to make informed decision-making. This technological revolution in agriculture addresses food security, sustainability, and resource efficiency, in addition to increasing output [4].

22.1.2 Importance of technology for agribusiness owners and farmers

Incorporating technology into agricultural processes is crucial for both agribusiness owners and farmers. Technology improves productivity and efficiency, enabling farmers to produce more with fewer resources. Automated technologies, precision agricultural procedures, and enhanced irrigation systems reduce labor costs and increase crop yields. Technology provides farmers with useful data and insights to improve crop management, pest control, and resource allocation. This data-driven approach decreases risks while increasing earnings. Third, technology promotes sustainable farming by maximizing water, fertilizer, and pesticide use,

DOI: 10.1201/9781003520733-26

reducing environmental impact. Digital platforms and mobile applications enable farmers to connect with markets, suppliers, and customers, improving supply chain management and market access. Furthermore, digital platforms and mobile applications enable farmers to communicate with markets, suppliers, and customers, hence improving supply chain management and market access. Technology allows for innovation and expansion, enabling agribusiness businesses to remain competitive in an ever-changing industry. Overall, current technology is crucial to improving agricultural efficiency, sustainability, and profitability.

22.2 HISTORICAL AGRICULTURAL PRACTICES AND EVOLUTION THROUGH TECHNOLOGICAL ADVANCEMENTS

Agriculture, the foundation of civilization, has evolved through ongoing innovation to improve efficiency, productivity, and sustainability. Agriculture has always been labor-intensive and reliant on natural elements such as soil fertility, rainfall, and manual labor. Early techniques for land management and food production included slash-and-burn, crop rotation, and primitive irrigation systems. These approaches were effective in their setting, but had limitations due to resource availability, climate, and ecosystem knowledge.

22.2.1 Early agricultural methods

Agriculture began circa 10,000 BCE, when people transitioned from hunter-gatherer to sedentary farming communities, primarily in the Fertile Crescent. Early agricultural practices included:

Slash-and-Burn Farming: It involves clearing forests and burning plants to provide fertile soil for agriculture. This was an early strategy utilized in many places of the world, including ancient Africa and the Americas, but it was not long-term viable because it depleted soil nutrients and required frequent relocation to fresh territory.

Crop Rotation: It is used by ancient civilizations such as the Romans and Greeks, involving planting different crops on the same ground to maintain soil fertility. This practice is still used today, even though modern agriculture incorporates it. Precision equipment allows for more accurate tracking of soil health.

Manual Irrigation: In regions where water was not naturally abundant, early farmers in Mesopotamia and Egypt built basic irrigation systems by diverting river water to their crops. However, these systems required constant maintenance and manual labor.

Animal-Powered Plowing: The domestication of animals like oxen and horses allowed farmers to till larger areas of land compared to manual plowing. This advancement led to increased food production and facilitated the growth of early cities and civilizations.

While these methods laid the groundwork for agricultural development, they were far from perfect. Yield per hectare was limited, and farming communities were highly vulnerable to natural disasters, pests, and diseases. This situation persisted for centuries until technological advancements began to reshape the agricultural landscape.

22.2.2 The industrial revolution and agricultural mechanization

The Industrial Revolution, which occurred between the 18th and 19th centuries, marked a significant turning point in agriculture. Innovations in machinery, transportation, and science transformed farming from a labor-intensive activity into a mechanized and increasingly efficient industry.

Mechanized Plowing and Harvesting: With the invention of steam-powered tractors and later gasoline-powered machinery, farmers could cultivate much larger areas of land with less physical labor. This significantly increased crop yields and allowed for the expansion of agricultural production into new regions.

Chemical Fertilizers and Pesticides: Jethro Tull's seed drill, invented in the 18th century, improved seed germination rates by allowing farmers to place seeds in precise rows at the appropriate depth. Similarly, automated threshers replaced hand grain harvesting, which accelerated the separation of seeds from stalks and husks.

Chemical Fertilizers and Insecticides: The introduction of chemical fertilizers in the 19th century gave a means of restoring nutrients to the soil, resulting in increased agricultural yields. Furthermore, pesticides became widely utilized to manage pests and illnesses that had previously destroyed crops. These advances not only enhanced productivity but also accelerated the establishment of agribusinesses, transforming agriculture from a local, subsistence activity to a commercial industry. However, this period also introduced challenges such as soil degradation, chemical runoff, and over-reliance on monoculture crops.

22.3 THE GREEN REVOLUTION

The mid-20th century saw one more extraordinary period known as the Green Insurgency. This worldwide agrarian development, especially conspicuous in agricultural nations, is meant to battle hunger and further develop food security through logical and mechanical advancements [5,6].

High-Return Crop Assortments: Researchers made illness safe, high-yielding cultivars of staple crop yields like wheat, rice, and maize. Norman Borlaug, known as the "father of the Green Upheaval," was instrumental in creating bantam wheat assortments that forestalled significant yearning in India and Mexico during the 1960s.

Manufactured Manures and Insect Poisons: The Green Insurgency saw a broad use of compound composts and manufactured insect poisons. While these data sources expanded efficiency, they likewise caused long-haul natural issues like soil debasement as well as water defilement.

Water System Foundation: Gigantic water system projects were executed to bring water to already parched districts, empowering all year cultivating in places that had been subject to occasional downpours.

While the Green Revolution exposed the limitations of industrial farming practices, it also helped prevent starvation in many countries and increased food production globally. Over the long haul, unnecessary utilization of synthetic manures, water bungle, and impractical land rehearses have prompted ecological corruption, requiring another rush of advancement in agribusiness [6].

22.3.1 Intersection of agriculture with biotechnology, AI, and IoT

In the 21st century, agriculture has merged with cutting-edge businesses like biotechnology, artificial intelligence (AI), and the Internet of Things (IoT). These advances not only increase agricultural output but also address challenges of sustainability, food security, and climate change [7].

22.3.1.1 Biotechnology and agriculture

Biotechnology has transformed agriculture by enabling scientists to modify genetic material to develop crops that are more resistant to pests, diseases, and environmental stressors.

Genetically Modified Organisms (GMOs): Biotechnology has greatly influenced the development of GMOs. Genetically modified crops, like Bt corn and herbicide-resistant soybeans, can reduce the need for chemical pesticides while increasing production potential.

Clustered Regularly Interspaced Short Palindromic Repeats (CRISPR) and Gene Editing: Advances in gene-editing technologies, such as CRISPR, have enabled researchers to accurately manipulate crop genomes. This method offers the potential to produce crops that thrive in extreme climatic conditions, such as drought and high salt levels, which are becoming more common as a result of climate change.

Biofertilizers and Biopesticides: Biotechnology has also allowed for the development of biofertilizers and biopesticides, which use natural organisms or their by-products to improve soil fertility or reduce pest populations. These alternatives to synthetic chemicals are part of a bigger push for sustainable agriculture.

22.3.1.2 Artificial Intelligence (AI) in agriculture

AI is increasingly vital in optimizing agricultural processes through predictive analytics, automation, and machine learning (ML) [7].

Crop and Soil Health Monitoring: AI-powered drones and sensors can detect crop and soil health issues early on. These gadgets assess data in real time and advise farmers to take corrective action, therefore decreasing the loss of crops.

Yield Prediction and Climate Modeling: AI systems can forecast future crop performance by analyzing historical weather patterns, soil data, and crop production. This allows farmers to make informed decisions about planting times, irrigation, and input usage. AI-powered climate models provide insights into how future climate change could impact specific regions and crops, allowing for more effective long-term planning.

Automation and Robotics: AI enables autonomous farming technology, including tractors, harvesters, and seeders. These technologies can do activities with minimum human involvement, resulting in decreased labor costs and greater efficiency. Robots are already being used for sensitive tasks such as fruit picking, which was formerly required human work.

22.3.1.3 Internet of Things (IoT) in agriculture

The IoT is transforming agriculture by connecting devices, sensors, and machinery to the internet for real-time monitoring and management.

Precision Agriculture: IoT devices such as soil moisture sensors, weather stations, and Global Positioning System (GPS)-enabled machinery help farmers practice precision agriculture. Using real-time data, farmers may adjust irrigation, fertilization, and pesticide application based on the specific needs of several divisions within a profession.

Smart Irrigation Systems: IoT-enabled irrigation systems can optimize water use by monitoring soil moisture and weather forecasts. This method conserves water and provides crops with optimal moisture at the proper moment, leading to improved growth and resilience to droughts.

Livestock Management: IoT devices are utilized to track cattle health and behaviors. Wearable sensors can monitor animal activity, feeding habits, and vital signs, alerting farmers of potential health risks early. This lowers disease spread and increases farm output [5,8].

22.4 OVERVIEW OF MODERN TECHNOLOGY IN VARIOUS SECTORS

Modern technology has brought about profound changes across numerous sectors, enhancing efficiency, productivity, and innovation:

Healthcare: The healthcare sector has seen major advancements, such as telemedicine, which allows remote consultations and treatment through digital platforms. Medical devices, like robotic surgery systems and AI-driven diagnostic tools, have significantly improved patient outcomes and reduced recovery times. Biotechnology continues to make strides in drug development, personalized medicine, and vaccine production [9–12].

Finance: Financial technology (fintech) has revolutionized the way financial transactions are conducted. Blockchain technology ensures secure transactions and has found applications in everything from banking to supply chain management. Digital banking and mobile payment platforms have made banking more accessible, especially in remote or underserved regions [13].

Manufacturing: Manufacturing automation includes robotics in assembly lines and 3D printing for rapid prototyping and production. These technologies have improved efficiency, shortened production times, and cut costs.

Energy: Renewable energy technologies such as solar panels and wind turbines are becoming more efficient and widely employed. Improved batteries allow renewable energy to be stored and used during peak demand [14].

Education: E-learning platforms and virtual classrooms have improved education accessibility and personalization. Global students can access high-quality educational resources and engage with teachers and peers via online platforms. Interactive digital technologies enhance engagement and provide for a more collaborative learning environment [15,16].

The pervasive influence of modern technology across these sectors underscores its capacity to drive significant progress and innovation. Agriculture, in particular, stands at the crossroads of several technological advancements, from AI and IoT to biotechnology, all of which promise to transform it into a more sustainable and efficient industry capable of feeding a growing global population [17].

22.5 FOOD SECURITY CHALLENGES

Food security is one of the most pressing global challenges of the 21st century. The growing global population, climate change, rapid urbanization, and the depletion of natural resources have exacerbated the food security crisis, particularly in developing nations. To meet the rising demand for food, significant improvements in agricultural productivity, sustainability, and efficiency are required.

Food security in India is hindered by poverty, migration, political instability, and environmental factors like droughts. India's Public Distribution System (PDS) addresses food insecurity by subsidizing key products. Despite its significance as a cornerstone of India's food and nutrition strategy, the government continues to battle with high levels of hunger and malnutrition. Although the PDS has potential, difficulties exist. Policy reforms should address concerns as corruption, mistargeting, and environmental concerns (including climate change). The review suggests that improving transparency, accountability, and efficiency of the system could improve the PDS's Contribution to food security in India. Originally a universal program, it has been modified to better target low-income populations through initiatives like the Targeted PDS and the Antyodaya Anna Yojana. The National Food Security Act provided legal backing to the PDS, covering 75% of the rural population and 50% of urban areas, with special focus on vulnerable groups such as children, pregnant women, and senior citizens. Despite improvements, the PDS is hampered by corruption, inefficiencies, and distribution issues that reduce its effectiveness in solving hunger and malnutrition.

22.5.1 Global food security issues

According to the United Nations, food security is a condition in which all people have physical and economic access to adequate, safe, and nutritious food to meet their dietary needs for an active and healthy life. However, fulfilling this goal remains a significant difficulty around the world, especially in developing countries. Several causes contribute to the current food security problem, which has immediate and long-term consequences for global well-being [18].

22.5.2 Population growth and urbanization

Rapid growth requires an increased global food supply to meet demand. Another global trend, urbanization, complicates this situation even more. Urbanization leads to limited arable land and increased demand for processed foods and agricultural products. Rural farming communities are declining, resulting in labor shortages and a loss of traditional agricultural knowledge. Rapid urbanization in nations like India and China puts pressure on agricultural sectors to fulfill the growing food demand. In Sub-Saharan Africa, rising urbanization poses similar issues for food production, but many countries lack the necessary resources to address them.

22.5.3 Climate change and environmental stressors

Climate change is a major cause of food insecurity worldwide. Climate change, altering weather patterns, and more frequent extreme events such as droughts, floods, and storms are causing disruptions to agricultural production. These changes affect farmers' capacity to produce enough food to meet demand.

Droughts in Africa: Droughts in Africa: Ethiopia, Somalia, and Kenya have recurring droughts, causing crop and livestock losses. Millions of people in these regions rely on rain-fed agriculture, rendering them sensitive to swings in rainfall patterns. South Asian countries such as Bangladesh, India, and Pakistan experience significant flooding during monsoon seasons, resulting in crop losses, soil erosion, and community displacement.

Desertification in the Sahel: The Sahel region in Africa, including Mali, Niger, and Chad, is experiencing desertification, which diminishes arable land and compels farmers to relocate to less productive locations. These environmental stressors not only affect crop production but also increase competition for scarce resources such as water and fertile land. Water shortage, for example, is a major issue in the Middle East and North Africa (MENA), and reliance on food imports has grown as local agriculture becomes less sustainable.

22.5.4 Vulnerable populations in developing nations

Due to a combination of factors including poverty, political instability, and weak agricultural infrastructure, developing nations, particularly those in Africa, Asia, and Latin

America, are at the greatest risk of food insecurity. The following examples highlight the extent of food security challenges in different regions:

Yemen: War-torn Yemen has one of the highest levels of food insecurity globally. Conflict has disrupted farming, destroyed infrastructure, and severely restricted access to food supplies, leaving millions of people reliant on humanitarian aid.

Venezuela: Economic collapse in Venezuela has resulted in widespread food shortages, leaving large segments of the population malnourished and struggling to find enough to eat.

Malawi and Zambia: Both countries have experienced food insecurity due to erratic rainfall patterns and the impact of pests like the fall armyworm, which can destroy entire fields of crops.

These examples illustrate that vulnerable populations often suffer the most from food security challenges. Climate change, economic crises, and political instability make it difficult for these countries to sustain traditional farming practices, emphasizing the need for innovative solutions to address these critical issues [19].

22.5.5 Population growth in developing countries

Developing nations in Africa and Asia are projected to account for the majority of global population growth. For example, Nigeria is expected to surpass the United States in population size by 2050, while India will become the world's most populous country. These population booms place immense pressure on already struggling agricultural systems to produce more food. In regions like Sub-Saharan Africa, agriculture is still largely reliant on traditional methods that are less productive and less efficient.

22.6 CURRENT AGRICULTURAL PRACTICES AND PRODUCTIVITY GAPS

In many parts of the world, agricultural productivity remains far below its potential. Smallholder farmers in developing nations often lack access to modern farming technologies, quality seeds, fertilizers, and irrigation systems, all of which limit their productivity. Without significant improvements in agricultural practices, it will be impossible to meet the food production targets set by the International Monetary Fund (IMF). Furthermore, substantial postharvest losses due to poor storage, transportation, and processing infrastructure limit the amount of food accessible for consumption.

The productivity gap is most visible in Sub-Saharan Africa, where staple crop yields such as maize, rice, and wheat are a fraction of those in rich countries. To close this gap, developing countries must invest in modern agricultural technologies, infrastructure, and education programs to provide farmers with the skills and knowledge they require to increase yields.

22.6.1 Role of technology in addressing food security

Modern agricultural technology is critical to meeting the expanding global food demand. Precision farming (PF) solutions and other smart agriculture technologies can help farmers optimize resource use, enhance productivity, and reduce environmental impact. These improvements are critical for reducing food insecurity and supporting sustainable agriculture practices in a constantly changing world.

22.6.2 Precision agriculture and resource optimization

PF and smart agriculture technology use real-time data from IoT devices, GPS, and satellite imaging to monitor and control farming operations, reducing waste and increasing yield. For instance, in Israel, a country with limited water resources, PF allows farmers to use inputs only where needed.

22.6.3 Climate resilience and adaptation

Smart farming technologies also help increase crop resilience to climate change by giving farmers the information and tools they need to adapt to changing climatic circumstances. Drones, satellite imagery, and AI-powered platforms can monitor weather patterns, soil health, and crop growth, allowing farmers to make better decisions about planting, irrigation, and harvesting. Farmers in Bangladesh, a country prone to frequent flooding, use satellite data and weather forecasting systems to predict floods and adjust planting dates accordingly. This has contributed to lower crop losses and increased food security in flood-prone areas. Similarly, in drought-prone parts of Sub-Saharan Africa, farmers are using drought-resistant crop varieties developed through biotechnology to withstand periods of low rainfall.

22.6.4 Reducing postharvest losses

One of the important domains where technology can make a significant difference is in reducing post-harvest losses. Many poor countries waste a considerable amount of food between harvest and consumption due to insufficient storage, transportation, and processing capabilities. Solar-powered cold storage, mobile market access apps, and blockchain-based supply chain transparency can all contribute to overcoming these problems. In India, mobile systems like National Agriculture Market (eNAM) connect farmers and buyers, eliminating the need for intermediaries and ensuring that farmers are appropriately compensated for their produce. This not only raises smallholder farmers' wages but also reduces food waste by ensuring that foods reach consumers more quickly.

22.7 INTERNATIONAL EFFORTS TO ADDRESS FOOD SECURITY

Several international organizations are working to address global food security challenges by promoting smart agriculture and sustainable farming practices [20].

The Food and Agriculture Organization (**FAO**), a United Nations specialized agency, oversees international efforts to maintain food security. The FAO assists countries in adopting innovative farming technologies, improving food systems, and strengthening resistance to climate change. The organization's efforts, such as the Save and Grow program, seek to promote sustainable agriculture approaches that increase productivity while minimizing environmental impact.

The World Food Programme (**WFP**): It is the world's largest humanitarian organization working to end hunger and improve food security. In addition to providing emergency food assistance during emergencies, the WFP is involved in projects that promote long-term food security through agricultural development. For example, the WFP aids smallholder farmers in Uganda by providing training, seeds, and advanced farming methods, enhancing production and resilience to climate shocks.

The Role of Public–Private Partnerships: These are widely seen as an important tool for tackling food security issues. Governments, nongovernmental organizations, and commercial enterprises are working to develop and implement smart agriculture solutions that benefit smallholder farmers. For example, the Gates Foundation collaborated with African governments and commercial enterprises to create and distribute drought-tolerant seeds to farmers across the continent. Similarly, corporations like John Deere are partnering with local farmers in Brazil to implement PF technology that boosts production and sustainability [20].

22.8 PRECISION FARMING (SMART AGRICULTURE)

PF is a type of agricultural management that focuses on monitoring, measuring, and adjusting to fluctuations in crop fields and within them. This technology goes by various other names, including site-specific crop management, on-demand farming, satellite agriculture, and precision agriculture. Precision agriculture makes use of information technology to ensure that crops and soil receive exactly what they need to thrive and be productive. This methodology not only increases profitability but also fosters sustainability and environmental conservation. It considers factors such as soil composition, topography, climatic conditions, and data related to plant growth and yield when managing agricultural practices [21]. The primary objective of this farming method, in contrast to conventional farming, is to use inputs in exact quantities to enhance crop yields and profitability. The International Society of Precision Agriculture (https://www.ispag.org/) adopted the following definition of precision agriculture in 2019:

> Precision Agriculture is a management strategy that gathers, processes and analyzes temporal, spatial and individual data and combines it with other information to support management decisions according to estimated variability for improved resource use efficiency, productivity, quality, profitability and sustainability of agricultural production

Today's technological advancements, particularly when integrated with the IoT, present significant opportunities to elevate PF to unprecedented levels, despite various challenges. The essential components driving the development of a comprehensive solution for PF include smart sensors, sensor networks, wireless communications, IoT, and AI. Several essential technologies employed in PF are as follows:

Global Positioning System: This technology enables the precise identification of farm equipment locations within the field, offering an accurate positioning system that is crucial for the application of variable rate technology (VRT) in the management of agricultural inputs. An effective agricultural management system that relies on remote sensing is significantly supported by the internet.

Geographic Information System (GIS): GIS software offers comprehensive maps that detail vegetation and productivity, encompassing crop data essential for informed decision-making. Agricultural GIS technologies can measure vegetation density in different portions of a field. This data can then be used by agricultural machinery to improve the seed, nutrition, herbicide, and fertilizer applications, which are adjusted to each plot's needs [22].

Grid Sampling: It involves dividing agricultural fields into smaller portions. Typical sizes range from 0.5 to 5 hectares. Soil samples are collected from these designated grids, which were then combined and subjected to laboratory investigation to determine the appropriate rates for applying agricultural inputs.

Internet of Things: Sensors can communicate with a mini-computer to monitor various characteristics including air temperature, humidity, and soil temperature.

Variable Rate Technology: VRT refers to agricultural machinery that correctly regulates crop inputs including fertilizers, irrigation, tillage, and pest management.

Table 22.1 Benefits of precision farming

S. No	*Benefit*	*Precision agriculture*
1	Enhanced agricultural productivity	Farmers have the ability to customize their management approaches to address the unique requirements of various sections within a field, potentially resulting in increased crop production.
2	Decreased expenses	It facilitates a decrease in the use of water, fertilizers, and pesticides among farmers, thereby assisting in the reduction of overall costs.
3	Diminished ecological footprint	Precision agriculture enhances resource efficiency, thereby minimizing the release of pollutants into the environment.
4	Mitigates the deterioration of soil	It has the potential to mitigate the excessive application of chemicals, thereby reducing the risk of detrimental substances infiltrating the soil.
5	Improved pest control strategies	Precision agriculture can mitigate the spread of pests and disease outbreaks through the implementation of early detection systems and proactive management strategies.
6	Labor effectiveness	Automated systems diminish the need for human labor.
7	Increased profits	Farmers have the potential to enhance their yields using new technology and, consequently, their profits.

Drones: They help farmers with precision agriculture by monitoring plant health, distributing seeds, applying pesticides, analyzing soil health, and managing livestock.

Plant Signal Induction: This technology enables precision agriculture by tailoring biostimulant treatments to crops' physiological needs at any time. This method enables focused interventions and optimizes productivity, nutrient absorption, and stress resilience, leading to more effective resource utilization and improved crop quality in various areas.

There are different technologies like yield monitors, yield maps, remote sensors, computer hardware and software, big data and analytics, farm management software, AI, ML, and cloud computing, all of which play important roles in improving the effectiveness of PF operations. Application of these technologies represents the most sophisticated agricultural methodology, effectively addressing both production challenges and significant social issues, including those of a global nature. Precision agriculture focuses on executing the appropriate actions in the correct location, utilizing the proper methods, and at the optimal moment. The application of precision technologies contributes to minimizing environmental harm and enhancing fuel efficiency, which ultimately leads to a reduction in carbon emissions. There are many other benefits of PF as described in Table 22.1.

22.9 TECHNOLOGICAL TOOLS IN SMART AGRICULTURE

Contemporary technological innovations, particularly when integrated with the IoT, present significant opportunities for advancing PF despite various obstacles. The essential components driving the development of a comprehensive PF solution include smart sensors, sensor networks, wireless communication, IoT, and AI [23,24].

Sensors: Sensors serve as the fundamental means for real-time data acquisition, while the communication interface facilitates connectivity to either local or cloud-based storage systems. Utilizing the data gathered, analytical tools, and decision-making systems enables the automation of agricultural processes. The adoption of PF, leveraging these advanced technologies, ensures that crops receive optimal input resources and care, thereby enhancing efficiency, quality, yield, and overall profitability [24]. Technological advancements are significantly influencing the evolution of crop and livestock farming, serving as potential catalysts for the sustainable intensification of agricultural systems. Notably, the emergence of compact, cost-effective, and high-performance sensors presents opportunities for integration into production systems. This integration facilitates a greater accumulation of data, ultimately enhancing the quality and quantity of information available. Sensors are instruments designed to identify, document, and quantify the radiation of electromagnetic energy, including both the energy they emit and the energy that is reflected. These devices can be classified based on their construction, the specific range of the electromagnetic spectrum they can register, and the methods they employ for detection, documentation, measurement, and representation of the energy they detect. Sensors can be classified as either passive or active based on the type of energy they detect. Active sensors create their own energy, which is sent to the target object, and they subsequently measure the radiation that is reflected from it. In contrast, passive sensors solely transmit and receive energy without generating any of their own. The primary attributes of sensors include sensitivity, linearity, precision, and signal resolution. Sensors transform a physical quantity into either analog electrical signals or digital data. They respond to various conditions, which they subsequently convert into output signals. Agricultural sensors are utilized in various applications, including soil testing devices, machinery for fertilization and crop protection, and equipment for harvesting crops. The most commonly employed types of sensors today are electromagnetic and ultrasonic sensors. Additionally, Light Detection and Ranging (LIDAR) and infrared sensors are currently under scientific research for agricultural applications.

Internet of Things: The future of PF, enhanced by IoT technology, holds significant promise. Sophisticated sensors can be utilized to track environmental conditions, including rainfall, moisture levels, and temperature, thereby optimizing agricultural productivity. IoT devices can also facilitate the assessment of water and nitrogen levels in the soil. Furthermore, by analyzing CO_2 concentrations in agricultural areas, it becomes possible to monitor evapotranspiration rates, leading to improved oversight of crop health. Additionally, the threat of pest infestations can be mitigated through the use of IoT-enabled traps that are equipped with high-resolution cameras and other advanced features [25].

Cloud Computing: On-demand computing involves the allocation of system resources and data to users based on their requests. This principle can be realized in various forms, such as Infrastructure as a Service (IaaS), Platform as a Service (PaaS), and Software as a Service (SaaS). In the agricultural sector, cloud computing can support farmers in adopting intelligent agricultural methods by utilizing cloud-based information and the IoT to enable data-informed decision-making concerning their crops.

Satellite Technology: It is revolutionizing agriculture by providing farmers with quick access to data and imagery, allowing them to make more educated decisions. Examining these photos allows for the early detection of numerous illnesses, stress indicators, drought situations, and nutrient shortages in crops. This enables a fast response to reduce crop losses and increase overall yield. Satellites collect data on soil moisture, composition, and temperature to evaluate soil health. This assessment helps you make informed decisions about soil management. It is especially useful in determining the appropriate soil additives, irrigation systems, and soil quality improvements. Weather forecasts are critical to agricultural production. Satellite technology is used to monitor weather conditions and make predictions based on the data collected.

Big Data Analytics: Integrating Big Data Analytics into agriculture offers tremendous benefits and is transforming farmers' crop production practices. Using information from sensors, satellites, and weather monitoring stations can help farmers make better decisions and improve their operations. Big Data enables predictive analytics to identify possible dangers, such as insect invasions or disease outbreaks. Recognizing dangers early on allows farmers to take proactive measures to protect their crops effectively [18].

22.10 APPLICATIONS OF SMART AGRICULTURE

Precision agricultural system is in a state of continuous evolution, enabling producers to enhance both production and resource utilization. Consequently, farmers are able to elevate soil quality and overall productivity. The objective of sustainable agriculture is to satisfy the present demands of society, particularly in terms of food production, while ensuring the protection of the environment as shown in Figure 22.1. Precision agriculture enhances resource conservation by enabling farmers to optimize the cultivation of crops and soil with the highest level of efficiency. Precision agriculture enables farmers to target the application of fertilizers and herbicides specifically to the areas of the field that require them, thereby reducing input costs and enhancing crop yields. It assists farmers in enhancing their irrigation methods to conserve both water and energy resources. When creating a field map, assess environmental factors like landscape and weather, utilizing GPS and satellite remote sensing. Crop monitoring helps farmers determine harvest readiness through vegetation indices, indicating maturity when leaves yellow and dry as nutrients are directed to the fruits. Quality ground mapping is essential for PF, allowing farmers to assess soil properties and nutrition. While soil mapping has a long history, modern technologies enhance accuracy, resulting in more efficient digital maps.

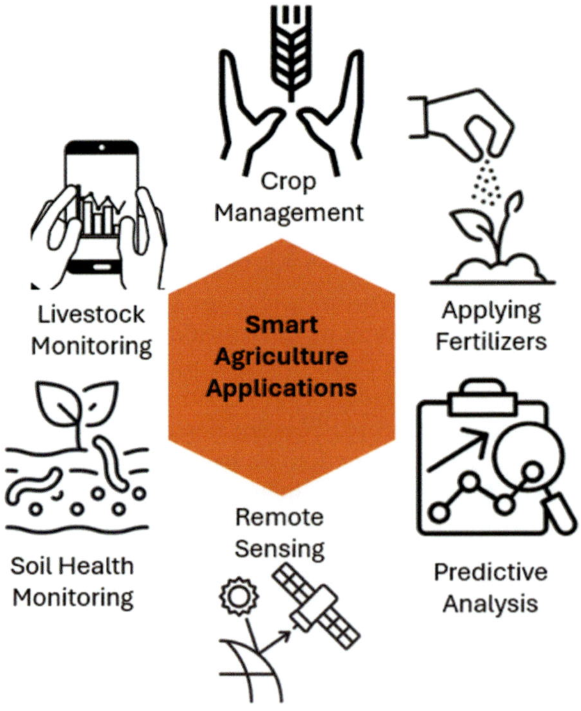

Figure 22.1 Applications of smart agriculture.

22.10.1 Summary of the case studies

A summary of the case studies regarding the smart farming implementation has been provided in Table 22.2.

Table 22.2 Smart farming implementations

Case study	*Technology used*	*Details*	*Impact*
John Deere's precision agriculture (global) [26]	GPS, IoT sensors, AI-driven analytics, autonomous tractors	John Deere's precision agriculture tools allow farmers to monitor soil conditions, optimize planting strategies, and automate tasks like sowing and harvesting.	Improved yield, reduced waste, and optimized input costs; promotes sustainable farming by minimizing water, fertilizer, and pesticide usage.
Smart greenhouse project—Netherlands [7]	IoT sensors, climate control systems, data analytics, AI	Advanced smart greenhouses in the Netherlands use IoT sensors to monitor environmental factors and adjust climate conditions to optimize crop growth.	Increased crop yields, enhanced energy efficiency, better use of water and nutrients, enabling higher food production on less land.
Agrivi farm management software—Europe [27]	Cloud computing, big data analytics, mobile apps	Agrivi offers farm management software that tracks field activities, weather, pest management, and crop yields, providing real-time insights for farmers.	Reduced risk of crop diseases and weather issues, improved resource management, and higher productivity through data-driven decisions.
Drone usage in smart farming—Japan [23]	Drones, AI, satellite imaging	Drones equipped with AI and cameras monitor crop health, soil conditions, and spray crops, while satellite imaging helps optimize farm operations.	Increased precision in crop management, minimized chemical usage, and reduced labor costs, addressing Japan's aging farming population.
LettUs Grow (vertical farming)—UK [28]	Aeroponics, IoT, cloud-based farm management systems	LettUs Grow uses aeroponics in vertical farms to grow plants without soil, using mist to deliver nutrients, with IoT monitoring and cloud-based management.	Efficient water usage (up to 95% less), increased urban food production, and reduced food miles, contributing to sustainable and local food systems.
Vineview—Napa Valley, California [29]	Aerial imaging, AI, data analytics	Vineview uses aerial imaging to monitor vineyard health, analyzing multispectral images to identify disease or irrigation issues early.	Improved grape quality, optimized irrigation, and reduced chemical usage, contributing to more sustainable vineyard management.
Cattle tracking with Moocall—Ireland [30]	IoT sensors, wearable devices, cloud computing	Moocall's wearable sensor monitors cows for signs of labor, sending alerts to farmers, enabling timely intervention.	Reduced calf mortality, improved animal welfare, and reduced labor for farmers during calving seasons.

22.10.2 Success stories from various regions, particularly developing nations

Case study	*Technology*	*Details*	*Success story*
Smallholder farmers in Kenya—using mobile technologies [31]	Mobile platforms, GPS, IoT sensors	Smallholder farmers in Kenya have adopted mobile-based technologies like *iCow* and *M-Farm* to receive real-time information on weather conditions, crop management, and market prices. IoT sensors and GPS systems help them manage farms more efficiently by monitoring soil moisture and optimizing irrigation.	Farmers using precision agriculture technologies have reported up to a 50% increase in yields, reduced water usage, and better pest management. Improved access to markets for crops.
Mexico's corn farming initiative [2]	Satellite imagery, data analytics, GPS-guided equipment	In Mexico, precision agriculture is employed in corn farming to optimize seed planting, fertilization, and irrigation. Satellite imagery is used to monitor crop health, and GPS-guided tractors help sow seeds with greater accuracy.	Corn farmers reported yield increases of up to 30%, reduced water consumption by 25%, and optimized fertilizer use, boosting productivity while ensuring sustainability.
Rice farming in Bangladesh—precision water management [25]	IoT sensors, mobile apps	Rice farmers in Bangladesh are utilizing precision irrigation technologies like IoT-based soil moisture sensors and mobile apps to manage water usage more efficiently. These tools provide real-time data on soil conditions and help optimize water application.	The initiative led to a 20%–30% reduction in water usage, while maintaining or increasing yields. Contributed to the sustainable use of water resources in a water-stressed region.
Smart farming in Zimbabwe—boosting maize yields [32]	Drones, data analytics, mobile-based advisory services	Maize farmers in Zimbabwe have adopted drones and data analytics to monitor crop health and detect issues like pest infestations or nutrient deficiencies early. Farmers also receive mobile-based advisory services offering tailored recommendations based on field data.	Maize yields have increased by 15%–20%, with fewer crop losses due to timely pest and disease detection. Optimized fertilizer use reduced costs.

(Continued)

(Continued)

Case study	*Technology*	*Details*	*Success story*
Precision irrigation in India—Tamil Nadu project [22]	Drip irrigation systems, IoT sensors, GIS (Geographic Information System)	The Tamil Nadu Precision Farming Project introduced drip irrigation systems and GIS mapping for crop monitoring. IoT sensors are used to optimize water and fertilizer use based on real-time soil data.	Farmers in Tamil Nadu have seen a 50%–60% increase in crop yields, reduced water use by up to 40%, and improved crop quality, empowering sustainable and profitable practices.
Brazil's soybean farmers—precision planting and harvesting [33]	GPS-guided machinery, satellite imagery, data analytics	Brazilian soybean farmers use GPS-guided machinery to optimize planting and harvesting. Satellite imagery and data analytics help monitor crop health and optimize input use, such as fertilizers and pesticides.	Precision agriculture has led to a 15%–20% increase in soybean yields, reduced input costs, and a decreased environmental impact by minimizing fertilizer and pesticide use.
Cocoa farming in Ghana—smart sensors and GIS [34]	GIS, IoT sensors, mobile-based advisory services	Ghana's cocoa growers use GIS technology to map farms and monitor soil conditions. IoT sensors monitor humidity and temperature, which are essential for cocoa growth, while mobile advisory services make crop management recommendations.	Cocoa yields have increased by 10%–15%, owing to higher crop quality. Technology has reduced the usage of chemical inputs, encouraging more environmentally friendly farming techniques.
Precision agriculture for wheat farmers in Pakistan [3]	IoT sensors, drone technology, data analytics	Wheat farmers in Pakistan use IoT sensors and drone technology to track crop health, soil conditions, and irrigation requirements. Data analytics provide informed judgments about fertilizer and pesticide applications.	Wheat yields have increased by 10%–20%, with significant reductions in water use. Farmers detect and address issues such as pests early, minimizing crop losses.
Precision agriculture for olive cultivation in Morocco [34]	IoT sensors, data analytics, weather forecasting tools	Moroccan olive producers use IoT devices to monitor soil moisture and optimize irrigation. Farmers may use data analytics and weather predictions to make more informed decisions regarding crop watering and harvesting.	Olive yields have increased by 15%–25%, with a 30% reduction in water usage. Improved olive quality has contributed to higher market prices.

(Continued)

(*Continued*)

Case study	*Technology*	*Details*	*Success story*
Precision agriculture for coffee farmers in Vietnam [35]	IoT sensors, drones, mobile apps	Coffee producers in Vietnam use IoT devices to monitor soil nutrients and water levels. Drones can monitor huge plantations to detect pest infestations and water stress. Mobile apps offer real-time crop health information.	Coffee yields have grown by 20%, while water and fertilizer usage have decreased by 15%–20%. The technology has enhanced sustainability and profitability.

22.11 ECONOMIC IMPACT OF SMART AGRICULTURE ON DEVELOPING COUNTRIES

The economic impact of smart agriculture on developing countries has been significant, offering both direct and indirect benefits. Here is an overview of the key areas where smart agriculture contributes to economic growth:

Aspect	*Economic Impact*
Increased crop yields	Precision farming, IoT sensors, and data analytics enable farmers to optimize planting, irrigation, and fertilization, resulting in higher yields. Yields in some locations have increased by 20%–50%, resulting in more income.
Cost efficiency	Cost efficiency is achieved by technologies such as GPS-guided machinery, precision irrigation, and automated systems that optimize water, fertilizer, and pesticide use. This helps farmers reduce production costs and increase profitability.
Reduction in crop losses	Drones, IoT sensors, and data analytics enable early identification of pests, illnesses, and nutritional deficiencies, reducing crop losses. This minimizes economic risk and income losses from crop failure or low quality.
Access to markets	Mobile platforms and digital marketplaces enable smallholder farmers to access market prices, sell crops at better rates, and reduce reliance on middlemen. This leads to better prices and improved bargaining power for farmers.
Employment opportunities	Smart farming requires skilled labor for managing technologies like drones, IoT systems, and data analytics, creating new job opportunities in rural areas and promoting entrepreneurship in agri-tech sectors.
Water and resource efficiency	Precision agriculture technologies optimize the use of resources like water and fertilizers, lowering costs and conserving rare resources. This is critical for developing countries with water scarcity and limited resource constraints.
Sustainable farming practices	Smart agriculture promotes sustainable practices by reducing chemical inputs and optimizing resource use, which improves long-term soil health and ensures sustainable economic returns for farmers and communities.
Access to financial services	Digital platforms offer farmers access to microloans, insurance, and advisory services, allowing them to invest in better farming technologies and practices, increasing their overall economic resilience.
Rural economic development	Smart agriculture promotes economic growth in rural regions by increasing production and opening up new business options, eliminating poverty, and improving living conditions.
Export potential	Smart farming produces higher quality and larger quantities, leading to increased agricultural exports, higher foreign exchange revenues, and economic growth in developing countries.

22.12 CURRENT STATE OF SMART AGRICULTURE IN INDIA

There is significant growth in the market of precision agriculture in India, which is supported by the more aware farmers and the latest growth in technology. This development in smart agriculture in India is expected to achieve a compound annual growth rate of 13.38% during the period 2022–2028. There are several key factors responsible for market expansion like more use of sustainable farming solutions and more encouragement to the adoption of modern agriculture by the government. At the same time, easy accessibility and less pricing of the advanced farming tools and software have enabled the farmers to adopt these more willingly [1]. This precision faring is majorly being adopted in the states of Maharashtra, Haryana, and Punjab. Infrastructure projects that promote smart and precision agriculture can be funded through the Agriculture Infrastructure Fund, which was established in response to the Covid-19 outbreak. The Indian government has provided Rs 450 crore budget for the Digital Agriculture Mission (DAM) with an additional Rs 600 crore with an aim to enhance technology usage in agriculture. DAM has also been launched by the Indian government, which improves the India Digital Ecosystem of Agriculture (IDEA), which provides financial support to states for new technologies (Next Generation Platform for Agriculture (NeGPA)), the creation of a Unified Farmers Service Interface, and the modernization of the Mahalanobis National Crop Forecast Centre along with the steps taken for soil health, fertility, and profile mapping. Similarly, online trading platforms like eNAM and "IDEA" strategy have supported the digital transformation within the agricultural sector. These steps have been taken to utilize the technological trends to optimize the trading procedure by empowering the farmers and developing a market for the agricultural goods. DiCRA (Data in Climate Resilient Agriculture) initiative gives crucial inputs on climate resilience, policies for enhanced agricultural sustainability and shaping public investments. Smart farming is also supported by the government initiative to invite foreign agricultural companies to invest in 100% Foreign Direct Investment (FDI) to aid farming activities. This includes vegetable cultivation and controlled aquaculture. ITC's launch of the 'Meta Market for Advanced Agricultural Services' app (ITCMAARS) is the result of this initiative. This helps the farmers in terms of PF, real-time soil testing, and market connections and tools for effective pest resolution and crop management. The implementation of micro irrigation technologies, like drip and sprinkler systems, is a step to improve water utilization efficiency under the scheme "'Per Drop More Crop' initiative under the Pradhan Mantri Krishi Sichai Yojana (PMKSY-PDMC)." This program was launched by Prime Minister Narendra Modi. The Indian Council of Agricultural Research (ICAR) fosters advancements, outreach, and educational initiatives within the agricultural sector. Apart from this, one more project named SENSAGRI, which stands for Sensor-based Smart Agriculture, is an innovative research initiative that employs drones and hyperspectral remote sensing technology to assess the condition of crops and soil. This project was developed by the ICAR in collaboration with the Indian Agricultural Research Institute. Figure 22.2 shows the general trends in smart agriculture adoption in India, comparing 2019 with projected adoption by 2024. This chart highlights the growing use of technologies like IoT, AI, drone technology, and smart irrigation, as well as government initiatives supporting the sector.

22.13 CHALLENGES IN SMART FARMING

PF appears to be a rational management strategy for agricultural producers, and numerous manufacturers and research organizations have dedicated significant resources to this technology. However, despite the considerable investment and numerous assurances from

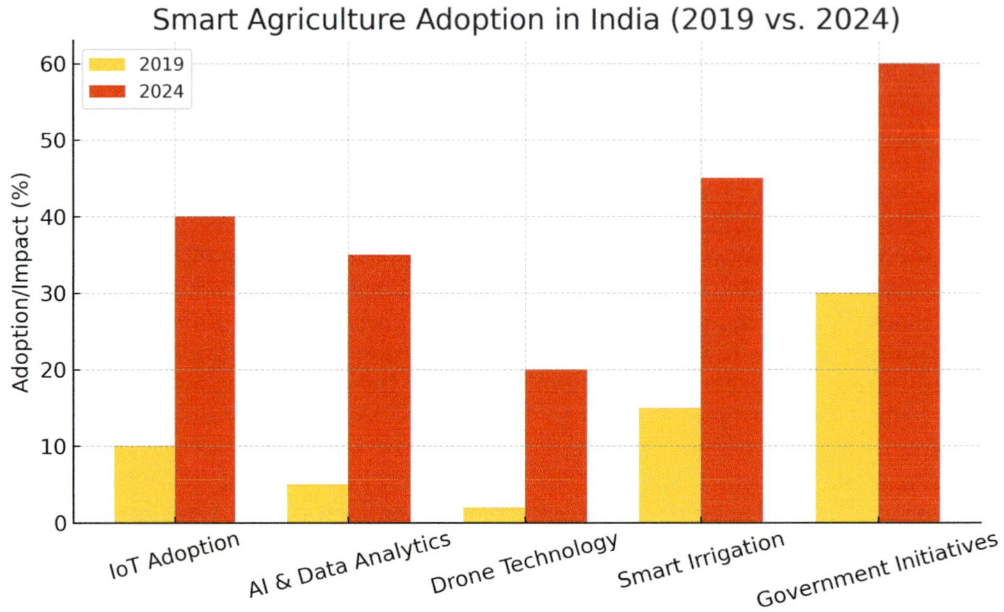

Figure 22.2 Trends in smart agriculture adoption in India.

manufacturers, many farmers remain hesitant for several reasons. To realize the full potential of precision agriculture, it is essential to address the following challenges:

Implementation of Cutting-Edge Technologies: Precision agriculture technologies like auto-guidance are widely adopted, while VRT faces slower uptake due to cost, complexity, and required expertise. Growers must assess the cost–benefit ratio of new technologies, with support from extension services aiding training efforts.

Establishing Uniformity in Technology across Various Platforms: The advancement of technology in agriculture fosters significant innovation among original equipment manufacturers. While these innovations present considerable benefits, they also pose challenges, particularly due to differing operational standards. Future efforts will need to focus on standardizing these technologies for greater integration.

Connectivity: To implement smart farming through IoT, it requires a strong 4G/5G or higher internet connectivity, majorly in rural areas.

Advanced Agricultural Data Management: Another challenge to implement precision agriculture is to manage the diverse data from equipment manufacturers and various technologies. Hence, there is a need to streamline the data by consolidating formats to enhance efficiency. This will make the data analysis easier and will remove the burden on growers and agronomists.

Management of Soil and Water Resources: Another need for precision agriculture is to manage the soil and water more effectively. However, it is difficult to implement due to varying climates and soil types. More advanced technologies can provide more efficient water distribution and real-time data for soil testing.

Proficiencies in Data Analysis: Precision agriculturalists face challenges in understanding and processing large amounts of data from daily, monthly, or seasonal data on a smaller farm unit. However, it is crucial to assess the value of these data points and focus on valuable insights.

Less Scalability: There is a need to enhance the usage of agricultural tool in the personal gardens and industrial agriculture. Moreover, these tools should be self-configurable, which will make them easy to use by the farmers.

Encountering the Challenges Associated with Vertical Farming: There is a need to integrate the precision agriculture tools in vertical farming and indoor operations, allowing climate control and smart farming without seasonal or weather-affecting arrangements.

Data Privacy and Information Sharing: Due to the increasing misuse of personal data of mobile users by hackers, farmer's community hesitates to share data that hampers collaboration. Hence, building trust and providing data security have to be ensured so that smart farming can be implemented more effectively.

The Literacy Levels of Farmers: One of the major challenges that developing countries are facing is low literacy levels of farmers and awareness about technology.

This makes information processing difficult for the farmers. Hence, the Agri-tech companies can implement the facilities to provide digital literacy to the farmers.

Insufficient Financial Resources: Small farmers are unable to access communication technologies like GPS and drones. Hence, PF becomes difficult for them.

Need of Technical Support: To implement effective precision agriculture, there is always a need of technical support and continuous maintenance. This will enhance the productivity of farmers using advanced equipment.

Difficult to Implement for Small Land Holdings: PF approach is not feasible to implement in case of small landholders as more investment is needed to deploy sensors on equipment.

22.14 RESEARCH GAPS IN SMART FARMING SYSTEMS

To implement the concept of smart farming, there are a lot of challenges like managing, network performance, sensors node's energy efficiency, and most importantly, security concerns. To mitigate these difficulties in IoT-driven agriculture, advancements in perception, communication, big data, security, and blockchain technologies are being leveraged. Such challenges restrict the practical deployment of smart farming in large-scale agricultural operations, revealing significant research deficiencies in the area. It is imperative to engage in research to effectively combine IoT, AI, cloud computing, and blockchain to improve agricultural productivity.

i. There are gaps in how to standardize protocols, calibrate devices, validate data, and integrate data from multiple sources.
ii. It is essential to gain a deeper understanding of security and data privacy concerns associated with smart farming.
iii. Designing wireless sensor networks that are both durable and energy-efficient presents significant challenges.
iv. Challenges exist concerning energy efficiency, network performance, and security.
v. Identifying superfluous data traffic is crucial, as is delivering timely and relevant information to farmers.

22.15 CONCLUSION

This study emphasizes the significant impact of modern technologies on agriculture, namely, the implementation of smart agriculture or PF. Advanced technologies like IoT, AI, big data analytics, and satellite imagery significantly improve agricultural productivity, according to key studies. These technologies help farmers monitor crops and soil conditions in real time, optimize resource use, minimize input costs (e.g., water and fertilizers), and boost crop yields. Smart agriculture is crucial for meeting global food security challenges, including population growth, climate change, and limited resources. However, several barriers were also identified, including high costs of technology implementation, data security concerns, and the low level of digital literacy among farmers, particularly in developing nations. Smart agriculture has a substantial impact on long-term food security. The IMF projects that by 2050, the world's population would exceed 10 billion, necessitating a 70% increase in food supply. PF is necessary to continue food supply, as traditional methods are insufficient. Smart agriculture has the ability to improve land-use efficiency, optimize input management, and reduce environmental impact, all of which are essential for sustainable food production. Drones, soil sensors, and cloud computing can increase output while conserving resources like water and fertilizer. This addresses the requirement for both increased yields and sustainable agriculture techniques. Precision agriculture has the ability to transform local economies by enabling farmers to produce more with fewer resources, hence reducing poverty and hunger. In due course, widespread implementation of smart agriculture can boost rural development, reduce the risk of food scarcity, and fortify global food systems against climate shocks.

22.16 RECOMMENDATIONS FOR POLICYMAKERS, RESEARCHERS, AND FARMERS

22.16.1 Policymakers

Need to Provide More Advanced Infrastructure: Governments need to prioritize infrastructure development in underdeveloped nations to enable smart agriculture (digital and physical). This includes improving rural internet connectivity so that farmers may adopt IoT and cloud-based crop management technologies.

Policymakers should create financial support mechanisms, such as subsidies, grants, or low-interest loans, to help smallholder farmers adopt precision agriculture technologies. This will make expensive technologies like drones and sensors more accessible to farmers with varying skill levels.

Data Security Regulations: With the increasing need of data in smart farming, a regulatory framework is necessary to protect farmers' data and prevent misuse. Secure and transparent data management systems will boost trust and encourage farmers to use them.

Educational Programs: Investment by the government should be done in digital literacy efforts and vocational training to help farmers understand and use PF technologies. Collaborations between educational institutions and technology companies could help with the creation of scalable learning models.

22.16.2 Researchers

Innovations to Provide Low-Cost Solution: Implementing smart farming in poor regions poses a significant barrier due to high technology costs. Researchers should build low-cost precision agricultural equipment and open-source platforms that small-scale farmers can easily embrace.

Data Management Solutions: Managing huge datasets from several sensors and devices is a significant challenge in precision agriculture.

Prioritizing Local Contexts: Research should focus on building technology solutions that are adaptable to local environments. PF methods must be tailored to the specific crops, weather patterns, and soil types found in different regions, particularly in developing countries. Solutions must also include local economic conditions and resource availability.

Energy-Efficient Systems: Using low-power sensors and wireless networks in smart farming can reduce expenses and environmental impact. This is particularly important for small landholders.

22.16.3 Farmers

Engagement with Digital Tools: Encourage farmers to actively use digital tools to improve their operations. Farmers who employ smartphone apps, GPS systems, and sensor-based technologies can improve decision-making and productivity while minimizing waste.

Collaborative Learning: Cooperative associations and local agricultural networks can assist farmers in sharing information about PF tools and techniques. Farmers can work together to obtain more expensive technologies, such as drones, and exchange data to improve decision-making.

Training and Skill Development: Farmers should take part in digital literacy and technical training programs, particularly in low-tech areas. This would enable people to not only understand the possibilities of smart agriculture but also to make educated decisions about technology investments.

Sustainable Practices: Farmers should use PF tools to optimize water usage with smart irrigation systems, reduce chemical inputs with targeted pest control, and improve soil health with accurate nutrient management. Although smart agriculture offers immense potential for improving food security, its successful implementation necessitates collaboration among policymakers, researchers, and farmers. Governments make policies to provide the necessary infrastructure and financial help, researchers must endeavor to make technology more accessible and efficient, and farmers must look into new learning opportunities to fully benefit from PF. With concentrated efforts, smart agriculture has the potential to become a crucial solution to the global food security concerns of the 21st century.

REFERENCES

1. U. Shafi, R. Mumtaz, J. García-Nieto, S. A. Hassan, S. A. R. Zaidi, and N. Iqbal, "Precision Agriculture Techniques and Practices: From Considerations to Applications," *Sensors*, vol. 19, no. 17, p. 3796, 2019. https://doi.org/10.3390/s19173796
2. M. R. Gonzalez and J. L. Martinez, "Precision Agriculture in Mexico: Advancing Corn Farming through Satellite Imagery and GPS Technology," *Journal of Agricultural Technology and Sustainability*, vol. 12, no. 4, pp. 198–206, 2023.
3. Z. A. Khan and A. Hussain, "Precision Agriculture for Wheat Farming in Pakistan: IoT and Drone Integration," *Pakistan Journal of Agricultural Research*, vol. 14, no. 3, pp. 112–121, 2023.
4. M. Videgain, J. A. Martínez-Casasnovas, A. Vigo-Morancho, et al., "On-Farm Experimentation of Precision Agriculture for Differential Seed and Fertilizer Management in Semi-Arid Rainfed Zones," *Precision Agriculture*, 2024. https://doi.org/10.1007/s11119-024-10189-y
5. D. Gollin, C. W. Hansen, and A. M. Wingender, "Two Blades of Grass: The Impact of the Green Revolution," *Journal of Political Economy*, vol. 129, no. 8, pp. 2344–2384, 2021. https://doi.org/10.1086/714444
6. R. J. Herrera and R. Garcia-Bertrand, "The Agricultural Revolutions," In *Ancestral DNA, Human Origins, and Migrations*, R. J. Herrera and R. Garcia-Bertrand (Eds.) Academic Press, pp. 475–509, 2018. https://doi.org/10.1016/B978-0-12-804124-6.00013-6
7. Smart Greenhouse Project, "IoT and AI in Greenhouse Farming - Netherlands," Available: https://www.smartgreenhouse.nl. [Accessed: Jan. 28, 2025].

8. S. Manasa, K. N. Anusha, S. Sharma, A. Thakur, and A. Sood, "An Overview of Precision Farming," *International Journal of Environment and Climate Change*, vol. 13, no. 12, pp. 441–456, 2023.
9. P. Nijalingappa, T. Gururaj, S. B. Goyal, V. Shukla, and A. Bruno (Eds.). *Examining the Metaverse in Healthcare: Opportunities, Challenges, and Future Directions: Opportunities, Challenges, and Future Directions*, IGI Global, 2024.
10. S. Vyas, A. Upadhyaya, D. Bhargava, and V. K. Shukla (Eds.). *Edge-AI in Healthcare: Trends and Future Perspectives*. London: CRC Press, 2023.
11. S. Dutta, V. K. Shukla, S. Nagpal, and M. Niranjanamurthy (Eds.). *Industry 4.0 and Intelligent Business Analytics for Healthcare*. Nova Science Publishers, 2022.
12. F. Columbus (Ed.). *Blockchain Technology for Healthcare Applications: Challenges, Privacy and Securing of Data*. Nova Science Publishers, 2024.
13. M. S. Thekkethil, V. K. Shukla, F. Beena and A. Chopra, "Robotic Process Automation in Banking and Finance Sector for Loan Processing and Fraud Detection." In *2021 9th International Conference on Reliability, Infocom Technologies and Optimization (Trends and Future Directions) (ICRITO)*, Noida, India, 2021, pp. 1–6. https://doi.org/10.1109/ICRITO51393.2021.9596076.
14. S. Vyas, S. Gupta, and V. K. Shukla, "Sustainable Energy Usage in Urban and Rural Context-A Study," In *Enabling Technologies for Effective Planning and Management in Sustainable Smart Cities*, pp. 389–402, 2023. Cham: Springer International Publishing.
15. M. Monfared, V. K. Shukla, S. Dutta, and A. Chaubey, "Reshaping Education through Augmented Reality and Virtual Reality," In *Cyber Intelligence and Information Retrieval: Proceedings of CIIR 2021*, pp. 619–629, 2021. Singapore: Springer Singapore.
16. S. Java, H. Mohammed, A. B. Bhardwaj, and V. K. Shukla, "Education 4.0 and Web 3.0 Applications in Enhancing Learning Management System: Post-lockdown Analysis in Covid-19 Pandemic," *Knowledge Management and Web*, vol. 3, p. 85, 2021.
17. R. Ratasuk, B. Vejlgaard, N. Mangalvedhe, and A. Ghosh, "IoT system for M2M Wireless sensor communication for smart farming," In *Proceedings of IEEE Wireless Communications and Networking Conference*, Doha, Qatar, April 3–6, 2016, pp. 1–5.
18. U. L. C. Baldos and T. W. Hertel, "Global food security in 2050: The role of agricultural productivity and climate change," *Australian Journal of Agricultural and Resource Economics*, vol. 58, no. 4, pp. 554–570, 2014. [Online]. Available: https://doi.org/10.1111/1467-8489.12048
19. V. Suma, "Internet of Things (IoT) based smart agriculture in India: An overview," *Journal of ISMAC*, vol. 3, no. 1, pp. 1–15, 2021. https://doi.org/10.36548/jismac.2021.1.001
20. N. A. George and F. H. McKay, "The Public Distribution System and food security in India," *International Journal of Environmental Research and Public Health*, vol. 16, no. 17, p. 3221, 2019. [Online]. Available: https://doi.org/10.3390/ijerph16173221
21. I. Cisternas, I. Velásquez, A. Caro, and A. Rodríguez, "Systematic literature review of implementations of precision agriculture," *Computers and Electronics in Agriculture*, vol. 176, p. 105626, 2020. https://doi.org/10.1016/j.compag.2020.105626
22. R. K. Sharma and S. S. Reddy, "Drip Irrigation and GIS Mapping in Tamil Nadu Precision Farming Project," *Indian Journal of Precision Agriculture*, vol. 11, no. 1, pp. 15–22, 2023.
23. Japan Smart Farming, "Drone Technology in Agriculture," Available: https://www.japan-agriculture.com. [Accessed: Jan. 28, 2025].
24. R. Karothia and M. K. Chattopadhyay, "Review of Various Technologies Involved in Precision Farming Automation." In *Precision Agriculture for Sustainability*, Narendra Khatri, Ajay Kumar Vyas, Celestine Iwendi, Prasenjit Chatterjee (Eds.), London: Apple Academic Press, pp. 3–20, 2024.
25. S. Ahmed, M. Rahman, and T. Hossain, "IoT-Based Precision Water Management for Rice Farming in Bangladesh," *Journal of Smart Agriculture and Environment*, vol. 8, no. 2, pp. 101–109, 2023.
26. John Deere, "Precision Agriculture," Available: https://www.deere.com. [Accessed: Jan. 28, 2025].

27. Agrivi, "Farm Management Software," Available: https://www.agrivi.com. [Accessed: Jan. 28, 2025].
28. LettUs Grow, "Vertical Farming with Aeroponics," Available: https://www.lettusgrow.com. [Accessed: Jan. 28, 2025].
29. VineView, "Aerial Imaging and AI for Vineyard Management," Available: https://www.vinev-iew.com. [Accessed: Jan. 28, 2025].
30. Moocall, "IoT-Based Cattle Tracking and Monitoring," Available: https://www.moocall.com. [Accessed: Jan. 28, 2025].
31. J. K. Mwangi and A. N. Wanjohi, "Empowering Smallholder Farmers in Kenya through Mobile Technologies: A Case of iCow and M-Farm," *International Journal of Agricultural Innovations and Research*, vol. 9, no. 3, pp. 45–52, 2023.
32. P. N. Chikanda, "Smart Farming in Zimbabwe: Enhancing Maize Yields with Drones and Data Analytics," *African Journal of Agricultural Research and Innovation*, vol. 10, no. 3, pp. 75–84, 2023.
33. L. S. Oliveira and F. A. Costa, "Precision Agriculture for Soybean Cultivation in Brazil: GPS-Guided Machinery and Data Analytics," *Journal of Agricultural Science and Technology*, vol. 15, no. 2, pp. 123–131, 2023.
34. M. E. Bensaid and A. L. Amrani, "IoT and Data Analytics for Olive Cultivation in Morocco: A Precision Agriculture Approach," *Journal of Mediterranean Agricultural Studies*, vol. 7, no. 2, pp. 89–97, 2023.
35. T. Nguyen and P. Tran, "Precision Agriculture for Coffee Farmers in Vietnam: IoT and Drone Applications," *Asian Journal of Agricultural Innovations*, vol. 13, no. 1, pp. 32–41, 2023.

Index

For Product Safety Concerns and Information please contact our
EU representative GPSR@taylorandfrancis.com Taylor & Francis
Verlag GmbH, Kaufingerstraße 24, 80331 München, Germany